CAMBRIDGE LIBRARY COLLECTION

Books of enduring scholarly value

Darwin

Two hundred years after his birth and 150 years after the publication of 'On the Origin of Species', Charles Darwin and his theories are still the focus of worldwide attention. This series offers not only works by Darwin, but also the writings of his mentors in Cambridge and elsewhere, and a survey of the impassioned scientific, philosophical and theological debates sparked by his 'dangerous idea'.

The Naturalist on the River Amazon

First published in 1863, this is a first-hand account of Henry Walter Bates' eleven-year expedition to the river Amazon in 1848, during which he discovered some eight thousand species unknown to the natural sciences. Written in the first person, it records the astonishing range of natural life in the regions traversed by the Amazon and its tributaries. Describing his adventures south of the equator, Bates takes the reader through Pará, Tocantins, Cametá, Marajó, Caripí, Obydos, Manos, Santarem, Tapajos, and Ega, descriptively cataloguing the rich vegetation, aboriginal population, and wondrous birds, animals and insects of these regions. More than just a scientist's log, the work that took Bates three years to complete was considered by Darwin to be 'the best work of natural history travels ever published in England'. This third edition of the book (1873) also contains numerous illustrations by the noted zoologist Joseph Wolf.

Cambridge University Press has long been a pioneer in the reissuing of out-of-print titles from its own backlist, producing digital reprints of books that are still sought after by scholars and students but could not be reprinted economically using traditional technology. The Cambridge Library Collection extends this activity to a wider range of books which are still of importance to researchers and professionals, either for the source material they contain, or as landmarks in the history of their academic discipline.

Drawing from the world-renowned collections in the Cambridge University Library, and guided by the advice of experts in each subject area, Cambridge University Press is using state-of-the-art scanning machines in its own Printing House to capture the content of each book selected for inclusion. The files are processed to give a consistently clear, crisp image, and the books finished to the high quality standard for which the Press is recognised around the world. The latest print-on-demand technology ensures that the books will remain available indefinitely, and that orders for single or multiple copies can quickly be supplied.

The Cambridge Library Collection will bring back to life books of enduring scholarly value (including out-of-copyright works originally issued by other publishers) across a wide range of disciplines in the humanities and social sciences and in science and technology.

The Naturalist on the River Amazon

A Record of Adventures, Habits of Animals, Sketches of Brazilian and Indian Life, and Aspects of Nature under the Equator, during Eleven Years of Travel

Henry Walter Bates

CAMBRIDGE UNIVERSITY PRESS

Cambridge, New York, Melbourne, Madrid, Cape Town, Singapore,
São Paolo, Delhi, Dubai, Tokyo

Published in the United States of America by Cambridge University Press, New York

www.cambridge.org
Information on this title: www.cambridge.org/9781108001632

This edition first published 1873
This digitally printed version 2009

ISBN 978-1-108-00163-2 Paperback

MOBBED BY CURL-CRESTED TOUCANS.

THE

NATURALIST ON THE RIVER AMAZONS.

A RECORD OF ADVENTURES, HABITS OF ANIMALS, SKETCHES OF BRAZILIAN AND INDIAN LIFE, AND ASPECTS OF NATURE UNDER THE EQUATOR, DURING ELEVEN YEARS OF TRAVEL.

BY HENRY WALTER BATES, F.L.S.,

(*Assistant-Secretary, Royal Geographical Society.*)

Pelopæus Wasp building nest.

THIRD EDITION.

WITH NUMEROUS ILLUSTRATIONS.

LONDON:

JOHN MURRAY, ALBEMARLE STREET.

1873.

CONTENTS.

CHAPTER I.

PARA.

CHAPTER II.

PARA

CHAPTER III.

PARÁ.

CHAPTER IV.

THE TOCANTINS AND CAMETA.

CHAPTER V.

CARIPÍ AND THE BAY OF MARAJÓ.

CHAPTER VI.

THE LOWER AMAZONS—PARÁ TO OBYDOS.

CHAPTER VII.

THE LOWER AMAZONS—OBYDOS TO MANAOS, OR THE BARRA OF THE RIO NEGRO.

CHAPTER VIII.

SANTAREM.

CHAPTER IX.

VOYAGE UP THE TAPAJOS.

LIST OF ILLUSTRATIONS.

LIST OF ILLUSTRATIONS.

THE

NATURALIST ON THE AMAZONS.

CHAPTER I.

PARÁ.

Arrival—Aspect of the country—The Pará river—First walk in the suburbs of Pará—Birds, Lizards and Insects of the suburbs—Leaf-carrying Ant—Sketch of the climate, history, and present condition of Pará.

I EMBARKED at Liverpool, with Mr. Wallace, in a small trading vessel, on the 26th of April, 1848; and, after a swift passage from the Irish Channel to the equator, arrived, on the 26th of May, off Salinas. This is the pilot-station for vessels bound to Pará, the only port of entry to the vast region watered by the Amazons. It is a small village, formerly a missionary settlement of the Jesuits, situated a few miles to the eastward of the Pará river. Here the ship anchored in the open sea, at a distance of six miles from the shore, the shallowness of the water far out around the mouth of the great river not permitting in safety a nearer approach; and the signal was hoisted for a pilot. It was with deep interest that my companion and myself, both now about to see and examine the beauties of a tropical country for the first time, gazed on the land where I, at least, eventually spent eleven of the best years of my life. To the eastward the country was not remarkable in appearance, being slightly undulating, with bare sandhills and scattered trees; but to the westward, stretching towards the mouth of the river, we could see through the captain's glass a long line of forest, rising apparently out of the water; a densely-packed mass of tall trees, broken into groups, and finally into single trees, as it dwindled away in the distance. This was the frontier, in this direction, of the great primæval forest characteristic of this region, which contains so many wonders in its recesses, and clothes the whole surface of the country for two thousand miles from this point to the foot of the Andes.

On the following day and night we sailed, with a light wind, partly aided by the tide, up the Pará river. Towards evening we passed Vigia and Colares, two fishing villages, and saw many native canoes, which seemed like toys beneath the lofty walls of dark forest. The air was excessively close, the sky overcast, and sheet lightning played almost incessantly around the horizon, an appropriate greeting on the threshold of a country lying close under the equator! The evening was calm, this being the season when the winds are not strong, so we glided along in a noiseless manner, which contrasted pleasantly with the unceasing turmoil to which we had been lately accustomed on the Atlantic. The immensity of the river struck us greatly, for although sailing sometimes at a distance of eight or nine miles from the eastern bank, the opposite shore was at no time visible. Indeed, the Pará river is 36 miles in breadth at its mouth; and at the city of Pará, nearly 70 miles from the sea, it is 20 miles wide; but at that point a series of islands commences, which contracts the river view in front of the port.

On the morning of the 28th of May we arrived at our destination. The appearance of the city at sunrise was pleasing in the highest degree. It is built on a low tract of land, having only one small rocky elevation at its southern extremity; it therefore affords no amphitheatral view from the river; but the white buildings roofed with red tiles, the numerous towers and cupolas of churches and convents, the crowds of palm trees reared above the buildings, all sharply defined against the clear blue sky, give an appearance of lightness and cheerfulness which is most exhilarating. The perpetual forest hems the city in on all sides landwards; and towards the suburbs, picturesque country houses are seen scattered about, half buried in luxuriant foliage. The port was full of native canoes and other vessels, large and small; and the ringing of bells and firing of rockets, announcing the dawn of some Roman Catholic festival day, showed that the population was astir at that early hour.

The impressions received during our first walk, on the evening of the day of our arrival, can never wholly fade from my mind.

After traversing the few streets of tall, gloomy, convent-looking buildings near the port, inhabited chiefly by merchants and shopkeepers; along which idle soldiers, dressed in shabby uniforms, carrying their muskets carelessly over their arms, priests, negresses with red water-jars on their heads, sad-looking Indian women carrying their naked children astride on their hips, and other samples of the motley life of the place, were seen; we passed down a long narrow street leading to the suburbs. Beyond this, our road lay across a grassy common into a picturesque lane leading to the virgin forest. The long street was inhabited by the poorer class of the population. The houses were of one story only, and had an irregular and mean appearance. The windows were without glass, having, instead, projecting lattice casements. The street was unpaved, and inches deep in loose sand. Groups of people were cooling themselves outside their doors—people of all shades in colour of skin, European, Negro and Indian, but chiefly an uncertain mixture of the three. Amongst them were several handsome women, dressed in a slovenly manner, barefoot or shod in loose slippers; but wearing richly decorated ear-rings, and around their necks strings of very large gold beads. They had dark expressive eyes, and remarkably rich heads of hair. It was a mere fancy, but I thought the mingled squalor, luxuriance and beauty of these women were pointedly in harmony with the rest of the scene; so striking, in the view, was the mixture of natural riches and human poverty. The houses were mostly in a dilapidated condition, and signs of indolence and neglect were everywhere visible. The wooden palings which surrounded the weed-grown gardens were strewn about, broken; and hogs, goats, and ill-fed poultry wandered in and out through the gaps. But amidst all, and compensating every defect, rose the overpowering beauty of the vegetation. The massive dark crowns of shady mangoes were seen everywhere amongst the dwellings, amidst fragrant blossoming orange, lemon, and many other tropical fruit trees; some in flower, others in fruit, at varying stages of ripeness. Here and there, shooting above the more dome-like and sombre trees, were the smooth columnar stems of palms, bearing aloft their magnificent crowns of finely-cut fronds. Amongst the latter the slim assai-palm was especially noticeable, growing in groups of four and five; its smooth, gently-curving stem, twenty to thirty feet high, terminating in a head of feathery foliage, inexpressibly light and elegant in outline. On the boughs of the taller and more ordinary-looking trees sat tufts of curiously-leaved parasites. Slender woody lianas hung in festoons from the branches, or were suspended in the form of cords and ribbons; whilst luxuriant creeping plants overran alike tree-trunks, roofs and walls, or toppled over palings in copious profusion of foliage. The superb banana (Musa paradisiaca), of which I had always read as forming one of the charms of tropical vegetation, here grew with great luxuriance: its glossy velvety-green leaves, twelve feet in length, curving over the roofs of verandahs in the rear of every house. The shape of the leaves, the varying shades of green which they present when lightly moved by the wind, and especially the contrast they afford in colour and form to the more sombre hues and more rounded outline of the other trees, are quite sufficient to account for the charm of this glorious tree. Strange forms of vegetation drew our attention at almost every step. Amongst them were the different kinds of Bromelia, or pine-apple plants, with their long, rigid, sword-shaped leaves, in some species jagged or toothed along their edges. Then there was the bread-fruit tree—an importation, it is true; but remarkable from its large, glossy, dark green, strongly digitated foliage, and its interesting history. Many other trees and plants, curious in leaf, stem, or manner of growth, grew on the borders of the thickets along which lay our road; they were all attractive to new-comers, whose last country ramble, of quite recent date, was over the bleak moors of Derbyshire on a sleety morning in April.

As we continued our walk the brief twilight commenced, and the sounds of multifarious life came from the vegetation around. The whirring of cicadas; the shrill stridulation of a vast number and variety of field crickets and grasshoppers, each species sounding its peculiar note; the plaintive hooting of tree frogs—all blended together in one continuous ringing sound,—the audible expression of the teeming profusion of Nature. As night came on, many species of frogs and toads in the marshy places joined in the chorus: their croaking and drumming, far louder than anything I had before heard in the same line, being added to the other noises, created an almost deafening din. This uproar of life, I afterwards found, never wholly ceased, night or day: in course of time I became, like other residents, accustomed to it. It is, however, one of the peculiarities of a tropical—at least, a Brazilian—climate which is most likely to surprise a stranger. After my return to England, the death-like stillness of summer days in the country appeared to me as strange as the ringing uproar did on my first arrival at Pará. The object of our visit being accomplished, we returned to the city. The fire-flies were then out in great numbers, flitting about the sombre woods, and even the frequented streets. We turned into our hammocks, well pleased with what we had seen, and full of anticipation with regard to the wealth of natural objects we had come to explore.

During the first few days we were employed in landing our baggage and arranging our extensive apparatus. We then accepted the invitation of the consignee of the vessel to make use of his rocinha, or country-house in the suburbs, until we finally decided on a residence. Upon this we made our first essay in housekeeping. We bought cotton hammocks, the universal substitute for beds in this country, cooking utensils and crockery, and engaged a free negro, named Isidoro, as cook and servant-of-all-work. Our first walks were in the immediate suburbs of Pará. The city lies on a corner of land formed by the junction of the river Guamá with the Pará. As I have said before, the forest, which covers the whole country, extends close up to the city streets; indeed, the town is built on a tract of cleared land, and is kept free from the jungle only by the constant care of the Government. The surface, though everywhere low, is slightly undulating, so that areas of dry land alternate throughout with areas of swampy ground, the vegetation and animal tenants of the two being widely different. Our residence lay on the side of the city nearest the Guamá, on the borders of one of the low and swampy areas which here extends over a portion of the suburbs. The tract of land is intersected by well-macadamized suburban roads, the chief of which, Estrada das Mongubeiras (the Monguba road), about a mile long, is a magnificent avenue of silk-cotton trees (Bombax monguba and B. ceiba), huge trees whose trunks taper rapidly from the ground upwards, and whose flowers before opening look like red balls studding the branches. This fine road was constructed under the governorship of the Count dos Arcos, about the year 1812. At right angles to it run a number of narrow green lanes, and the whole district is drained by a system of small canals or trenches through which the tide ebbs and flows, showing the lowness of the site. Before I left the country, other enterprising presidents had formed a number of avenues lined with coco-nut palms, almond and other trees, in continuation of the Monguba road., over the more elevated and drier ground to the north-east of the city. On the high ground the vegetation has an aspect quite different from that which it presents in the swampy parts. Indeed, with the exception of the palm trees, the suburbs here have an aspect like that of a village green at home. The soil is sandy, and the open commons are covered with a short grassy and shrubby vegetation. Beyond this, the land again descends to a marshy tract, where, at the bottom of the moist hollows, the public wells are situated. Here all the linen of the city is washed by hosts of noisy negresses, and here also the water-carts are filled—painted hogsheads on wheels, drawn by bullocks. In early morning, when the sun sometimes shines through a light mist, and everything is dripping with moisture, this part of the city is full of life: vociferous negroes and wrangling Gallegos,* the proprietors of the water-carts, are gathered about, jabbering continually, and taking their morning drams in dirty wine-shops at the street corners.

Along these beautiful roads we found much to interest us during the first few days. Suburbs of towns, and open, sunny, cultivated places in Brazil, are tenanted by species of animals and plants which are mostly different from those of the dense primæval forests. I will, therefore, give an account of what we observed of the animal world, during our explorations in the immediate neighbourhood of Pará.

The number and beauty of the birds and insects did not at first equal our expectations. The majority of the birds we saw were small and obscurely coloured; they were indeed similar, in general appearance, to such as are met with in country places in England. Occasionally a flock of small parroquets, green, with a patch of yellow on the forehead, would come at early morning to the trees near the Estrada. They would feed quietly, sometimes chattering in subdued tones, but setting up a harsh scream, and flying off, on being disturbed. Humming-birds we did not see at this time, although I afterwards found them by hundreds when certain trees were in flower. Vultures we only saw at a distance, sweeping round at a great height, over the public slaughter-houses. Several flycatchers, finches, ant-thrushes, a tribe of plainly-coloured birds, intermediate in structure between flycatchers and thrushes, some of which startle the new-comer by their extraordinary notes emitted from their places of concealment in the dense thickets; and also

* Natives of Galicia, in Spain, who follow this occupation in Lisbon and Oporto, as well as at Pará.

tanagers, and other small birds, inhabited the neighbourhood. None of these had a pleasing song, except a little brown wren (Troglodytes furvus), whose voice and melody resemble those of our English robin. It is often seen, hopping and climbing about the walls and roofs of houses and on trees in their vicinity. Its song is more frequently heard in the rainy season, when the Monguba trees shed their leaves. At those times the Estrada das Mongubeiras has an appearance quite unusual in a tropical country. The tree is one of the few in the Amazons region which sheds all its foliage before any of the new leaf-buds expand. The naked branches, the sodden ground matted with dead leaves, the grey mist veiling the surrounding vegetation, and the cool atmosphere soon after sunrise; all combine to remind one of autumnal mornings in England. Whilst loitering about at such times in a half-oblivious mood, thinking of home, the song of this bird would create for the moment a perfect illusion. Numbers of tanagers frequented the fruit and other trees in our garden. The two principal kinds which attracted our attention were the Rhamphocœlus jacapa and the Tanagra episcopus. The females of both are dull in colour, but the male of Jacapa has a beautiful velvety purple and black plumage, the beak being partly white, whilst the same sex in Episcopus is of a pale blue colour, with white spots on the wings. In their habits they both resemble the common house-sparrow of Europe, which does not exist in South America, its place being in some measure filled by these familiar tanagers. They are just as lively, restless, bold, and wary; their notes are very similar, chirping and inharmonious, and they seem to be almost as fond of the neighbourhood of man. They do not, however, build their nests on houses.

Another interesting and common bird was the Japím, a species of Cassicus (C. icteronotus). It belongs to the same family of birds as our starling, magpie, and rook, and has a rich yellow and black plumage, remarkably compact and velvety in texture. The shape of its head and its physiognomy are very similar to those of the magpie; it has light grey eyes, which give it the same knowing expression. It is social in its habits; and builds its nest, like the English rook, on trees in the neighbourhood of habitations. But the nests are quite differently constructed, being shaped like purses, two feet in length, and suspended from the slender branches all round the tree, some of them very near the ground. The entrance is on the side near the bottom of the nest. The bird is a great favourite with the Brazilians of Pará: it is a noisy, stirring babbling creature, passing constantly to and fro, chattering to its comrades, and is very ready at imitating other birds, especially the domestic poultry of the vicinity. There was at one time a weekly newspaper published at Pará, called "The Japím;" the name being chosen, I suppose, on account of the babbling propensities of the bird. Its eggs are nearly round, and of a bluish-white colour, speckled with brown.

Of other vertebrate animals we saw very little, except of the lizards. They are sure to attract the attention of the new-comer from Northern Europe, by reason of their strange appearance, great numbers, and variety. The species which are seen crawling over the walls of buildings in the city, are different from those found in the forest or in the interior of houses. They are unpleasant-looking animals, with colours assimilated to those of the dilapidated stone and mud walls on which they are seen. The house lizards belong to a peculiar family, the Geckos, and are found even in the best-kept chambers, most frequently on the walls and ceilings, to which they cling motionless by day, being active only at night. They are of speckled grey or ashy colours. The structure of their feet is beautifully adapted for clinging to and running over smooth surfaces; the underside of their toes being expanded into cushions, beneath which folds of skin form a series of flexible plates. By means of this apparatus they can walk or run across a smooth ceiling with their backs downwards; the plated soles, by quick muscular action, exhausting and admitting air alternately. The Geckos are very repulsive in appearance. The Brazilians give them the name of Osgas, and firmly believe them to be poisonous; they are, however, harmless creatures. Those found in houses are small; but I have seen others of great size, in crevices of tree trunks in the forest. Sometimes Geckos are found with forked tails; this results from the budding of a rudimentary tail at the side, from an injury done to the member. A slight rap will cause their tails to snap off; the loss being afterwards partially repaired by a new growth. The tails of lizards seem to be almost useless appendages to the animals. I used often to amuse myself in the suburbs, whilst resting in the verandah of our house during the heat of mid-day, by watching the variegated green, brown, and yellow ground-lizards. They would come nimbly forward, and commence grubbing with their fore feet and snouts around the roots of herbage, searching for insect larvæ. On the slightest alarm they would scamper off; their tails cocked up in the air as they waddled awkwardly away, evidently an incumbrance to them in their flight.

Next to the birds and lizards, the insects of the suburbs of Pará deserve a few remarks. I will pass over the many other orders and families of this class, and proceed at once to the ants. These were in great numbers everywhere, but I will mention here only two kinds. We were amazed at seeing ants an inch and a quarter in length, and stout in proportion, marching in single file through the thickets. These belonged to the species called Dinoponera grandis. Its colonies consist

Saüba or Leaf-carrying Ant.—1. Worker-minor; 2. Worker-major; 3. Subterranean worker.

of a small number of individuals, and are established about the roots of slender trees. It is a stinging species, but the sting is not so severe as in many of the smaller kinds. There was nothing peculiar or attractive in the habits of this giant among the ants. Another far more interesting species was the Saüba (Œcodoma cephalotes). This ant is seen everywhere about the suburbs, marching to and fro in broad columns. From its habit of despoiling the most valuable cultivated trees of their foliage, it is a great scourge to the Brazilians. In some districts it is so abundant that agriculture is almost impossible, and everywhere complaints are heard of the terrible pest.

The workers of this species are of three orders, and vary in size from two to seven lines; some idea of them may be obtained from the accompanying woodcut. The true working-class of a colony is formed by the small-sized order of workers, the worker-minors as they are called (Fig. 1). The two other kinds, whose functions, as we shall see, are not yet properly understood, have enormously swollen and massive heads; in one (Fig. 2), the head is highly polished; in the other (Fig. 3), it is opaque and hairy. The worker-minors vary greatly in size, some being double the bulk of others. The entire body is of very solid consistence, and of a pale reddish-brown colour. The thorax or middle segment is armed with three pairs of sharp spines; the head, also, has a pair of similar spines proceeding from the cheeks behind.

In our first walks we were puzzled to account for large mounds of earth, of a different colour from the surrounding soil, which were thrown up in the plantations and woods. Some of them were very extensive, being forty yards in circumference, but not more than two feet in height. We soon ascertained that these were the work of the Saüba, being the outworks, or domes, which overlie and protect the entrances to their vast subterranean galleries. On close examination, I found the earth of which they are composed to consist of very minute granules, agglomerated without cement, and forming many rows of little ridges and turrets. The difference in colour from the superficial soil of the vicinity is owing to their being formed of the undersoil, brought up from a considerable depth. It is very rarely that the ants are seen at work on these mounds; the entrances seem to be generally closed; only now and then, when some particular work is going on, are the galleries opened. The entrances are small and numerous; in the large hillocks it would require a great amount of excavation to get at the main galleries; but I succeeded in removing portions of the dome in smaller hillocks, and then I found that the minor entrances converged, at the depth of about two feet, to one broad elaborately-worked gallery or mine, which was four or five inches in diameter.

This habit in the Saüba ant of clipping and carrying away immense quantities of leaves has long been recorded in books on natural history. When employed on this work, their processions look like a multitude of animated leaves on the march. In some places I found an accumulation of such leaves, all circular pieces, about the size of a sixpence, lying on the pathway, unattended by ants, and at some distance from any colony. Such heaps are always found to be removed when the place is revisited the next day. In course of time I had plenty of opportunities of seeing them at work. They mount the tree in multitudes, the individuals being all worker-minors. Each one places itself on the surface of a leaf, and cuts with its sharp scissor-like jaws a nearly semicircular incision on the upper side; it then takes the edge between its jaws, and by a sharp jerk detaches the piece. Sometimes they let the leaf drop to the ground, where a little heap accumulates, until carried off by another relay of workers; but, generally, each marches off with the piece it has operated upon, and as all take the same road to their colony, the path they follow becomes in a short time smooth and bare, looking like the impression of a cart-wheel through the herbage.

It is a most interesting sight to see the vast host of busy diminutive labourers occupied on this work. Unfortunately they choose cultivated trees for their purpose. This ant is quite peculiar to Tropical America, as is the entire genus to which it belongs; it sometimes despoils the young trees of species growing wild in its native forests; but seems to prefer, when within reach, plants imported from other countries, such as the coffee and orange trees. It has not hitherto been shown satisfactorily to what use it applies the leaves. I discovered this only after much time spent in investigation. The leaves are used to thatch the domes which cover the entrances to their subterranean dwellings, thereby protecting from the deluging rains the young broods in the nests beneath. The larger mounds, already described, are so extensive that few persons would attempt to remove them for the purpose of examining their interior; but smaller hillocks, covering other entrances to the same system of tunnels and chambers, may be found in sheltered places, and these are always thatched with leaves, mingled with granules of earth. The heavily-laden workers, each carrying its segment of leaf vertically, the lower edge secured in its mandibles, troop up and cast their burthens on the hillock; another relay of labourers place the leaves in position, covering them with a layer of earthy granules, which are brought one by one from the soil beneath.

The underground abodes of this wonderful ant are known to be very extensive. The Rev. Hamlet Clark has related that the Saüba of Rio de Janeiro, a species closely allied to ours, has excavated a tunnel under the bed of the river Parahyba, at a place where it is as broad as the Thames at London Bridge. At the Magoary rice mills, near Pará, these ants once pierced the embankment of a large reservoir: the great body of water which it contained escaped before the damage could be repaired. In the Botanic Gardens, at Pará, an enterprising French gardener tried all he could think of to extirpate the Saüba. With this object he made fires over some of the main entrances to their colonies, and blew the fumes of sulphur down the galleries by means of bellows. I saw the smoke issue from a great number of outlets, one of which was 70 yards distant from the place where the bellows were used. This shows how extensively the underground galleries are ramified.

Besides injuring and destroying young trees by despoiling them of their foliage, the Saüba ant is troublesome to the inhabitants from its habit of plundering the stores of provisions in houses at night, for it is even more active by night than in the day-time. At first I was inclined to discredit the stories of their entering habitations and carrying off grain by grain the farinha or mandioca meal, the bread of the poorer classes of Brazil. At length, while residing at an Indian village on the Tapajos, I had ample proof of the fact. One night my servant woke me three or four hours before sunrise by calling out that the rats were robbing the farinha baskets; the article at that time being scarce and dear. I got up, listened, and found the noise was very unlike that made by rats. So I took the light and went into the store-room, which was close to my sleeping-place. I there found a broad column of Saüba ants, consisting of thousands of individuals, as busy as possible, passing to and fro between the door and my precious baskets. Most of those passing outwards were laden each with a grain of farinha, which was, in some cases, larger and many times heavier than the bodies of the carriers. Farinha consists of grains of similar size and appearance to the tapioca of our shops; both are products of the same root, tapioca being the pure starch, and farinha the starch mixed with woody fibre, the latter ingredient giving it a yellowish colour. It was amusing to see some of the dwarfs, the smallest members of their family, staggering along, completely hidden under their load. The baskets, which were on a high table, were entirely covered with ants, many hundreds of whom were employed in snipping the dry leaves which served as lining. This produced the rustling sound which had at first disturbed us. My servant told me that they would carry off the whole contents of the two baskets (about two bushels) in the course of the night, if they were not driven off; so we tried to exterminate them by killing them with our wooden clogs. It was impossible, however, to prevent fresh hosts coming in as fast as we killed their companions. They returned the next night; and I was then obliged to lay trains of gunpowder along their line, and blow them up. This, repeated many times, at last seemed to intimidate them, for we were free from their visits during the remainder of my residence at the place. What they did with the hard dry grains of mandioca I was never able to ascertain, and cannot even conjecture. The meal contains no gluten, and therefore would be useless as cement. It contains only a small relative portion of starch, and, when mixed with water, it separates and falls away like so much earthy matter. It may serve as food for the subterranean workers. But the young or larvæ of ants are usually fed by juices secreted by the worker nurses.

Ants, it is scarcely necessary to observe, consist, in each species, of three sets of individuals, or, as some express it, of three sexes—namely, males, females, and workers; the last-mentioned being undeveloped females. The perfect sexes are winged on their first attaining the adult state; they alone propagate their kind, flying away, previous to the act of reproduction, from the nest in which they have been reared. This winged state of the perfect males and females, and the habit of flying abroad before pairing, are very important points in the economy of ants; for they are thus enabled to intercross with members of distant colonies which swarm at the same time, and thereby increase the vigour of the race, a proceeding essential to the prosperity of any species. In many ants, especially those of tropical climates, the workers, again, are of two classes, whose structure and functions are widely different. In some species they are wonderfully unlike each other, and constitute two well-defined forms of workers. In others, there is a gradation of individuals between the two extremes. The curious differences in structure and habits between these two classes form an interesting, but very difficult, study. It is one of the great peculiarities of the Saüba ant to possess *three* classes of workers. My investigations regarding them were far from complete; I will relate, however, what I have observed on the subject.

When engaged in leaf-cutting, plundering farinha, and other operations, two classes of workers are always seen (Figs. 1 and 2, page 9). They are not, it is true, very sharply defined in structure, for individuals of intermediate grades occur. All the work, however, is done by the individuals which have small heads (Fig. 1), whilst those which have enormously large heads, the worker-majors (Fig. 2), are observed to be simply walking about. I could never satisfy myself as to the function of these worker-majors. They are not the soldiers or defenders of the working portion of the community, like the armed class in the Termites, or white ants; for they never fight. The species has no sting, and does not display active resistance when interfered with. I once imagined they exercised a sort of superintendence over the others; but this function is entirely unnecessary in a community where all work with a precision and regularity resembling the subordinate parts of a piece of machinery. I came to the conclusion, at last, that they have no very precisely defined function. They cannot, however, be entirely useless to the community, for the sustenance of an idle class of such bulky individuals would be too heavy a charge for the species to sustain. I think they serve, in some sort, as passive instruments of protection to the real workers. Their enormously large, hard, and indestructible heads may be of use in protecting them against the attacks of insectivorous animals. They would be, on this view, a kind of "pièces de resistance," serving as a foil against onslaughts made on the main body of workers.

The third order of workers is the most curious of all. If the top of a small fresh hillock, one in which the thatching process is going on, be taken off, a broad cylindrical shaft is disclosed, at a depth of about two feet from the surface. If this be probed with a stick, which may be done to the extent of three or four feet without touching bottom, a small number of colossal fellows (Fig. 3) will slowly begin to make their way up the smooth sides of the mine. Their heads are of the same size as those of the class Fig. 2; but the front is clothed with hairs, instead of being polished, and they have in the middle of the forehead a twin ocellus, or simple eye, of quite different structure from the ordinary compound eyes, on the sides of the head. This frontal eye is totally wanting in the other workers, and is not known in any other kind of ant. The apparition of these strange creatures from the cavernous depths of the mine reminded me, when I first observed them, of the Cyclopes of Homeric fable. They were not very pugnacious, as I feared they would be, and I had no difficulty in securing a few with my fingers. I never saw them under any other circumstances than those here related, and what their special functions may be I cannot divine.

The whole arrangement of a Formicarium, or ant-colony, and all the varied activity of ant-life, are directed to one main purpose — the perpetuation and dissemination of the species. Most of the labour which we see performed by the workers has for its end the sustenance and welfare of the young brood, which are helpless grubs. The true females are incapable of attending to the wants of their offspring; and it is on the poor sterile workers, who are denied all the other pleasures of maternity, that the entire care devolves. The workers are also the chief agents in carrying out the different migrations of the colonies, which are of vast importance to the dispersal and consequent prosperity of the species. The successful *début* of

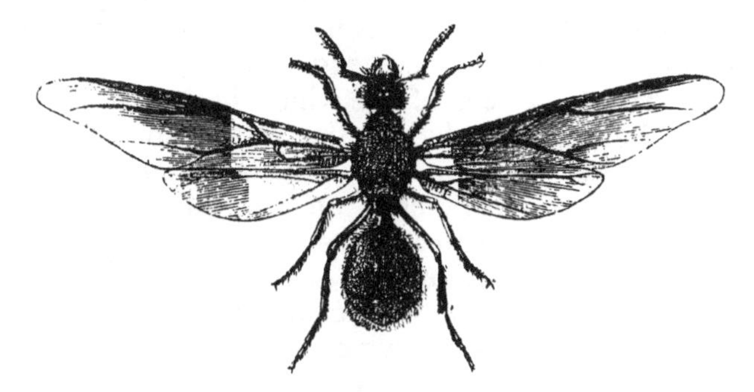

Saüba Ant.—Female.

the winged males and females depends likewise on the workers. It is amusing to see the activity and excitement which reign in an ant's nest when the exodus of the winged individuals is taking place. The workers clear the roads of exit, and show the most lively interest in their departure, although it is highly improbable that any of them will return to the same colony. The swarming or exodus of the winged males and females of the Saüba ant takes place in January and February, that is, at the commencement of the rainy season. They come out in the evening in vast numbers, causing quite a commotion in the streets and lanes. They are of very large size, the female measuring no less than two inches and a quarter in expanse of wing; the male is not much more than half this size. They are so eagerly preyed upon by insectivorous animals, that on the morning after their flight not an individual is to be seen, a few impregnated females alone escaping the slaughter to found new colonies.

At the time of our arrival, Pará had not quite recovered from the effects of a series of revolutions, brought about by the hatred which existed between the native Brazilians and the Portuguese; the former, in the end, calling to their aid the Indian and mixed coloured population. The number of inhabitants of the city had decreased, in consequence of these disorders, from 24,500 in 1819, to 15,000 in 1848. Although the public peace had not been broken for twelve years before the date of our visit, confidence was not yet completely restored, and the Portuguese merchants and tradesmen would not trust themselves to live at their beautiful country-houses or rocinhas, which lie embosomed in the luxuriant shady gardens around the city. No progress had been made in clearing the second-growth forest, which had grown over the once cultivated grounds and now reached the end of all the suburban streets. The place had the aspect of one which had seen better days; the public buildings, including the palaces of the President and Bishop, the cathedral, the principal churches and convents, all seemed constructed on a scale of grandeur far beyond the present requirements of the city. Streets full of extensive private residences, built in the Italian style of architecture, were in a neglected condition, weeds and flourishing young trees growing from large cracks in the masonry. The large public squares were overgrown with weeds, and impassable on account of the swampy places which occupied portions of their areas. Commerce, however, was now beginning to revive, and before I left the country I saw great improvements, as I shall have to relate towards the conclusion of this narrative.

The province of which Pará is the capital was, at the time I allude to, the most extensive in the Brazilian empire, being about 1560 miles in length from east to west, and about 600 in breadth. Since that date—namely, in 1853—it has been divided into two by the separation of the Upper Amazons as a distinct province. It formerly constituted a section, capitania, or governorship of the Portuguese colony. Originally it was well peopled by Indians, varying much in social condition according to their tribe, but all exhibiting the same general physical characters, which are those of the American red man, somewhat modified by long residence in an equatorial forest country. Most of the tribes are now extinct or forgotten, at least those which originally peopled the banks of the main river, their descendants having amalgamated with the white and negro immigrants:* many still exist, however, in their original state on the Upper Amazons and most of the branch rivers. On this account Indians in this province are far more numerous than elsewhere in Brazil, and the Indian element may be said to prevail in the mongrel population, the negro proportion being much smaller than in South Brazil.

The city is built on the best available site for a port of entry to the Amazons region, and must in time become a vast emporium; for the northern shore of the main river, where alone a rival capital could be founded, is much more difficult of access to vessels, and is besides extremely unhealthy. Although lying so near the equator (1° 28′ S. lat.) the climate is not excessively hot. The temperature during three years only once reached 95° of Fahrenheit. The greatest heat of the day, about 2 p.m., ranges generally between 89° and 94°; but on the other hand, the air is never cooler than 73°, so that a uniformly high temperature exists, and the mean of the year is 81°. North American residents say that the hea is not so oppressive as it is in summer in New York and Philadelphia. The humidity is, of course, excessive, but the rains are not so heavy and continuous in the wet season as in many other tropical climates. The country had for a long time a reputation for extreme salubrity. Since the small-pox in 1819, which attacked chiefly the Indians, no serious epidemic had visited the province. We were agreeably surprised to find no danger from exposure to the night air or residence in the low swampy lands. A few English residents, who had been established here for twenty or thirty years, looked almost as fresh in colour as if they had never left their native country. The native women, too, seemed to preserve their good looks and plump condition until late in life. I nowhere observed that early decay of appearance in Brazilian ladies, which is said to be so general in the women of North America. Up to 1848 the salubrity of Pará was quite remarkable for a city lying in the delta of a great river in the middle of the tropics and half

* The mixed breeds which now form, probably, the greater part of the population, have each a distinguishing name. Mameluco denotes the offspring of White with Indian; Mulatto, that of White with Negro; Cafuzo, the mixture of the Indian and Negro; Curiboco, the cross between the Cafuzo and the Indian; Xibaro, that between the Cafuzo and Negro. These are seldom, however, well-demarcated, and all shades of colour exist; the names are generally applied only approximatively. The term Creole is confined to negroes born in the country. The civilised Indian is called Tapuyo or Caboclo.

surrounded by swamps. It did not much longer enjoy its immunity from epidemics. In 1850 the yellow fever visited the province for the first time, and carried off in a few weeks more than four per cent. of the population.

The province of Pará, or as we may now say, the two provinces of Pará and the Amazons, contain an area of 800,000 square miles; the population of which is only about 230,000, or in the ratio of one person to four square miles! The country is covered with forests, and the soil fertile in the extreme, even for a tropical country. It is intersected throughout by broad and deep navigable rivers. It is the pride of the Paraenses to call the Amazons the Mediterranean of South America. The colossal stream perhaps deserves the name, for not only have the main river and its principal tributaries an immense expanse of water, bathing the shores of extensive and varied regions, but there is also throughout a system of back channels, connected with the main rivers by narrow outlets, and linking together a series of lakes, some of which are fifteen, twenty, and thirty miles in length. The whole Amazons valley is thus covered by a network of navigable waters, forming a vast inland freshwater sea with endless ramifications, rather than a river.

I resided at Pará nearly a year and a half altogether, returning thither and making a stay of a few months after each of my shorter excursions into the interior; until the 6th of November, 1851, when I started on my long voyage to the Tapajos and the Upper Amazons, which occupied me seven years and a half.

CHAPTER II.

PARÁ.

The swampy forests of Pará—A Portuguese landed proprietor—Country house at Nazareth—Life of a Naturalist under the equator—The drier virgin forests—Magoary—Retired creeks—Aborigines.

After having resided about a fortnight at Mr. Miller's rocinha, we heard of another similar country-house to be let, much better situated for our purpose, in the village of Nazareth, a mile and a half from the city, and close to the forest. The owner was an old Portuguese gentleman named Danin, who lived at his tile manufactory at the mouth of the Una, a small river lying two miles below Pará. We resolved to walk to his place through the forest, a distance of three miles, although the road was said to be scarcely passable at this season of the year, and the Una much more easily accessible by boat. We were glad, however, of this early opportunity of traversing the rich swampy forest, which we had admired so much from the deck of the ship; so, about eleven o'clock one sunny morning, after procuring the necessary information about the road, we set off in that direction. This part of the forest afterwards became one of my best hunting-grounds. I will narrate the incidents of the walk, giving my first impressions and some remarks on the wonderful vegetation. The forest is very similar on most of the low lands, and therefore one description will do for all.

On leaving the town, we walked along a straight suburban road, constructed above the level of the surrounding land. It had low swampy ground on each side, built upon, however, and containing several spacious rocinhas, which were embowered in magnificent foliage. Leaving the last of these, we arrived at a part where the lofty forest towered up like a wall, five or six yards from the edge of the path, to the height of, probably,

100 feet. The tree trunks were only seen partially here and there, nearly the whole frontage from ground to summit being covered with a diversified drapery of creeping plants, all of the most vivid shades of green; scarcely a flower to be seen, except in some places a solitary scarlet passion-flower, set in the green mantle like a star. The low ground on the borders, between the forest wall and the road, was encumbered with a tangled mass of bushy and shrubby vegetation, amongst which prickly mimosas were very numerous, covering the other bushes in the same way as brambles do in England. Other dwarf mimosas trailed along the ground close to the edge of the road, shrinking at the slightest touch of the feet as we passed by. Cassia trees, with their elegant pinnate foliage and conspicuous yellow flowers, formed a great proportion of the lower trees, and arborescent aruns grew in groups around the swampy hollows. Over the whole fluttered a larger number of brilliantly-coloured butterflies than we had yet seen; some wholly orange or yellow (Callidryas), others with excessively elongated wings, sailing horizontally through the air, coloured black, and varied with blue, red, and yellow (Heliconii). One magnificent grassy-green species (Colænis Dido) especially attracted our attention. Near the ground hovered many other smaller species very similar in appearance to those found at home, attracted by the flowers of numerous leguminous and other shrubs. Besides butterflies, there were few other insects except dragonflies, which were in great numbers, similar in shape to English species, but some of them looking conspicuously different on account of their fiery red colours.

After stopping a long time to examine and admire, we at length walked onward. The road then ascended slightly, and the soil and vegetation became suddenly altered in character. The shrubs here were grasses, low sedges and other plants, smaller in foliage than those growing in moist grounds. The forest was second growth, low, consisting of trees which had the general aspect of laurels and other evergreens in our gardens at home: the leaves glossy and dark green. Some of them were elegantly veined and hairy (Melastomæ). whilst many, scattered amongst the rest had smaller foliage (Myrtles), but these were not sufficient to subtract much from the general character of the whole.

The sun, now, for we had loitered long on the road, was exceedingly powerful. The day was most brilliant; the sky without a cloud. In fact, it was one of those glorious days which announce the commencement of the dry season. The radiation of heat from the sandy ground was visible by the quivering motion of the air above it. We saw or heard no mammals or birds; a few cattle belonging to an estate down a shady lane were congregated, panting, under a cluster of wide-spreading trees. The very. soil was hot to our feet; and we hastened onward to the shade of the forest which we could see not far ahead. At length, on entering it, what a relief! We found ourselves in a moderately broad pathway or alley, where the branches of the trees crossed overhead and produced a delightful shade. The woods were at first of recent growth, dense, and utterly impenetrable; the ground, instead of being clothed with grass and shrubs as in the woods of Europe, was everywhere carpeted with Lycopodiums (fern-shaped mosses). Gradually the scene became changed. We descended slightly from an elevated, dry, and sandy area to a low and swampy one; a cool air breathed on our faces, and a mouldy smell of rotting vegetation greeted us. The trees were now taller, the underwood less dense, and we could obtain glimpses into the wilderness on all sides. The leafy crowns of the trees, scarcely two of which could be seen together of the same kind, were now far away above us, in another world as it were. We could only see at times, where there was a break above, the tracery of the foliage against the clear blue sky. Sometimes the leaves were palmate, or of the shape of large outstretched hands; at others, finely cut or feathery, like the leaves of Mimosæ. Below, the tree trunks were everywhere linked together by sipós; the woody flexible stems of climbing and creeping trees, whose foliage is far away above, mingled with that of the taller independent trees. Some were twisted in strands like cables, others had thick stems contorted in every variety of shape, entwining snake-like round the tree trunks, or forming gigantic loops and coils among the larger branches; others, again, were of zigzag shape, or indented like the steps of a staircase, sweeping from the ground to a giddy height.

It interested me much afterwards to find that these climbing trees do not form any particular family. There is no distinct group of plants whose especial habit is to climb, but species of many and the most diverse families, the bulk of whose members are not climbers, seem to have been driven by circumstances to adopt this habit. There is even a climbing genus of palms (Desmoncus), the species of which are called, in the Tupí language, Jacitára. These have slender, thickly-spined, and

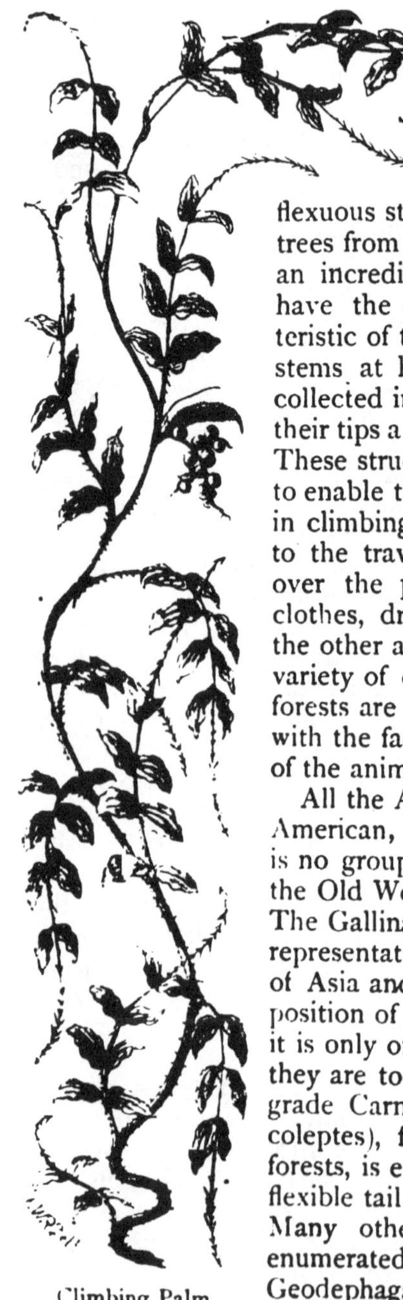
Climbing Palm (Desmoncus).

flexuous stems, which twine about the taller trees from one to the other, and grow to an incredible length. The leaves, which have the ordinary pinnate shape characteristic of the family, are emitted from the stems at long intervals, instead of being collected into a dense crown, and have at their tips a number of long recurved spines. These structures are excellent contrivances to enable the trees to secure themselves by in climbing, but they are a great nuisance to the traveller, for they sometimes hang over the pathway, and catch the hat or clothes, dragging off the one or tearing the other as he passes. The number and variety of climbing trees in the Amazons forests are interesting, taken in connection with the fact of the very general tendency of the animals also to become climbers.

All the Amazonian, and in fact all South American, monkeys are climbers. There is no group answering to the baboons of the Old World, which live on the ground. The Gallinaceous birds of the country, the representatives of the fowls and pheasants of Asia and Africa, are all adapted by the position of the toes to perch on trees, and it is only on trees, at a great height, that they are to be seen. A genus of Plantigrade Carnivora, allied to the bears (Cercoleptes), found only in the Amazonian forests, is entirely arboreal, and has a long flexible tail like that of certain monkeys. Many other similar instances could be enumerated, but I will mention only the Geodephaga, or carnivorous ground beetles, a great proportion of whose genera and species in these forest regions are, by the structure of their feet, fitted to live exclusively on the branches and leaves of trees.

Many of the woody lianas suspended from trees are not climbers, but the air-roots of epiphytous plants (Aroideæ), which sit on the stronger boughs of the trees above, and hang down straight as plumb-lines. Some are suspended singly, others in clusters; some reach half-way to the ground and others touch it, striking their rootlets into the earth. The underwood in this part of the forest was composed partly of younger trees of the same species as their taller neighbours, and partly of palms of many species, some of them twenty to thirty feet in height, others small and delicate, with stems no thicker than a finger. These latter (different kinds of Bactris) bore small bunches of fruit, red or black, often containing a sweet grape-like juice.

Further on the ground became more swampy, and we had some difficulty in picking our way. The wild banana (Urania Amazonica) here began to appear, and, as it grew in masses, imparted a new aspect to the scene. The leaves of this beautiful plant are like broad sword-blades, eight feet in length and a foot broad; they rise straight upwards, alternately, from the top of a stem five or six feet high. Numerous kinds of plants with leaves similar in shape to these, but smaller, clothed the ground. Amongst them were species of Marantaceæ, some of which had broad glossy leaves, with long leaf-stalks radiating from joints in a reed-like stem. The trunks of the trees were clothed with climbing ferns, and Pothos plants with large, fleshy, heart-shaped leaves. Bamboos and other tall grass and reed-like plants arched over the pathway. The appearance of this part of the forest was strange in the extreme; description can convey no adequate idea of it. The reader who has visited Kew may form some notion by conceiving a vegetation like that in the great palm-house spread over a large tract of swampy ground, but he must fancy it mingled with large exogenous trees similar to our oaks and elms covered with creepers and parasites, and figure to himself the ground encumbered with fallen and rotten trunks, branches, and leaves; the whole illuminated by a glowing vertical sun, and reeking with moisture.

We at length emerged from the forest, on the banks of the Una, near its mouth. It was here about one hundred yards wide. The residence of Senhor Danin stood on the opposite shore; a large building, whitewashed and red-tiled as usual, raised on wooden piles above the humid ground. The second story was the part occupied by the family, and along it was an open verandah, where people, male and female, were at work. Below were several negroes employed carrying clay on their heads. We called out for a boat, and one of them crossed over to fetch us. Senhor Danin received us with the usual formal politeness of the Portuguese; he spoke English very well, and after we had arranged our business we remained conversing with him on various subjects connected with the country. Like all employers in this province, he was full of one topic—the scarcity of hands. It appeared that he had made great exertions to introduce white labour, but had failed, after having brought numbers of men from Portugal and other countries under engagement to work for him. They all left him one by one soon after their arrival. The abundance of unoccupied land, the liberty that exists, a state of things produced by the half-wild canoe-life of the people, and the ease with which a mere subsistence can be obtained with moderate work, tempt even the best-disposed to quit regular labour as soon as they can.

Shortly afterwards we took possession of our new residence. The house was a square building, consisting of four equal-sized rooms; the tiled roof projected all round, so as to form a broad verandah, cool and pleasant to sit and work in. The cultivated ground, which appeared as if newly cleared from the forest, was planted with fruit trees and small plots of coffee and mandioca. The entrance to the grounds was by an iron-grille gateway from a grassy square, around which were built the few houses and palm-thatched huts which then constituted the village. The most important building was the chapel of our Lady of Nazareth, which stood opposite our place. The saint here enshrined was a great favourite with all orthodox Paraenses, who attributed to her the performance of many miracles. The image was to be seen on the altar, a handsome doll about four feet high, wearing a silver crown and a garment of blue silk, studded with golden stars. In and about the chapel were the offerings that had been made to her, proofs of the miracles which she had performed. There were models of legs, arms, breasts, and so forth, which she had cured. But most curious of all was a ship's boat, deposited here by the crew of a Portuguese vessel which had foundered, a year or two before our arrival, in a squall off Cayenne; part of them having been saved in the boat, after invoking the protection of the saint here enshrined. The annual festival in honour of our Lady of Nazareth is the greatest of the Pará holidays; many persons come to it from the neighbouring city of Maranham, 300 miles distant. Once the president ordered the mail steamer to be delayed two days at Pará for the convenience of these visitors. The popularity of the festa is partly owing to the beautiful weather that prevails when it takes place, namely, in the middle of the fine season, on the ten days preceding the full moon in October or November. Pará is then seen at its best. The weather is not too dry, for three weeks never follow in succession without a shower: so that all the glory of verdure and flowers can be enjoyed with clear skies. The moonlit nights are then especially beautiful; the atmosphere is transparently clear, and the light sea-breeze produces an agreeable coolness.

We now settled ourselves for a few months' regular work. We had the forest on three sides of us; it was the end of the wet season; most species of birds had finished moulting, and every day the insects increased in number and variety. Behind the rocinha, after several days' exploration, I found a series of pathways through the woods, which led to the Una road; about half-way was the house in which the celebrated travellers Spix and Martius resided during their stay at Pará, in 1819. It was now in a neglected condition, and the plantations were overgrown with bushes. The paths hereabout were very productive of insects, and being entirely under shade were very pleasant for strolling. Close to our doors began the main forest road. It was broad enough for two horsemen abreast, and branched off in three directions; the main line going to the village of Ourem, a distance of 50 miles. This road formerly extended to Maranham, but it had been long in disuse, and was now grown up, being scarcely passable between Pará and Ourem.

Our researches were made in various directions along these paths, and every day produced us a number of new and interesting species. Collecting, preparing our specimens, and making notes, kept us well occupied. One day was so much like another, that a general description of the diurnal round of incidents, including the sequence of natural phenomena, will be sufficient to give an idea of how days pass to naturalists under the equator.

We used to rise soon after dawn, when Isidoro would go down to the city, after supplying us with a cup of coffee, to purchase the fresh provisions for the day. The two hours before breakfast were devoted to ornithology. At that early period of the day the sky was invariably cloudless (the thermometer marking 72° or 73° Fahr.): the heavy dew or the previous night's rain, which lay on the moist foliage, becoming quickly dissipated by the glowing sun, which rising straight out of the east, mounted rapidly towards the zenith. All nature was fresh, new leaf and flower-buds expanding rapidly. Some mornings a single tree would appear in flower amidst what was the preceding evening a uniform green mass of forest—a dome of blossom suddenly created as if by magic. The birds were all active; from the wild-fruit trees, not far off, we often heard the shrill yelping of the Toucans (Ramphastos vitellinus). Small flocks of parrots flew over on most mornings, at a great height, appearing in distinct relief against the blue sky, always two by two, chattering to each other, the pairs being separated by regular intervals; their bright colours, however, were not apparent at that height. After breakfast we devoted the hours from 10 a.m. to 2 or 3 p.m. to entomology; the best time for insects in the forest being a little before the greatest heat of the day.

The heat increased rapidly towards two o'clock (92° and 93° Fahr.), by which time every voice of bird or mammal was hushed; only in the trees was heard at intervals the harsh whirr of a cicada. The leaves, which were so moist and fresh in early morning, now become lax and drooping; the flowers shed their petals. Our neighbours, the Indian and Mulatto inhabitants of the open palm-thatched huts, as we returned home fatigued with our ramble, were either asleep in their hammocks or seated on mats in the shade, too languid even to talk. On most days in June and July a heavy shower would fall some time in the afternoon, producing a most welcome coolness. The approach of the rain-clouds was after a uniform fashion very interesting to observe. First, the cool sea-breeze, which commenced to blow about 10 o'clock, and which had increased in force with the increasing power of the sun, would flag and finally die away. The heat and electric tension of the atmosphere would then become almost insupportable. Languor and uneasiness would seize on every one; even the denizens of the forest betraying it by their motions. White clouds would appear in the east and gather into cumuli, with an increasing blackness along their lower portions. The whole eastern horizon would become almost suddenly black, and this would spread upwards, the sun at length becoming obscured. Then the rush of a mighty wind is heard through the forest, swaying the tree-tops; a vivid flash of lightning bursts forth, then a crash of thunder, and down streams the deluging rain. Such storms soon cease, leaving bluish-black motionless clouds in the sky until night. Meantime all nature is refreshed; but heaps of flower-petals and fallen leaves are seen under the trees. Towards evening life revives again, and the ringing uproar is resumed from bush and tree. The following morning the sun again rises in a cloudless sky, and so the cycle is completed; spring, summer, and autumn, as it were, in one tropical day. The days are more or less like this throughout the year in this country. A little difference exists between the dry and wet seasons; but generally, the dry season, which lasts from July to December, is varied with showers, and the wet, from January to June, with sunny days. It results from this, that the periodical phenomena of plants and animals do not take place at about the same time in all species, or in the individuals of any given species, as they do in temperate countries. Of course there is no hybernation; nor, as the dry season is not excessive, is there any summer torpidity as in some tropical countries. Plants do not flower or shed their leaves, nor do birds moult, pair, or breed simultaneously. In Europe, a woodland scene has its spring, its summer, its autumnal, and its winter aspects. In the equatorial forests the aspect is the same or nearly so every day in the year: budding, flowering, fruiting, and leaf-shedding are always going on in one species or other. The activity of birds and insects proceeds without interruption, each species having its own separate times; the colonies of wasps, for instance, do not die off annually, leaving only the queens, as in cold climates; but the succession of generations and colonies goes on incessantly. It is never either spring, summer, or autumn, but each day is a combination of all three. With the day and night always of equal length, the atmospheric disturbances of each day neutralising themselves before each succeeding morn; with the sun in its course proceeding mid-way across the sky, and the daily temperature the same within two or three degrees throughout the year—how grand in its perfect equilibrium and simplicity is the march of Nature under the equator!

Our evenings were generally fully employed preserving our collections, and making notes. We dined at four, and took tea about seven o'clock. Sometimes we walked to the city to see Brazilian life or enjoy the pleasures of European and American society. And so the time passed away from June 15th to August 26th. During this period we made two excursions of greater length to the rice and saw-mills of Magoary, an establishment owned by an American gentleman, Mr. Upton, situated on the banks of a creek in the heart of the forest, about twelve miles from Pará. I will narrate some of the incidents of these excursions, and give an account of the more interesting observations made on the Natural History and inhabitants of these interior creeks and forests.

Our first trip to the mills was by land. The creek on whose banks they stand, the Iritirí, communicates with the river Pará through another larger creek, the Magoary; so that there is a passage by water, but this is about 20 miles round. We started at sunrise, taking Isidoro with us. The road plunged at once into the forest after leaving Nazareth, so that in a few minutes we were enveloped in shade. For some distance the woods were of second growth, the original forest near the town having been formerly cleared or thinned. They were dense and impenetrable on account of the close growth of the young trees and the mass of thorny shrubs and creepers. These thickets swarmed with ants and ant-thrushes : they were also frequented by a species of puff-throated manikin, a little bird which flies occasionally across the road, emitting a strange noise, made, I believe, with the wings, and resembling the clatter of a small wooden rattle.

A mile or a mile and a half further on, the character of the woods began to change, and we then found ourselves in the primæval forest. The appearance was greatly different from that of the swampy tract I have already described. The land was rather more elevated and undulating; the many swamp plants with their long and broad leaves were wanting, and there was less underwood, although the trees were wider apart. Through this wilderness the road continued for seven or eight miles. The same unbroken forest extends all the way to Maranham and in other directions, as we were told, a distance of about 300 miles southward and eastward of Pará. In almost every hollow part the road was crossed by a brook, whose cold, dark, leaf-stained waters were bridged over by tree trunks. The ground was carpeted, as usual, by Lycopodiums, but it was also encumbered with masses of vegetable *débris* and a thick coating of dead leaves. Fruits of many kinds were scattered about, amongst which were many sorts of beans, some of the pods a foot long, flat and leathery in texture, others hard as stone. In one place there was a quantity of large empty wooden vessels, which Isidoro told us fell from the Sapucaya tree. They are called monkeys' drinking-cups (Cuyas de Macaco), and are the capsules which contain the nuts sold under the name just mentioned, in Covent Garden Market. At the top of the vessel is a circular hole, in which a natural lid fits neatly. When the nuts are ripe, this lid becomes loosened, and the heavy cup falls with a crash, scattering the nuts over the ground. The tree which yields the nut (Lecythis ollaria), is of immense height. It is closely allied to the Brazil-nut tree (Bertholletia excelsa), whose seeds are also enclosed in large woody vessels; but these have no lid, and fall entire to the ground. This is the reason why the one kind of nut is so much dearer than the other. The Sapucaya is not less abundant, probably, than the Bertholletia, but its nuts in falling are scattered about and eaten by wild animals; whilst the full capsules of Brazil-nuts are collected entire by the natives.

What attracted us chiefly were the colossal trees. The general run of trees had not remarkably thick stems; the great and uniform height to which they grow without emitting a branch, was a much more noticeable feature than their thickness; but at intervals of a furlong or so a veritable giant towered up. Only one of these monstrous trees can grow within a given space; it monopolises the domain, and none but individuals of much inferior size can find a footing near it. The cylindrical trunks of these larger trees were generally about 20 to 25 feet in circumference. Von Martius mentions having measured trees in the Pará district, belonging to various species (Symphonia coccinea, Lecythis sp. and Cratæva Tapia), which were 50 to 60 feet in girth at the point where they become cylindrical. The height of the vast column-like stems could not be less than 100 feet from the ground to their lowest branch. Mr. Leavens, at the saw-mills, told me they frequently squared logs for sawing 100 feet long, of the Pao d'Arco and the Massaranduba. The total height of these trees, stem and crown together, may be estimated at from 180 to 200 feet: where one of them stands, the vast dome of foliage rises above the other forest trees as a domed cathedral does above the other buildings in a city.

A very remarkable feature in these trees is the growth of buttress-shaped projections around the lower part of their stems. The spaces between these buttresses, which are generally thin walls of wood, form spacious chambers, and may be compared to stalls in a stable: some of them are large enough to hold half a dozen persons. The purpose of these structures is as obvious, at the first glance, as that of the similar props of brick-work which support a high wall. They are not peculiar to one species, but are common to most of the larger forest trees. Their nature and manner of growth are explained when a series of young trees of different ages is examined. It is then seen that they are the roots which have raised themselves ridge-like out of the earth; growing gradually upwards as the increasing height of the tree required augmented support. Thus they are plainly intended to sustain the massive crown and trunk in these crowded forests, where lateral growth of the roots in the earth is rendered difficult by the multitude of competitors.

The other grand forest trees whose native names we learnt, were the Moira-tinga (the White or King-tree), probably the same as, or allied to, the Mora excelsa, which Sir Robert Schomburgk discovered in British Guiana; the Samaüma (Eriodendron Samauma) and the Massaranduba, or Cow-tree. The last-mentioned is the most remarkable. We had already heard a good deal about this tree, and about its producing from its bark a copious supply of milk as pleasant to drink as that of the cow. We had also eaten its fruit in Pará, where it is sold in the streets by negro market women; and had heard a good deal of the durableness in water of its timber. We were glad, therefore, to see this wonderful tree growing in its native wilds. It is one of the largest of the forest monarchs, and is peculiar in appearance on account of its deeply-scored, reddish, and ragged bark. A decoction of the bark, I was told, is used as a red dye for cloth. A few days afterwards we tasted its milk, which was drawn from dry logs that had been standing many days in the hot sun, at the saw-mills. It was pleasant with coffee, but had a slight rankness when drank pure; it soon thickens to a glue, which is excessively tenacious, and is often used to cement broken crockery. I was told that it was not safe to drink much of it, for a slave had recently nearly lost his life through taking it too freely.

In some parts of the road ferns were conspicuous objects. But I afterwards found them much more numerous on the Maranham road, especially in one place where the whole forest glade formed a vast fernery; the ground was covered with terrestrial species, and the tree trunks clothed with climbing and epiphytous kinds. I saw no tree ferns in the Pará district; they belong to hilly regions; some occur, however, on the Upper Amazons.

Such were the principal features in the vegetation of the wilderness; but where were the flowers? To our great disappointment we saw none, or only such as were insignificant in appearance. Orchids are very rare in the dense forests of the low lands. I believe it is now tolerably well ascertained that the majority of forest trees in equatorial Brazil have small and inconspicuous flowers. Flower-frequenting insects are also rare in the forest. Of course they would not be found where their favourite food was wanting, but I always noticed that even where flowers occurred in the forest, few or no insects were seen upon them. In the open country or campos of Santarem, on the Lower Amazons, flowering trees and bushes are more abundant, and there a large number of floral insects are attracted. The forest bees of South America belonging to the genera Melipona and Euglossa are more frequently seen feeding on the sweet sap which exudes from the trees, or on the excrement of birds on leaves, than on flowers.

We were disappointed also in not meeting with any of the larger animals in the forest. There was no tumultuous movement, or sound of life. We did not see or hear monkeys, and no tapir or jaguar crossed our path. Birds, also, appeared to be exceedingly scarce. We heard, however, occasionally the long-drawn, wailing note of the Inambú, a kind of partridge (Crypturus cinereus ?); and, also, in the hollows on the banks of the rivulets, the noisy notes of another bird, which seemed to go in pairs, amongst the tree-tops, calling to each other as they went. These notes resounded through the wilderness. Another solitary bird had a most sweet and melancholy song; it consisted simply of a few notes, uttered in a plaintive key, commencing high, and descending by harmonic intervals. It was probably a species of warbler of the genus Trichas. All these notes of birds are very striking and characteristic of the forest.

I afterwards saw reason to modify my opinion, founded on these first impressions, with regard to the amount and variety of animal life in this and other parts of the Amazonian forests. There is, in fact, a great variety of mammals, birds, and reptiles, but they are widely scattered, and all excessively shy of man. The region is so extensive, and uniform in the forest clothing of its surface, that it is only at long intervals that animals are seen in abundance, where some particular spot is found which is more attractive than others. Brazil, moreover, is throughout poor in terrestrial mammals, and the species are of small size; they do not, therefore, form a conspicuous feature in its forests. The huntsman would be disappointed who expected to find here flocks of animals similar to the buffalo herds of North America, or the swarms of antelopes and herds of ponderous pachyderms of Southern Africa. The largest and most interesting portion of the Brazilian mammal fauna is arboreal in its habits; this feature of the animal denizens of these forests I have already alluded to. The most *intensely* arboreal animals in the world are the South American monkeys of the family Cebidæ, many of which have a fifth hand for climbing in their prehensile tails, adapted for this function by their strong muscular development, and the naked palms under their tips. This seems to teach us that the South American fauna has been slowly adapted to a forest life, and, therefore, that extensive forests must have always existed since the region was first peopled by mammalia. But to this subject, and to the natural history of the monkeys, of which thirty-eight species inhabit the Amazon region, I shall have to return.

We often read, in books of travels, of the silence and gloom of the Brazilian forests. They are realities, and the impression deepens on a longer acquaintance. The few sounds of birds are of that pensive or mysterious character which intensifies the feeling of solitude rather than imparts a sense of life and cheerfulness. Sometimes, in the midst of the stillness, a sudden yell or scream will startle one; this comes from some defenceless fruit-eating animal, which is pounced upon by a tiger-cat or stealthy boa-constrictor. Morning and evening the howling monkeys make a most fearful and harrowing noise, under which it is difficult to keep up one's buoyancy of spirit. The feeling of inhospitable wildness which the forest is calculated to inspire, is increased tenfold under this fearful uproar. Often, even in the still hours of midday, a sudden crash will be heard resounding afar through the wilderness, as some great bough or entire tree falls to the ground. There are, besides, many sounds which it is impossible to account for. I found the natives generally as much at a loss in this respect as myself. Sometimes a sound is heard like the clang of an iron bar against a hard, hollow tree, or a piercing cry rends the air: these are not repeated, and the succeeding silence tends to heighten the unpleasant impression which they make on the mind. With the native it is always the Curupíra, the wild man or spirit of the forest, which produces all noises they are unable to explain. For myths are the rude theories which

INTERIOR OF PRIMÆVAL FOREST ON THE AMAZONS.

mankind, in the infancy of knowledge, invent to explain natural phenomena. The Curupíra is a mysterious being, whose attributes are uncertain, for they vary according to locality. Sometimes he is described as a kind of orang-otang, being covered with long shaggy hair, and living in trees. At others he is said to have cloven feet and a bright red face. He has a wife and children, and sometimes comes down to the roças to steal the mandioca. At one time I had a mameluco youth in my service, whose head was full of the legends and superstitions of the country. He always went with me into the forest; in fact, I could not get him to go alone, and whenever we heard any of the strange noises mentioned above, he used to tremble with fear. He would crouch down behind me, and beg of me to turn back; his alarm ceasing only after he had made a charm to protect us from the Curupíra. For this purpose he took a young palm leaf, plaited it, and formed it into a ring, which he hung to a branch on our track.

At length, after a six hours' walk, we arrived at our destination, the last mile or two having been again through second-growth forest. The mills formed a large pile of buildings, pleasantly situated in a cleared tract of land, many acres in extent, and everywhere surrounded by the perpetual forest. We were received in the kindest manner by the overseer, Mr. Leavens, who showed us all that was interesting about the place, and took us to the best spots in the neighbourhood for birds and insects. The mills were built a long time ago by a wealthy Brazilian. They had belonged to Mr. Upton for many years. I was told that when the dark-skinned revolutionists were preparing for their attack on Pará, they occupied the place, but not the slightest injury was done to the machinery or building, for the leaders said it was against the Portuguese and their party that they were at war, not against the other foreigners.

The creek Iritirí at the mills is only a few yards wide; it winds about between two lofty walls of forest for some distance, then becomes much broader, and finally joins the Magoary. There are many other ramifications, creeks or channels, which lead to retired hamlets and scattered houses, inhabited by people of mixed white, Indian, and negro descent. Many of them did business with Mr. Leavens, bringing for sale their little harvests of rice, or a few logs of timber. It was interesting to see them in their little heavily-laden montarias. Sometimes the boats were managed by handsome, healthy young lads, loosely clad in straw hat, white shirt, and dark blue trousers, turned up to the knee. They steered, paddled, and managed the varejaõ (the boating pole) with much grace and dexterity.

We made many excursions down the Iritirí, and saw much of these creeks; besides, our second visit to the mills was by water. The Magoary is a magnificent channel; the different branches form quite a labyrinth, and the land is everywhere of little elevation. All these smaller rivers throughout the Pará estuary are of the nature of creeks. The land is so level, that the short local rivers have no sources and downward currents, like rivers as we generally understand them. They serve the purpose of draining the land, but instead of having a constant current one way, they have a regular ebb and flow with the tide. The natives call them, in the Tupí language, Igarapés, or canoe-paths. The igarapés and furos or channels, which are infinite in number in this great river delta, are characteristic of the country. The land is everywhere covered with impenetrable forests; the houses and villages are all on the waterside, and nearly all communication is by water. This semi-aquatic life of the people is one of the most interesting features of the country. For short excursions, and for fishing in still waters, a small boat, called montaria, is universally used. It is made of five planks; a broad one for the bottom, bent into the proper shape by the action of heat, two narrow ones for the sides, and two small triangular pieces for stem and stern. It has no rudder; the paddle serves for both steering and propelling. The montaria takes here the place of the horse, mule, or camel of other regions. Besides one or more montarias, almost every family has a larger canoe, called Igarité. This is fitted with two masts, a rudder, and keel, and has an arched awning or cabin near the stern, made of a framework of tough lianas, thatched with palm leaves. In the igarité they will cross stormy rivers fifteen or twenty miles broad. The natives are all boat-builders. It is often remarked, by white residents, that an Indian is a carpenter and shipwright by intuition. It is astonishing to see in what crazy vessels these people will risk themselves. I have seen Indians cross rivers in a leaky montaria, when it required the nicest equilibrium to keep the leak just above water; a movement of a hair's breadth would send all to the bottom, but they managed to cross in safety. They are especially careful when they have strangers under their charge, and it is the custom of Brazilian and Portuguese travellers to leave the whole management to them. When they are alone, they are more reckless, and often have to swim for their lives. If a squall overtakes them as they are crossing in a heavily-laden canoe, they all jump overboard and swim about until the heavy sea subsides, when they re-embark.

A few words on the aboriginal population of the Pará estuary will here not be out of place. The banks of the Pará were originally inhabited by a number of distinct tribes, who, in their habits, resembled very much the natives of the sea-coast from Maranham to Bahia. It is related that one large tribe, the Tupinambas, migrated from Pernambuco to the Amazons. One fact seems to be well established, namely, that all the coast tribes were far more advanced in civilisation, and milder in their manners, than the savages who inhabited the interior lands of Brazil. They were settled in villages, and addicted to agriculture. They navigated the rivers in large canoes, called ubás, made of immense hollowed-out tree trunks; in these they used to go on war expeditions, carrying in the prows their trophies and calabash rattles, whose clatter was meant to intimidate their enemies. They were gentle in disposition, and received the early Portuguese settlers with great friendliness. The inland savages, on the other hand, led a wandering life, as they do at the present time, only coming down occasionally to rob the plantations of the coast tribes, who always entertained the greatest enmity towards them.

The original Indian tribes of the district are now either civilised, or have amalgamated with the white and negro immigrants. Their distinguished tribal names have long been forgotten, and the race bears now the general appellation of Tapuyo, which seems to be one of the names of the ancient Tupinambas. The Indians of the interior, still remaining in the savage state, are called by the Brazilians Indios, or Gentios (Heathens). All the semi-civilised Tapuyos of the villages, and in fact the inhabitants of retired places generally, speak the Lingoa geral, a language adapted by the Jesuit missionaries from the original idiom of the Tupinambas. The language of the Guaranis, a nation living on the banks of the Paraguay, is a dialect of it, and hence it is called by philologists the Tupi-Guarani language; printed grammars of it are always on sale at the shops of the Pará booksellers. The fact of one language having been spoken over so wide an extent of country as that from the Amazons to Paraguay, is quite an isolated one in this country, and points to considerable migrations of the Indian tribes in former times. At present the languages spoken by neighbouring tribes on the banks of the interior rivers are totally distinct; on the Juruá, even scattered hordes belonging to the same tribe are not able to understand each other.

The civilised Tapuyo of Pará differs in no essential point, in physical or moral qualities, from the Indian of the interior. He is more stoutly built, being better fed than some of them; but in this respect there are great differences amongst the tribes themselves. He presents all the chief characteristics of the American red man. The skin of a coppery brown colour, the features of the face broad, and the hair black, thick, and straight. He is generally about the middle height, thick-set, has a broad muscular chest, well-shaped but somewhat thick legs and arms, and small hands and feet. The cheek bones are not generally prominent; the eyes are black, and seldom oblique like those of the Tartar races of Eastern Asia, which are supposed to have sprung from the same original stock as the American red man. The features exhibit scarcely any mobility of expression; this is connected with the excessively apathetic and undemonstrative character of the race. They never betray, in fact they do not feel keenly, the emotions of joy, grief, wonder, fear, and so forth. They can never be excited to enthusiasm; but they have strong affections, especially those connected with family. It is commonly stated by the whites and negroes that the Tapuyo is ungrateful. Brazilian mistresses of households, who have much experience of Indians, have always a long list of instances to relate to the stranger, showing their base ingratitude. They certainly do not appear to remember or think of repaying benefits, but this is probably because they did not require, and do not value, such benefits as their would-be masters confer upon them. I have known instances of attachment and fidelity on the part of Indians towards their masters, but these are exceptional cases. All the actions of the Indian show that his ruling desire is to be let alone; he is attached to his home, his quiet monotonous forest and river life; he likes to go to towns occasionally, to see the wonders introduced by the white man, but he has a great repugnance to living in the midst of the crowd; he prefers handicraft to field labour, and especially dislikes binding himself to regular labour for hire. He is shy and uneasy before strangers, but if they visit his abode, he treats them well, for he has a rooted appreciation of the duty of hospitality; there is a pride about him, and being naturally formal and polite, he acts the host with great dignity. He withdraws from towns as soon as the stir of civilisation begins to make itself felt. When we first arrived at Pará, many Indian families resided there, for the mode of living at that time was more like that of a large village than a city; but as soon as river steamers and more business activity were introduced, they all gradually took themselves away.

These characteristics of the Pará Indians are applicable, of course, to some extent, to the Mamelucos, who now constitute a great proportion of the population. The inflexibility of character of the Indian, and his total inability to accommodate himself to new arrangements, will infallibly lead to his extinction, as immigrants, endowed with more supple organisations, increase, and civilisation advances in the Amazon region. But, as the different races amalgamate readily, and the offspring of white and Indian often become distinguished Brazilian citizens, there is little reason to regret the fate of the race. Formerly the Indian was harshly treated, and even now he is so in many parts of the interior. But, according to the laws of Brazil, he is a free citizen, having equal privileges with the whites; and there are very strong enactments providing against the enslaving and ill-treatment of the Indians. The residents of the interior, who have no higher principles to counteract instinctive selfishness or antipathy of race, cannot comprehend why they are not allowed to compel Indians to work for them, seeing that they will not do it of their own accord. The inevitable result of the conflict of interests between a European and a weaker indigenous race, when the two come in contact, is the sacrifice of the latter. In the Pará district, the Indians are no longer enslaved, but they are deprived of their lands, and this they feel bitterly, as one of them, an industrious and worthy man, related to me.

On our second visit to the mills, we stayed ten days. There is a large reservoir and also a natural lake near the place, both containing aquatic plants, whose leaves rest on the surface like our water lilies, but they are not so elegant as our nymphæa, either in leaf or flower. On the banks of these pools grow quantities of a species of fan-leaved palm-tree, the Caraná, whose stems are surrounded by whorls of strong spines. I sometimes took a montaria, and paddled myself alone down the creek. One day I got upset, and had to land on a grassy slope leading to an old plantation, where I ran about naked whilst my clothes were being dried on a bush. The creek Iritirí is not so picturesque as many others which I subsequently explored. Towards the Magoary the banks at the edge of the water are clothed with mangrove bushes, and beneath them the muddy banks, into which the long roots that hang down from the fruit before it leaves the branches strike their fibres, swarm with crabs. On the lower branches the beautiful bird, Ardea helias, is found. This is a small heron of exquisitely graceful shape and mien; its plumage is minutely variegated with bars and spots of many colours, like the wings of certain kinds of moths. It is difficult to see the bird in the woods, on account of its sombre colours and the shadiness of its dwelling-places; but its note, a soft long-drawn whistle, often betrays its hiding-place. I was told by the Indians that it builds in trees, and that the nest, which is made of clay, is beautifully constructed. It is a favourite pet-bird of the Brazilians, who call it Pavaõ (pronounced pavaong), or peacock. I often had opportunities of observing its habits. It soon becomes tame, and walks about the floors of houses, picking up scraps of food, or catching insects, which it secures by walking gently to the place where they settle, and spearing them with its long slender beak. It allows itself to be handled by children, and will answer to its name "Pavaõ! Pavaõ!" walking up with a dainty, circumspect gait, and taking a fly or beetle from the hand.

During these rambles by land and water we increased our collections considerably. Before we left the mills we arranged a joint excursion to the Tocantins. Mr. Leavens wished to ascend that river to ascertain if the reports were true, that cedar grew abundantly between the lowermost cataract and the mouth of the Araguaya, and we agreed to accompany him. Whilst we were at the mills, a Portuguese trader arrived with a quantity of worm-eaten logs of this cedar, which he had gathered from the floating timber in the current of the main Amazons. The tree producing this wood, which is named cedar on account of the similarity of its aroma to that of the true cedars, is not, of course, a coniferous tree, as no member of that class is found in equatorial America, at least in the Amazons region. It is, according to Von Martius, the Cedrela odorata, an exogen belonging to the same order as the mahogany tree. The wood is light, and the tree is therefore, on falling into the water, floated down with the river currents. It must grow in great quantities somewhere in the interior, to judge from the number of uprooted trees annually carried to the sea; and as the wood is much esteemed for cabinet work and canoe-building, it is of some importance to learn where a regular supply can be obtained. We were glad of course to arrange with Mr. Leavens, who was familiar with the language, and an adept in river-navigation; so we returned to Pará to ship our collections for England, and prepare for the journey to a new region.

CHAPTER III.

PARÁ.

Religious holidays—Marmoset monkeys—Serpents—Insects.

BEFORE leaving the subject of Pará, where I resided, as already stated, in all eighteen months, it will be necessary to give a more detailed account of several matters connected with the customs of the people and the Natural History of the neighbourhood, which have hitherto been only briefly mentioned. I reserve an account of the trade and improved condition of Pará in 1859 for the end of this narrative.

During the first few weeks of our stay many of those religious festivals took place, which occupied so large a share of the time and thoughts of the people. These were splendid affairs, wherein artistically-arranged processions through the streets, accompanied by thousands of people, military displays, the clatter of fireworks, and the clang of military music, were superadded to pompous religious services in the churches. To those who had witnessed similar ceremonies in the southern countries of Europe, there would be nothing remarkable perhaps in these doings, except their taking place amidst the splendours of tropical nature; but to me they were full of novelty, and were besides interesting as exhibiting much that was peculiar in the manners of the people. The festivals celebrate either the anniversaries of events concerning saints, or those of the more important transactions in the life of Christ. To them have been added, since the Independence, many gala days connected with the events in the Brazilian national history; but these have all a semi-religious character. The holidays had become so numerous, and interfered so much with trade and industry towards the year 1852, that the Brazilian Government was obliged to reduce them; obtaining the necessary permission from Rome to abolish several which were of minor importance. Many of those which have been retained are declining in importance since the introduction of railways and steamboats, and the increased devotion of the people to commerce; at the time of our arrival, however, they were in full glory. The way they were managed was in this fashion. A general manager or "Juiz" for each festa was elected by lot every year in the vestry of the church, and to him were handed over all the paraphernalia pertaining to the particular festival which he was chosen to manage; the image of the saint, the banners, silver crowns, and so forth. He then employed a number of people to go the round of the parish and collect alms towards defraying the expenses. It was considered that the greater the amount of money spent in wax-candles, fireworks, music and feasting, the greater the honour done to the saint. If the Juiz was a rich man, he seldom sent out alms-gatherers, but celebrated the whole affair at his own expense, which was sometimes to the extent of several hundred pounds. Each festival lasted nine days (a *novena*), and in many cases refreshments for the public were provided every evening. In the smaller towns a ball took place two or three evenings during the novena, and on the last day there was a grand dinner. The priest, of course, had to be paid very liberally, especially for the sermon delivered on the Saint's-day or termination of the festival, sermons being extra duty in Brazil.

There was much difference as to the accessories of these festivals between the interior towns and villages and the capital; but little or no work was done anywhere whilst they lasted, and they tended much to demoralise the people. It is soon perceived that religion is rather the amusement of the Paraenses than their serious exercise. The ideas of the majority evidently do not reach beyond the belief that all the proceedings are, in each case, in honour of the particular wooden image enshrined at the church. The uneducated Portuguese immigrants seemed to me to have very degrading notions of religion. I have often travelled in the company of these shining examples of European enlightenment. They generally carry with them, wherever they go, a small image of some favourite saint in their trunks; and when a squall or any other danger arises, their first impulse is to rush to the cabin, take out the image and clasp it to their lips, whilst uttering a prayer for protection. The negroes and mulattos are similar in this respect to the low Portuguese, but I think they show a purer devotional feeling; and in conversation I have always found them to be more rational in religious views than the lower orders of Portuguese. As to the Indians; with the exception of the more civilised families residing near the large towns, they exhibit no religious sentiment at all. They have their own patron saint, St. Thomé, and celebrate his anniversary in the orthodox way, for they are fond of observing all the formalities; but they think the feasting to be of equal importance with the church ceremonies. At some of the festivals, masquerading forms a large part of the proceedings, and then the Indians really shine. They get up capital imitations of wild animals, dress themselves to represent the Caypór and other fabulous creatures of the forest, and act their parts throughout with great cleverness. When St. Thomé's festival takes place, every employer of Indians knows that all his men will get drunk. The Indian, generally too shy to ask directly for cashaça (rum), is then very bold; he asks for a frasco at once (two bottles and a half), and says, if interrogated, that he is going to fuddle in honour of St. Thomé.

In the city of Pará, the provincial government assists to augment the splendour of the religious holidays. The processions which traverse the principal streets consist, in the first place, of the image of the saint, and those of several other subordinate ones belonging to the same church; these are borne on the shoulders of respectable householders, who volunteer for the purpose: sometimes you will see your neighbour the grocer or the carpenter groaning under the load. The priest and his crowd of attendants precede the images, arrayed in embroidered robes, and protected by magnificent sunshades—no useless ornament here, for the heat is very great when the sun is not obscured. On each side of the long line the citizens walk, clad in crimson silk cloaks, and holding each a large lighted wax candle. Behind follows a regiment or two of foot soldiers with their bands of music, and last of all the crowd, the coloured people being cleanly dressed and preserving a grave demeanour. The women are always in great force, their luxuriant black hair decorated with jasmines, white orchids, and other tropical flowers. They are dressed in their usual holiday attire, gauze chemises and black silk petticoats; their necks are adorned with links of gold beads, which when they are slaves are generally the property of their mistresses, who love thus to display their wealth.

At night, when festivals are going on in the grassy squares around the suburban churches, there is really much to admire. A great deal that is peculiar in the land and the life of its inhabitants can be seen best at those times. The cheerful white church is brilliantly lighted up, and the music, not of a very solemn description, peals forth from the open windows and doors. Numbers of young gaudily-dressed negresses line the path to the church doors with stands of liqueurs, sweet-meats, and cigarettes, which they sell to the outsiders. A short distance off is heard the rattle of dice-boxes and roulette at the open-air gambling stalls. When the festival happens on moon-lit nights, the whole scene is very striking to a new-comer. Around the square are groups of tall palm trees, and beyond it, over the illuminated houses, appear the thick groves of mangoes near the suburban avenues, from which comes the perpetual ringing din of insect life. The soft tropical moon-light lends a wonderful charm to the whole. The inhabitants are all out, dressed in their best. The upper classes, who come to enjoy the fine evening and the general cheerfulness, are seated on chairs around the doors of friendly houses. There is no boisterous conviviality, but a quiet enjoyment seems to be felt everywhere, and a gentle courtesy rules amongst all classes and colours. I have seen a splendidly-dressed colonel, from the President's palace, walk up to a mulatto, and politely ask his permission to take a light from his cigar. When the service is over, the church bells are set ringing, a shower of rockets mounts upwards, the bands strike up, and parties of coloured people in the booths begin their dances. About ten o'clock the Brazilian national air is played, and all disperse quietly and soberly to their homes.

At the festival of Corpus Christi there was a very pretty arrangement. The large green square of the Trinidade was lighted up all round with bonfires. On one side a fine pavilion was erected, the upright posts consisting of real fan-leaved palm trees, the Mauritia flexuosa, which had been brought from the forest, stems and heads entire, and fixed in the ground. The booth was illuminated with coloured lamps, and lined with red and white cloth. In it were seated the ladies, not all of pure Caucasian blood, but presenting a fine sample of Pará beauty and fashion.

The grandest of all these festivals is that held in honour of Our Lady of Nazareth: it is, I believe, peculiar to Pará. As I have said before, it falls in the second quarter of the moon, about the middle of the dry season—that is, in October or November—and lasts, like the others, nine days. On the first day a very extensive procession takes place, starting from the

Cathedral, whither the image of the saint has been conveyed some days previously; and terminating at the chapel or hermitage, as it is called, of the saint at Nazareth, a distance of more than two miles. The whole population turns out on this occasion. All the soldiers, both of the line and the National Guard, take part in it, each battalion accompanied by its band of music. The civil authorities also, with the President at their head, and the principal citizens, including many of the foreign residents, join in the line. The boat of the shipwrecked Portuguese vessel is carried after the saint on the shoulders of officers or men of the Brazilian navy, and along with it are borne the other symbols of the miracles which Our Lady is supposed to have performed. The procession starts soon after the sun's heat begins to moderate—that is, about half-past four o'clock in the afternoon. When the image is deposited in the chapel, the festival is considered to be inaugurated, and the village every evening becomes the resort of the pleasure-loving population, the holiday portion of the programme being preceded, of course, by a religious service in the chapel. The aspect of the place is then that of a fair; without the humour and fun, but, at the same time, without the noise and coarseness, of similar holidays in England. Large rooms are set apart for panoramic and other exhibitions, to which the public are admitted gratis. In the course of each evening, large displays of fireworks take place, all arranged according to a published programme of the festival.

The various ceremonies which take place during Lent seemed to me the most impressive, and some of them were exceedingly well arranged. The people, both performers and spectators, conduct themselves with more gravity on these occasions, and there is no holiday-making. Performances, representing the last events in the life of Christ, are enacted in the churches or streets, in such a way as to remind one of the old miracle plays or mysteries. A few days before Good Friday, a torchlight procession takes place by night from one church to another, in which is carried a large wooden image of Christ bent under the weight of the cross. The chief members of the government assist, and the whole slowly moves to the sound of muffled drums. A double procession is managed a few days afterwards. The image of St. Mary is carried in one direction, and that of the Saviour in another. The two images meet in the middle of one of the most beautiful churches, which is previously filled to excess with the multitudes anxious to witness the affecting meeting of mother and son a few days before the crucifixion. The images are brought face to face in the middle of the church, the crowd falls prostrate, and a lachrymose sermon is delivered from the pulpit. The whole thing, as well as many other spectacles arranged during the few succeeding days, is highly theatrical, and well calculated to excite the religious emotions of the people, although, perhaps, only temporarily. On Good Friday the bells do not ring, all musical sounds are interdicted, and the hours, night and day, are announced by the dismal noise of wooden clappers, wielded by negroes stationed near the different churches. A sermon is delivered in each church. In the middle of it, a scroll is suddenly unfolded from the pulpit, on which is an exaggerated picture of the bleeding Christ. This act is accompanied by loud groans which come from stout-lunged individuals concealed in the vestry and engaged for the purpose. The priest becomes greatly excited, and actually sheds tears. On one of these occasions I squeezed myself into the crowd, and watched the effect of the spectacle on the audience. Old Portuguese men and Brazilian women seemed very much affected—sobbing, beating their breasts, and telling their beads. The negroes behaved themselves with great propriety, but seemed moved more particularly by the pomp, the gilding, the dresses, and the general display. Young Brazilians laughed. Several aborigines were there, coolly looking on. One old Indian, who was standing near me, said, in a derisive manner, when the sermon was over, " It's all very good ; better it could not be " (Está todo bom ; melhor naõ pude ser).

The negroes of Pará are very devout. They have built, by slow degrees, a fine church, as I was told, by their own unaided exertions. It is called Nossa Senhora do Rosario, or Our Lady of the Rosary. During the first weeks of our residence at Pará, I frequently observed a line of negroes and negresses, late at night, marching along the streets, singing a chorus. Each carried on his or her head a quantity of building materials—stones, bricks, mortar, or planks. I found they were chiefly slaves, who, after their hard day's work, were contributing a little towards the construction of their church. The materials had all been purchased by their own savings. The interior was finished about a year afterwards, and is decorated, I thought, quite as superbly as the other churches which were constructed, with far larger means, by the old religious orders more than a century ago. Annually, the negroes celebrate the festival of Nossa Senhora do Rosario, and generally make it a complete success.

I will now add a few more notes which I have accumulated on the subject of the natural history, and then we shall have done. for the present, with Pará and its neighbourhood.

I have already mentioned that monkeys were rare in the immediate vicinity of Pará. I met with three species only in the forest near the city; they are shy animals, and avoid the neighbourhood of towns, where they are subject to much persecution by the inhabitants, who kill them for food. The only kind which I saw frequently was the little Midas ursulus, one of the Marmosets, a family peculiar to tropical America, and differing in many essential points of structure and habits from all other apes. They are small in size, and more like squirrels than true monkeys in their manner of climbing. The nails, except those of the hind thumbs, are long and claw-shaped like those of squirrels, and the thumbs of the fore extremities, or hands, are not opposable to the other fingers. I do not mean to imply that they have a near relationship to squirrels, which belong to the Rodents, an inferior order of mammals; their resemblance to those animals is merely a superficial one. They have two molar teeth less in each jaw than the Cebidæ, the other family of American monkeys; they agree with them, however, in the sideway position of the nostrils, a character which distinguishes both from all the monkeys of the old world.. The body is long and slender, clothed with soft hairs, and the tail, which is nearly twice the length of the trunk, is not prehensile. The hind limbs are much larger in volume than the anterior pair. The Midas ursulus is never seen in large flocks; three or four are the greatest number observed together. It seems to be less afraid of the neighbourhood of man than any other monkey. I sometimes saw it in the woods which border the suburban streets, and once I espied two individuals in a thicket behind the English consul's house at Nazareth. Its mode of progression along the main boughs of the lofty trees is like that of the squirrel; it does not ascend to the slender branches, or take those wonderful flying leaps which the Cebidæ do, whose prehensile tails and flexible hands fit them for such headlong travelling. It confines itself to the larger boughs and trunks of trees, the long nails being of great assistance to the creature, enabling it to cling securely to the bark; and it is often seen passing rapidly round the perpendicular cylindrical trunks. It is a quick, restless, timid little creature, and has a great share of curiosity, for when a person passes by under the trees along which a flock is running, they always stop for a few moments to have a stare at the intruder. In Pará, Midas ursulus is often seen in a tame state in the houses of the inhabitants. When full grown, it is about nine inches long, independently of the tail, which measures fifteen inches. The fur is thick, and black in colour, with the exception of a reddish-brown streak down the middle of the back. When first taken, or when kept tied up, it is very timid and irritable. It will not allow itself to be approached, but keeps retreating backwards when any one attempts to coax it. It is always in a querulous humour, uttering a twittering, complaining noise; its dark, watchful eyes, expressive of distrust, observant of every movement which takes place near it. When treated kindly, however, as it generally is in the houses of the natives, it becomes very tame and familiar. I once saw one as playful as a kitten, running about the house after the negro children, who fondled it to their hearts' content. It acted somewhat differently towards strangers, and seemed not to like them to sit in the hammock which was slung in the room, leaping up, trying to bite, and otherwise annoying them. It is generally fed on sweet fruits, such as the banana; but it is also fond of insects, especially soft-bodied spiders and grasshoppers, which it will snap up with eagerness when within reach. The expression of countenance in these small monkeys is intelligent and pleasing. This is partly owing to the open facial angle, which is given as one of 60°; but the quick movements of the head, and the way they have of inclining it to one side when their curiosity is excited, contribute very much to give them a knowing expression.

On the Upper Amazons I once saw a tame individual of the Midas leoninus, a species first described by Humboldt, which was still more playful and intelligent than the one just described. This rare and beautiful little monkey is only seven inches in length, exclusive of the tail. It is named leoninus on account of the long brown mane which depends from the neck, and which gives it very much the appearance of a diminutive lion. In the house where it was kept, it was familiar with every one; its greatest pleasure seeming to be to climb about the bodies of different persons who entered. The first time I went in, it ran across the room straightway to the chair on which I had sat down, and climbed up to my shoulder; arrived there, it turned round and looked into my face, showing its little teeth, and chattering, as though it would say, "Well, and how do *you* do?" It showed more affection towards its master than towards strangers, and would climb up to his head a dozen times in the course of an hour, making a great show every time of searching there for certain animalcula. Isidore Geoffroy St. Hilaire relates of a species of this genus, that it distinguished between different objects depicted on an engraving. M. Audouin showed it the portraits of a cat and a wasp; at these it became much terrified: whereas, at the sight of a figure of a grasshopper or beetle, it precipitated itself on the picture, as if to seize the objects there represented.

Although monkeys are now rare in a wild state near Pará, a great number may be seen semi-domesticated in the city. The Brazilians are fond of pet animals. Monkeys, however, have not been known to breed in captivity in this country. I counted, in a short time, thirteen different species, whilst walking about the Pará streets, either at the doors or windows of houses, or in the native canoes. Two of them I did not meet with afterwards in any other part of the country. One of these was the well-known Hapale Jacchus, a little creature resembling a kitten, banded with black and gray all over the body and tail, and having a fringe of long white hairs surrounding the ears. It was seated on the shoulder of a young mulatto girl, as she was walking along the street, and I was told had been captured in the island of Marajo. The other was a species of Cebus, with a remarkably large head. It had ruddy-brown fur, paler on the face, but presenting a blackish tuft on the top of the forehead.

In the wet season serpents are common in the neighbourhood of Pará. One morning, in April, 1849, after a night of deluging rain, the lamplighter, on his rounds to extinguish the lamps, knocked me up to show me a boa-constrictor he had just killed in the Rua St. Antonio, not far from my door. He had cut it nearly in two with a large knife, as it was making its way down the sandy street. Sometimes the native hunters capture boa-constrictors alive in the forest near the city. We bought one which had been taken in this way, and kept it for some time in a large box under our verandah. This is not, however, the largest or most formidable serpent found in the Amazons region. It is far inferior, in these respects, to the hideous Sucurujú, or Water Boa (Eunectes murinus), which sometimes attacks man; but of this I shall have to give an account in a subsequent chapter.

It frequently happened, in passing through the thickets, that a snake would fall from the boughs close to me. Once I got for a few moments completely entangled in the folds of one, a wonderfully slender kind, being nearly six feet in length, and not more than half an inch in diameter at its broadest part. It was a species of Dryophis. The majority of the snakes seen were innocuous. One day, however, I trod on the tail of a young serpent belonging to a very poisonous kind, the Jararaca (Craspedocephalus atrox). It turned round and bit my trousers; and a young Indian lad, who was behind me, dexterously cut it through with his knife before it had time to free itself. In some seasons snakes are very abundant, and it often struck me as strange that accidents did not occur more frequently than was the case.

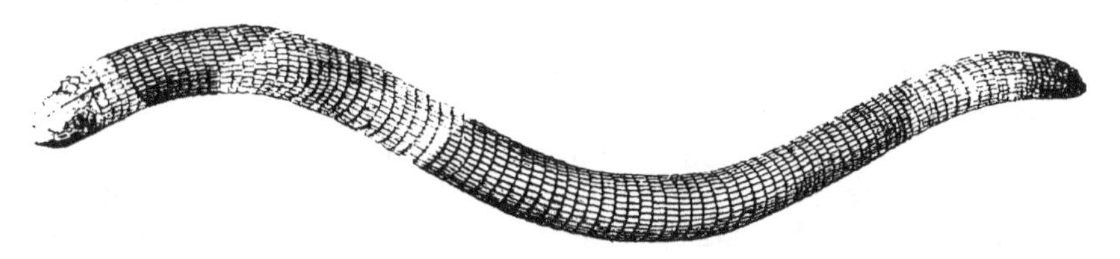

Amphisbæna.

Amongst the most curious snakes found here were the Amphisbænæ, a genus allied to the slow-worm of Europe. Several species occur at Pará. Those brought to me were generally not much more than a foot in length. They are of cylindrical shape, having, properly speaking, no neck, and the blunt tail, which is only about an inch in length, is of the same shape as the head. This peculiar form, added to their habit of wriggling backwards as well as forwards, has given rise to the fable that they have two heads, one at each extremity. They are extremely sluggish in their motions, and are clothed with scales that have the form of small imbedded plates arranged in rings round the body. The eye is so small as to be scarcely perceptible. They live habitually in the subterranean chambers of the Saüba ant; only coming out of their abodes occasionally in the night time. The natives call the Amphisbæna the "Mai das Saübas," or Mother of the Saübas, and believe it to be poisonous, although it is perfectly harmless. It is one of the many curious animals which have become the subject of mythical stories with the natives. They say the ants treat it with great affection, and that if the snake be taken away from a nest, the Saübas will forsake the spot. I once took one quite whole out of the body of a young Jararaca, the poisonous species already alluded to, whose body was so distended with its contents that the skin was stretched out to a film over the contained Amphisbæna. I was, unfortunately, not able to ascertain the exact relation which subsists between these curious snakes and the Saüba ants. I believe, however, they feed upon the Saübas, for I once found remains of ants in the stomach of one of them. Their motions are quite peculiar; the undilatable jaws, small eyes and curious plated integument also distinguish them from other snakes. These properties have evidently some relation to their residence in the subterranean abodes of ants. It is now well ascertained by naturalists, that some of the most anomalous forms amongst Coleopterous insects are those which live solely in the nests of ants, and it is curious that an abnormal form of snakes should also be found in the society of these insects.

The neighbourhood of Pará is rich in insects. I do not speak of the number of individuals, which is probably less than one meets with, excepting ants and Termites, in summer days in temperate latitudes; but the variety, or in other words, the number of species, is very great. It will convey some idea of the diversity of butterflies when I mention that about 700 species of that tribe are found within an hour's walk of the town; whilst the total number found in the British Islands does not exceed 66, and the whole of Europe supports only 321. Some of the most showy species, such as the swallow-tailed kinds, Papilio Polycaon, Thoas, Torquatus, and others, are seen flying about the streets and gardens; sometimes they come through the open windows, attracted by flowers in the partments. Those species of Papilio which are most characteristic of the country, so conspicuous in their velvety-black, green, and rose-coloured hues, which Linnæus, in pursuance of his elegant system of nomenclature,—naming the different kinds after the heroes of Greek mythology,—called Trojans, never leave the shades of the forest. The splendid metallic blue Morphos, some of which measure seven inches in expanse, are generally confined to the shady alleys of the forest. They sometimes come forth into the broad sunlight. When we first went to look at our new residence in Nazareth, a Morpho Menelaus, one of the most beautiful kinds, was seen flapping its huge wings like a bird along the verandah. This species, however, although much admired, looks dull in colour by the side of its congener, the Morpho Rhetenor, whose wings, on the upper face, are of quite a dazzling lustre. Rhetenor usually prefers the broad sunny roads in the forest, and is an almost unattainable prize, on account of its lofty flight; for it very rarely descends nearer the ground than about twenty feet. When it comes sailing along, it occasionally flaps its wings, and then the blue surface flashes in the sunlight, so that it is visible a quarter of a mile off. There is another species of this genus, of a satiny-white hue, the Morpho Uraneis; this is equally difficult to obtain; the male only has the satiny lustre, the female being of a pale-lavender colour. It is in the height of the dry season that the greatest number and variety of butterflies are found in the woods; especially when a shower falls at intervals of a few days. An infinite number of curious and rare species may then be taken, most diversified in habits, mode of flight, colours, and markings: some yellow, others bright red, green, purple, and blue, and many bordered or spangled with metallic lines and spots of a silvery or golden lustre. Some have wings transparent as glass; one of these clear-wings is especially beautiful, namely, the Hetæra Esmeralda; it has one spot only of opaque colouring on its wings, which is of a violet and rose hue; this is the only part visible when the insect is flying low over dead leaves, in the gloomy shades where alone it is found, and it then looks like a wandering petal of a flower.

Bees and wasps are not especially numerous near Pará, and I will reserve an account of their habits for a future chapter. Many species of Mygale, those monstrous hairy spiders, half a foot in expanse, which attract the attention so much in museums, are found in sandy places at Nazareth. The different kinds have the most diversified habits. Some construct, amongst the tiles or thatch of houses, dens of closely-woven web, which, in texture, very much resembles fine muslin; these are often seen crawling over the walls of apartments. Others build similar nests in trees, and are known to attack birds. One very robust fellow, the Mygale Blondii, burrows into the earth, forming a broad slanting gallery, about two feet long, the sides of which he lines beautifully with silk. He is nocturnal in his habits. Just before sunset he may be seen keeping watch within the mouth of his tunnel, disappearing suddenly when he hears a heavy foot-tread near his hiding-place. The number of spiders ornamented with showy colours was somewhat remarkable. Some double themselves up at the base of leaf-stalks, so as to resemble flower-buds, and thus deceive the insects on which they prey. The most extraordinary-looking spider was a species of Acrosoma, which had two curved bronze-coloured spines, an inch and a half in length, proceeding from the tip of its abdomen. It spins a large web, the monstrous appendages being apparently no impediment to it in its work; but what their use can be I am unable to divine.

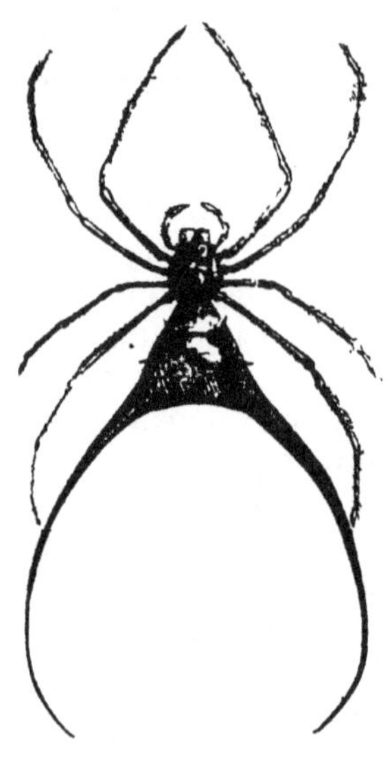

Acrosoma arcuatum.

Coleoptera, or beetles, at first seemed to be very scarce. This apparent scarcity has been noticed in other equatorial countries, and arises, probably, from the great heat of the sun not permitting them to exist in exposed situations, where they form such conspicuous objects in Europe. Many hundred species of the different families can be found, when they are patiently searched for in the shady places to which they are confined. It is vain to look for the Geodephaga, or carnivorous beetles, under stones, or anywhere. indeed, in open, sunny places. The terrestrial forms of this interesting family, which abound in England and temperate countries generally, are scare in the neighbourhood of Para, in fact I met with only four or five species: on the other hand the purely arboreal kinds were rather numerous. The contrary of this happens in northern latitudes, where the great majority of the species and genera are exclusively terrestrial. The arboreal forms are distinguished by the structure of the feet, which have broad spongy soles and toothed claws enabling them to climb over and cling to branches and leaves. The remarkable scarcity of ground beetles is, doubtless, attributable to the number of ants and Termites which people every inch of surface in all shady places, and which would most likely destroy the larvæ of Coleoptera. Moreover these active creatures have the same functions as Coleoptera, and thus render their existence unnecessary. The large proportion of climbing forms of carnivorous beetles is an interesting fact, because it affords another instance of the arboreal character which animal forms tend to assume in equinoctial America, a circumstance which points to the slow adaptation of the Fauna to a forest-clad country, throughout an immense lapse of geological time.

CHAPTER IV.

THE TOCANTINS AND CAMETÁ.

Preparations for the journey—The bay of Goajará—Grove of fan-leaved palms—The lower Tocantins—Sketch of the river—Vista alegre—Baiao—Rapids—Boat journey to the Guariba falls—Native life on the Tocantins—Second journey to Cametá.

August 26th. 1848.—Mr. Wallace and I started to-day on the excursion which I have already mentioned as having been planned with Mr. Leavens, up the river Tocantins, whose mouth lies about forty-five miles in a straight line, but eighty miles following the bends of the river channels, to the south-west of Pará. This river, as before stated, has a course of 1,600 miles, and stands third in rank amongst the streams which form the Amazons system. The preparations for the journey took a great deal of time and trouble. We had first to hire a proper vessel, a two-masted *vigilinga* twenty-seven feet long, with a flat prow and great breadth of beam, and fitted to live in heavy seas; for, although our voyage was only a river trip, there were vast sea-like expanses of water to traverse. It was not decked over, but had two arched awnings formed of strong wickerwork, and thatched with palm leaves. We had then to store it with provisions for three months, the time we at first intended to be away; procure the necessary passports; and, lastly, engage a crew. Mr. Leavens, having had much experience in the country, managed all these matters. He brought two Indians from the rice-mills, and these induced another to enrol himself. We, on our parts, took our cook, Isidoro, and a young Indian lad, named Antonio, who had attached himself to us in the course of our residence at Nazareth. Our principal man was Alexandro, one of Mr. Leavens's Indians. He was an intelligent and well-disposed young Tapuyo, an expert sailor and an indefatigable hunter. To his fidelity we were indebted for being enabled to carry out any of the objects of our voyage. Being a native of a district near the capital, Alexandro was a civilised Tapuyo, a citizen as free as his white neighbours. He spoke only Portuguese. He was a spare-built man, rather under the middle height, with fine regular features, and, what was unusual in Indians, the upper lip decorated with a moustache. Three years afterwards I saw him at Pará in the uniform of the National Guard, and he called on me often to talk about old times. I esteemed him as a quiet, sensible, manly young fellow.

We set sail in the evening, after waiting several hours in vain for one of our crew. It was soon dark, the wind blew stiffly, and the tide rushed along with great rapidity, carrying us swiftly past the crowd of vessels which were anchored in the port. The canoe rolled a good deal. After we had made five or six miles of way the tide turned, and we were obliged to cast anchor. Not long after, we laid ourselves down all three together on the mat, which was spread over the floor of our cabin, and soon fell asleep.

On awaking at sunrise the next morning, we found ourselves gliding upwards with the tide, along the Bahia or Bay, as it is called, of Goajará. This is a broad channel lying between the mainland and a line of islands which extend some distance beyond the city. Into it three large rivers discharge their waters, namely, the Guamá, the Acará, and the Mojú; so that it forms a kind of sub-estuary within the grand estuary of Pará. It is nearly four miles broad. The left bank, along which we were now sailing, was beautiful in the extreme; not an inch of soil was to be seen; the water frontage presented a compact wall of rich and varied forest, resting on the surface of the stream. It seemed to form a finished border to the water scene, where the dome-like, rounded shapes of exogenous trees which constituted the mass formed the groundwork, and the endless diversity of broad-leaved Heliconiæ and Palms—each kind differing in stem, crown, and fronds—the rich embroidery. The morning was calm and cloudless; and the slanting beams of the early sun, striking full on the front of the forest, lighted up the whole most gloriously. The only sound of life which reached us was the call of the Serracúra (Gallinula Cayennensis), a kind of wild fowl; all else was so still that the voices of boatmen could be plainly heard, from canoes passing a mile or two distant from us. The sun soon gains great power on the water, but with it the sea-breeze increases in strength, moderating the heat which would otherwise be almost insupportable. We reached the end of the Goajará about midday, and then entered the narrower channel of the Mojú. Up this we travelled, partly rowing and partly sailing, between the same unbroken walls of forest, until the morning of the 28th.

August 29th.—The Mojú, a stream little inferior to the Thames in size, is connected about twenty miles from its mouth, by means of a short artificial canal, with a small stream, the Igarapé-mirim, which flows the opposite way into the water-system of the Tocantins. Small vessels like ours take this route in preference to the stormy passage by way of the main river, although the distance is considerably greater. We passed through the canal yesterday, and to-day have been threading our way through a labyrinth of narrow channels, their banks all clothed with the same magnificent forest; but agreeably varied by houses of planters and settlers. We passed many quite large establishments, besides one pretty little village called Santa Anna. All these channels are washed through by the tides,—the ebb, contrary to what takes place in the short canal, setting towards the Tocantins. The water is almost tepid (77° Fahr.), and the rank vegetation all around seems reeking with moisture. The country, however, as we were told, is perfectly healthy. Some of the houses are built on wooden piles driven into the mud of the swamp.

In the afternoon we reached the end of the last channel, called the Murutipucú, which runs for several miles between two unbroken lines of fan-leaved palms, forming with their straight stems colossal palisades. On rounding a point of land we came in full view of the Tocantins. The event was announced by one of our Indians, who was on the look-out at the prow, shouting, "La está o Paraná-uassú!" "Behold the great river!" It was a grand sight—a broad expanse of dark waters dancing merrily to the breeze; the opposite shore, a narrow blue line, miles away.

We went ashore on an island covered with palm-trees, to make a fire and boil our kettle for tea. I wandered a short way inland, and was astounded at the prospect. The land lay below the upper level of the daily tides, so that there was no underwood, and the ground was bare. The trees were almost all of one species of Palm, the gigantic fan-leaved Mauritia flexuosa: on the borders only was there a small number of a second kind, the equally remarkable Ubussú palm

(Manicaria saccifera). The Ubussú has erect, uncut leaves, twenty-five feet long, and six feet wide, all arranged round the top of a four-feet high stem, so as to form a figure like that of a colossal shuttlecock. The fan-leaved palms, which clothed nearly the entire islet, had huge cylindrical smooth stems, three feet in diameter, and about a hundred feet high. The crowns were formed of enormous clusters of fan-shaped leaves, the stalks alone of which measured seven to ten feet in length. Nothing in the vegetable world could be more imposing than this grove of palms. There was no underwood to obstruct the view of the long perspective of towering columns. The crowns, which were densely packed together at an immense height overhead, shut out the rays of the sun; and the gloomy solitude beneath, through which the sound of our voices seemed to reverberate, could be compared to nothing so well as a solemn temple. The fruits of the two palms were scattered over the ground; those of the Ubussú adhere together by twos and threes, and have a rough, brown-coloured shell; the fruit of the Mauritia, on the contrary, is of a bright red hue, and the skin is impressed with deep crossing lines, which give it a resemblance to a quilted cricket-ball.

About midnight, the tide being favourable, and the breeze strong, we crossed the river, taking it in a slanting direction, a distance of sixteen miles, and arrived at eight o'clock the following morning at Cametá. This is a town of some importance, pleasantly situated on the somewhat high terra firma of the left bank of the Tocantins. I will defer giving an account of the place till the end of this narrative of our Tocantins voyage. We lost here another of our men, who got drinking with some old companions ashore, and were obliged to start on the difficult journey up the river with two hands only, and they in a very dissatisfied humour with the prospect.

The river view from Cametá is magnificent. The town is situated, as already mentioned, on a high bank, which forms quite a considerable elevation for this flat country, and the broad expanse of dark-green waters is studded with low, palm-clad islands; the prospect down river, however, being clear, or bounded only by a sea-like horizon of water and sky. The shores are washed by the breeze-tossed waters into little bays and creeks, fringed with sandy beaches. The Tocantins has been likened, by Prince Adalbert of Prussia, who crossed its mouth in 1846, to the Ganges. It is upwards of ten miles in breadth at its mouth; opposite Cametá it is five miles broad.

Mr. Burchell, the well-known English traveller, descended the river from the mining provinces of interior Brazil some years before our visit. Unfortunately, the utility of this fine stream is impaired by the numerous obstructions to its navigation in the shape of cataracts and rapids, which commence, in ascending, at about 120 miles above Cametá, as will be seen in the sequel.

August 30th.—Arrived, in company with Senhor Laroque, an intelligent Portuguese merchant, at Vista Alegre, fifteen miles above Cametá. This was the residence of Senhor Antonio Ferreira Gomez, and was a fair sample of a Brazilian planter's establishment in this part of the country. The buildings covered a wide space, the dwelling-house being separated from the place of business, and as both were built on low, flooded ground, the communication between the two was by means of a long wooden bridge. From the office and visitors' apartments a wooden pier extended into the river. The whole was raised on piles above high-water mark. There was a rude mill for grinding sugar-cane, worked by bullocks; but cashaça, or rum, was the only article manufactured from the juice. Behind the buildings was a small piece of ground cleared from the forest, and planted with fruit trees, orange, lemon, genipapa, goyava, and others: and beyond this, a broad path through a neglected plantation of coffee and cacao, led to several large sheds, where the farinha, or mandioca meal was manufactured. The plantations of mandioca are always scattered about in the forest, some of them being on islands in the middle of the river. Land being plentiful, and the plough, as well as, indeed, nearly all other agricultural implements, unknown, the same ground is not planted three years together; but a new piece of forest is cleared every alternate year, and the old clearing suffered to relapse into jungle.

We stayed here two days, sleeping ashore in the apartment devoted to strangers. As usual in Brazilian houses of the middle class, we were not introduced to the female members of the family, and, indeed, saw nothing of them except at a distance. In the forest and thickets about the place we were tolerably successful in collecting, finding a number of birds and insects which do not occur at Pará. I saw here, for the first time, the sky-blue Chatterer (Ampelis cotinga). It was on the topmost bough of a very lofty tree, and completely out of the reach of an ordinary fowling-piece. The beautiful light-blue colour of its plumage was plainly discernible at that distance.

It is a dull, quiet bird. A much commoner species was the Cigana or Gipsy (Opisthocomus cristatus), a bird belonging to the same order (Gallinacea) as our domestic fowl. It is about the size of a pheasant; the plumage is dark brown, varied with reddish, and the head is adorned with a crest of long feathers. It is a remarkable bird in many respects. The hind toe is not placed high above the level of the other toes, as it is in the fowl order generally, but lies on the same plane with them; the shape of the foot becomes thus suited to the purely arboreal habits of the bird, enabling it to grasp firmly the branches of trees. This is a distinguishing character of all the birds in equinoctial America which represent the fowl and pheasant tribes of the old world, and affords another proof of the adaptation of the Fauna to a forest region. The Cigana lives in considerable flocks on the lower trees and bushes bordering the streams and lagoons, and feeds on various wild fruits, especially the sour Goyava (Psidium sp.). The natives say it devours the fruit of arborescent Arums (Caladium arborescens), which grow in crowded masses around the swampy banks of lagoons. Its voice is a harsh, grating hiss; it makes the noise when alarmed, all the individuals sibilating as they fly heavily away from tree to tree, when disturbed by passing canoes. It is polygamous, like other members of the same order. It is never, however, by any chance, seen on the ground, and is nowhere domesticated. The flesh has an unpleasant odour of musk combined with wet hides—a smell called by the Brazilians catinga; it is, therefore, uneatable. If it be as unpalateable to carnivorous animals as it is to man, the immunity from persecution which it would thereby enjoy would account for its existing in such great numbers throughout the country.

We lost here another of our crew; and thus, at the commencement of our voyage, had before us the prospect of being forced to return, from sheer want of hands to manage the canoe. Senhor Gomez, to whom we had brought letters of introduction from Senhor Joaõ Augusto Correia, a Brazilian gentleman of high standing at Pará, tried what he could do to induce the canoe-men of his neighbourhood to engage with us, but it was a vain endeavour. The people of these parts seemed to be above working for wages. They are naturally indolent, and besides, have all some little business or plantation of their own, which gives them a livelihood with independence. It is difficult to obtain hands under any circumstances, but it was particularly so in our case, from being foreigners, and suspected, as was natural amongst ignorant people, of being strange in our habits. At length, our host lent us two of his slaves to help us on another stage, namely, to the village of Baiaõ, where we had great hopes of having this, our urgent want, supplied by the military commandant of the district.

September 2nd.—The distance from Vista Alegre to Baiao is about twenty-five miles. We had but little wind, and our men were therefore obliged to row the greater part of the way. The oars used in such canoes as ours are made by tying a stout paddle to the end of a long pole by means of woody lianas. The men take their stand on a raised deck, formed by a few rough planks placed over the arched covering in the fore part of the vessel, and pull with their backs to the stern. We started at 6 a.m., and about sunset reached a point where the west channel of the river, along which we had been travelling since we left Cametá, joined a broader middle one, and formed with it a great expanse of water. The islands here seem to form two pretty regular lines, dividing the great river into three channels. As we progressed slowly, we took the montaria, and went ashore, from time to time, to the houses, which were numerous on the river banks as well as on the larger islands. In low situations they had a very unfinished appearance, being mere frameworks raised high on wooden piles, and thatched with the leaves of the Ubussú palm. In their construction another palm-tree is made much use of, viz., the Assai (Euterpe oleracea). The outer part of the stem of this species is hard and tough as horn; it is split into narrow planks, and these form a great portion of the walls and flooring. The residents told us that the western channel becomes nearly dry in the middle of the fine season, but that at high water, in April and May, the river rises to the level of the house-floors. The river bottom is everywhere sandy, and the country perfectly healthy. The people seemed to be all contented and happy, but idleness and poverty were exhibited by many unmistakeable signs. As to the flooding of their island abodes, they did not seem to care about that at all. They seem to be almost amphibious, or as much at home on the water as on land. It was really alarming to see men and women and children, in little leaky canoes laden to the water-level with bag and baggage, crossing broad reaches of river. Most of them have houses also on the terra firma, and reside in the cool palm-swamps of the Ygapó islands, as they are called, only in the hot and dry season. They live chiefly on fish, shellfish (amongst which were large Ampullariæ, whose flesh I found, on trial, to be a very tough morsel), the never-failing farinha, and the fruits of the forest. Amongst the latter the fruits of palm-tree occupied the chief place. The Assai is the most in use, but this forms a universal article of diet in all parts of the country. The fruit, which is perfectly round, and about the size of a cherry, contains but a small portion of pulp lying between the skin and the hard kernel. This is made, with the addition of water, into a thick, violet-coloured beverage, which stains the lips like blackberries. The fruit of the Mirití is also a common article of food, although the pulp is sour and unpalateable, at least to European tastes. It is boiled, and then eaten with farinha. The Tucumá (Astrocaryum tucuma), and the Mucujá (Acrocomia lasiospatha), grow only on the mainland. Their fruits yield a yellowish, fibrous pulp, which the natives eat in the same way as the Mirití. They contain so much fatty matter, that vultures and dogs devour thèm greedily.

Assai Palm
(Euterpe oleracea).

Early on the morning of September 3rd we reached the right or eastern bank, which is here from forty to sixty feet high. The houses were more substantially built than those we had hitherto seen. We succeeded in buying a small turtle; most of the inhabitants had a few of these animals, which they kept in little enclosures made with stakes. The people were of the same class everywhere, Mamelucos. They were very civil; we were not able, however, to purchase much fresh food from them. I think this was owing to their really not having more than was absolutely required to satisfy their own needs. In these districts, where the people depend for animal food solely on fishing, there is a period of the year when they suffer hunger, so that they are disposed to prize highly a small stock when they have it. They generally answered in the negative when we asked, money in hand, whether they had fowls, turtles, or eggs to sell, "Naõ ha, sinto que naõ posso lhe ser bom;" or, "Naõ ha, meu coracaõ." "We have none; I am sorry I cannot oblige you;" or, "There is none, my heart.

Sept. 3rd to *7th.*—At half-past eight a.m. we arrived at Baiaõ, which is built on a very high bank, and contains about 400 inhabitants. We had to climb to the village up a ladder, which is fixed against the bank, and on arriving at the top, took possession of a room which Senhor Seixas had given orders to be prepared for us. He himself was away at his sitio, and would not be here till the next day. We were now quite dependent on him for men to enable us to continue our voyage, and so had no remedy but to wait his leisure. The situation of the place, and the nature of the woods around it, promised well for novelties in birds and insects; so we had no reason to be vexed at the delay, but brought our apparatus and store-boxes up from the canoe, and set to work.

The easy, lounging life of the people amused us very much. I afterwards had plenty of time to become used to tropical village life. There is a free, familiar, *pro bono publico* style of living in these small places, which requires some time for a European to fall into. No sooner were we established in our rooms, than a number of lazy young fellows came to look on and make remarks, and we had to answer all sorts of questions. The houses have their doors and windows open to the street, and people walk in and out as they please; there is always, however, a more secluded apartment, where the female members of the families reside. In their familiarity there is nothing intentionally offensive, and it is practised simply in the desire to be civil and sociable. A young Mameluco, named Soares, an Escrivaõ, or public clerk, took me into his house to show me his library. I was rather surprised to see a number of well-thumbed Latin classics, Virgil, Terence, Cicero's Epistles, and Livy. I was not familiar enough, at this early period of my residence in the country, with Portuguese to converse freely with Senhor Soares, or ascertain what use he made of these books;

it was an unexpected sight, a classical library in a mud-plastered and palm-thatched hut on the banks of the Tocantins.

The prospect from the village was magnificent, over the green wooded islands, far away to the grey line of forest on the opposite shore of the Tocantins. We were now well out of the low alluvial country of the Amazons proper, and the climate was evidently much drier than it is near Pará. They had had no rain here for many weeks, and the atmosphere was hazy around the horizon; so much so that the sun, before setting, glared like a blood-red globe. At Pará this never happens; the stars and sun are as clear and sharply defined when they peep above the distant tree-tops as they are at the zenith. This beautiful transparency of the air arises, doubtless, from the equal distribution through it of invisible vapour. I shall ever remember, in one of my voyages along the Pará river, the grand spectacle that was once presented at sunrise. Our vessel was a large schooner, and we were bounding along before a spanking breeze, which tossed the waters into foam, when the day dawned. So clear was the air, that the lower rim of the full moon remained sharply defined until it touched the western horizon, whilst, at the same time, the sun rose in the east. The two great orbs were visible at the same time, and the passage from the moonlit night to day was so gentle, that it seemed to be only the brightening of dull weather. The woods around Baiaõ were of second growth, the ground having been formerly cultivated. A great number of coffee and cotton trees grew amongst the thickets. A fine woodland pathway extends for miles over the high, undulating bank, leading from one house to another along the edge of the cliff. I went into several of them, and talked to their inmates. They were all poor people. The men were out fishing, some far away, a distance of many days' journey; the women plant mandioca, make the farinha, spin and weave cotton, manufacture soap of burnt cacao shells and andiroba oil, and follow various other domestic employments. I asked why they allowed their plantations to run to waste. They said that it was useless trying to plant anything hereabout; the Saüba ant devoured the young coffee trees, and every one who attempted to contend against this universal ravager was sure to be defeated. The country, for many miles along the banks of the river, seemed to be well peopled. The inhabitants were nearly all of the tawny-white mameluco class. I saw a good many mulattos, but very few negroes and Indians, and none that could be called pure whites.

When Senhor Seixas arrived, he acted very kindly. He provided us at once with two men, killed an ox in our honour, and treated us altogether with great consideration. We were not, however, introduced to his family. I caught a glimpse once of his wife, a prettly little mameluco woman, as she was tripping with a young girl, whom I supposed to be her daughter, across the back yard. Both wore long dressing-gowns, made of bright-coloured calico print, and had long wooden tobacco-pipes in their mouths. The room in which we slept and worked had formerly served as a storeroom for cacao, and at night I was kept awake for hours by rats and cockroaches, which swarm in all such places. The latter were running about all over the walls; now and then one would come suddenly with a whirr full at my face, and get under my shirt if I attempted to jerk it off. As to the rats, they were chasing one another by dozens all night long, over the floor, up and down the edges of the doors, and along the rafters of the open roof.

September 7th.—We started from Baiaõ at an early hour. One of our new men was a good-humoured, willing young mulatto, named José; the other was a sulky Indian, called Manoel, who seemed to have been pressed into our service against his will. Senhor Seixas, on parting, sent a quantity of fresh provisions on board. A few miles above Baiao the channel became very shallow; we got aground several times, and the men had to disembark and shove the vessel off. Alexandro here shot several fine fish, with bow and arrow. It was the first time I had seen fish captured in this way. The arrow is a reed, with a steel barbed point, which is fixed in a hole at the end, and secured by fine twine made from the fibres of pine-apple leaves. It is only in the clearest water that fish can be thus shot: and the only skill required is to make, in taking aim, the proper allowance for refraction.

The next day before sunrise a fine breeze sprang up, and the men awoke and set the sails. We glided all day through channels between islands with long white sandy beaches, over which, now and then, aquatic and wading birds were seen running. The forest was low, and had a harsh, dry aspect. Several palm trees grew here which we had not before seen. On low bushes, near the water, pretty red-headed tanagers (Tanagra gularis) were numerous, flitting about and chirping like sparrows. About half-past four p.m. we brought to at the mouth of a creek or channel, where there was a great extent of sandy beach. The sand had been blown by the wind into ridges and undulations, and over the moister parts large flocks of sandpipers were running about. Alexandro and I had a long ramble over the rolling plain, which came as an agreeable change after the monotonous forest scenery amid which we had been so long travelling. He pointed out to me the tracks of a huge jaguar on the sand. We found here, also, our first turtle's nest, and obtained 120 eggs from it, which were laid at a depth of nearly two feet from the surface, the mother first excavating a hole, and afterwards covering it up with sand. The place is discoverable only by following the tracks of the turtle from the water. I saw here an alligator for the first time, which reared its head and shoulders above the water just after I had taken a bath near the spot. The night was calm and cloudless, and we employed the hours before bed-time in angling by moonlight.

On the 10th we reached a small settlement called Patos, consisting of about a dozen houses, and built on a high rocky bank, on the eastern shore. The rock is the same nodular conglomerate which is found at so many places, from the sea-coast to a distance of 600 miles up the Amazons. Mr. Leavens made a last attempt here to engage men to accompany us to the Araguaya; but it was in vain; not a soul could be induced by any amount of wages to go on such an expedition. The reports as to the existence of cedar were very vague. All said that the tree was plentiful somewhere, but no one could fix on the precise locality. I believe that the cedar grows, like all other forest trees, in a scattered way, and not in masses anywhere. The fact of its being the principal tree observed floating down with the current of the Amazons, is to be explained by its wood being much lighter than that of the majority of trees. When the banks are washed away by currents, trees of all species fall into the river; but the heavier ones, which are the most numerous, sink, and the lighter, such as the cedar, alone float down to the sea.

Mr. Leavens was told that there were cedar trees at Trocará, on the opposite side of the river, near some fine rounded hills covered with forest, visible from Patos; so there we went. We found here several families encamped in a delightful spot. The shore sloped gradually down to the water, and was shaded by a few wide-spreading trees. There was no underwood. A great number of hammocks were seen slung between the tree trunks, and the litter of a numerous household lay scattered about. Women, old and young, some of the latter very good-looking, and a large number of children, besides pet animals, enlivened the encampment. They were all half-breeds, simple, well-disposed people, and explained to us that they were inhabitants of Cametá, who had come thus far, eighty miles, to spend the summer months. The only motive they could give for coming was, "that it was so hot in the town in the veraõ (summer), and they were all so fond of fresh fish." Thus these simple folks think nothing of leaving home and business to come on a three months' pic-nic. It is the annual custom of this class of people, throughout the province, to spend a few months of the fine season in the wilder parts of the country. They carry with them all the farinha they can scrape together, this being the only article of food necessary to provide. The men hunt and fish for the day's wants, and sometimes collect a little india-rubber, salsaparilla, or copaiba oil, to sell to traders on their return; the women assist in paddling the canoes, do the cooking, and sometimes fish with rod and line. The weather is enjoyable the whole time, and so days and weeks pass happily away.

One of the men volunteered to walk with us into the forest, and show us a few cedar trees. We passed through a mile or two of spiny thickets, and at length came upon the banks of the rivulet Trocará, which flows over a stony bed, and, about a mile above its mouth, falls over a ledge of rocks, thus forming a very pretty cascade. In the neighbourhood we found a number of specimens of a curious land-shell, a large flat Helix, with a labyrinthine mouth (Anastoma). We learnt afterwards that it was a species which had been discovered a few years previously by Dr. Gardner, the botanist, on the upper part of the Tocantins.

We saw here, for the first time, the splendid hyacinthine macaw (Macrocercus hyacinthinus, Lath., the Araruna of the natives), one of the finest and rarest species of the Parrot family. It only occurs in the interior of Brazil, from 16° S. lat. to the southern border of the Amazons valley. It is three feet long from the beak to the tip of the tail, and is entirely of a soft hyacinthine blue colour, except round the eyes, where the skin is naked and white. It flies in pairs, and feeds on the hard nuts of several palms, but especially of the Mucuja (Acrocomia lasiospatha). These nuts, which are so hard as to be difficult to break with a heavy hammer, are crushed to a pulp by the powerful beak of this macaw.

Being unable to obtain men, Mr. Leavens now gave up his

project of ascending the river as far as the Araguaya. He assented to our request, however, to ascend to the cataracts near Arroyos. We started therefore from Patos with a more definite aim before us than we had hitherto had. The river became more picturesque as we advanced. The water was very low, it being now the height of the dry season; the islands were smaller than those further down, and some of them were high and rocky. Bold wooded bluffs projected into the stream, and all the shores were fringed with beaches of glistening white sand. On one side of the river there was an extensive grassy plain or campo with isolated patches of trees scattered over it. On the 14th and following day we stopped several times to ramble ashore. Our longest excursion was to a large shallow lagoon, choked up with aquatic plants, which lay about two miles across the campo. At a place called Juquerapuá we engaged a pilot to conduct us to Arroyos, and a few miles above the pilot's house, arrived at a point where it was not possible to advance further in our large canoe, on account of the rapids.

September 16th.—Embarked at six a.m. in a large montaria which had been lent to us for this part of the voyage by Senhor Seixas, leaving the vigilinga anchored close to a rocky islet, named Santa Anna, to await our return. At ten a.m. we arrived at the first rapids, which are called Tapaiunaquára. The river, which was here about a mile wide, was choked up with rocks, a broken ridge passing completely across it. Between these confused piles of stone the currents were fearfully strong, and formed numerous eddies and whirlpools. We were obliged to get out occasionally and walk from rock to rock, whilst the men dragged the canoe over the obstacles. Beyond Tapaiunaquára the stream became again broad and deep, and the river scenery was beautiful in the extreme. The water was clear, and of a bluish-green colour. On both sides of the stream stretched ranges of wooded hills, and in the middle picturesque islets rested on the smooth water, whose brilliant green woods fringed with palms formed charming bits of foreground to the perspective of sombre hills fading into grey in the distance. Joaquim pointed out to us grove after grove of Brazil-nut trees (Bertholletia excelsa) on the mainland. This is one of the chief collecting grounds for this nut. The tree is one of the loftiest in the forest, towering far above its fellows; we could see the woody fruits, large and round as cannon-balls, dotted over the branches. The currents were very strong in some places, so that during the greater part of the way the men preferred to travel near the shore, and propel the boat by means of long poles.

We arrived at Arroyos about four o'clock in the afternoon, after ten hours' hard pull. The place consists simply of a few houses built on a high bank, and forms a station where canoemen from the mining countries of the interior of Brazil stop to rest themselves, before or after surmounting the dreaded falls and rapids of Guaribas, situated a couple of miles further up. We dined ashore, and in the evening again embarked to visit the falls. The vigorous and successful way in which our men battled with the terrific currents excited our astonishment. The bed of the river, here about a mile wide, is strewn with blocks of various sizes, which lie in the most irregular manner, and between them rush currents of more or less rapidity. With an accurate knowledge of the place and skilful management, the falls can be approached in small canoes by threading the less dangerous channels. The main fall is about a quarter of a mile wide: we climbed to an elevation overlooking it, and had a good view of the cataract. A body of water rushes with terrific force down a steep slope, and boils up with deafening roar around the boulders which obstruct its course. The wildness of the whole scene was very impressive. As far as the eye could reach, stretched range after range of wooded hills, scores of miles of beautiful wilderness, inhabited only by scanty tribes of wild Indians. In the midst of such a solitude the roar of the cataract seemed fitting music.

September 17th.—We commenced early in the morning our downward voyage. Arroyos is situated in about 4° 10′ S. lat.; and lies, therefore, about 130 miles from the mouth of the Tocantins. Fifteen miles above Guaribas another similar cataract, called Tabocas, lies across the river. We were told that there were in all fifteen of these obstructions to navigation between Arroyos and the mouth of the Araguaya. The worst was the Inferno, the Guaribas standing second to it in evil reputation. Many canoes and lives have been lost here, most of the accidents arising through the vessels being hurled against an enormous cubical mass of rock called the Guaribinha, which we, on our trip to the falls in the small canoe, passed round with the greatest ease about a quarter of a mile below the main falls. This, however, was the dry season; in the time of full waters a tremendous current sets against it. We descended the river rapidly, and found it excellent fun shooting the rapids. The men seemed to delight in choosing the swiftest parts of the current; they sang and yelled in the greatest excitement, working the paddles with great force, and throwing clouds of spray above us as we bounded downwards. We stopped to rest at the mouth of a rivulet named Caganxa. The pilot told us that gold had been found in the bed of this brook; so we had the curiosity to wade several hundred yards through the icy cold waters in search of it. Mr. Leavens seemed very much interested in the matter; he picked up all the shining stones he could espy in the pebbly bottom, in hopes of finding diamonds also. There is, in fact, no reason why both gold and diamonds should not be found here, the hills being a continuation of those of the mining countries of interior Brazil, and the brooks flowing through the narrow valleys between them.

On arriving at the place where we had left our canoe, we stayed all night and part of the following day, and I had a stroll along a delightful pathway, which led over hill and dale, two or three miles through the forest. I was surprised at the number and variety of brilliantly-coloured butterflies; they were all of small size, and started forth at every step I took, from the low bushes which bordered the road. I first heard here the notes of a trogon; it was seated alone on a branch, at no great elevation; a beautiful bird, with glossy-green back and rose-coloured breast (probably Trogon melanurus). At intervals it uttered, in a complaining tone, a sound resembling the words "quá, quá." It is a dull inactive bird, and not very ready to take flight when approached. In this respect, however, the trogons are not equal to the jacamars, whose stupidity in remaining at their posts, seated on low branches in the gloomiest shades of the forest, is somewhat remarkable in a country where all other birds are exceedingly wary. One species of jacamar was not uncommon here (Galbula viridis); I sometimes saw two or three together, seated on a slender branch, silent and motionless with the exception of a slight movement of the head; when an insect flew past within a short distance, one of the birds would dart off, seize it, and return again to its sitting-place. The trogons are found in the tropics of both hemispheres; the jacamars, which are clothed in plumage of the most beautiful golden-bronze and steel colours, are peculiar to tropical America.

At night I slept ashore as a change from the confinement of the canoe, having obtained permission from Senhor Joaquim to sling my hammock under his roof. The house, like all others in these out-of-the-way parts of the country, was a large, open, palm-thatched shed, having one end inclosed by means of partitions also made of palm-leaves, so as to form a private apartment. Under the shed were placed all the household utensils; earthenware jars, pots, and kettles, hunting and fishing implements, paddles, bows and arrows, harpoons, and so forth. One or two common wooden chests serve to contain the holiday clothing of the females; there is no other furniture, except a few stools and the hammock, which answers the purposes of chair and sofa. When a visitor enters, he is asked to sit down in a hammock: persons who are on intimate terms with each other recline together in the same hammock, one at each end; this is a very convenient arrangement for friendly conversation. There are neither tables nor chairs; the cloth for meals is spread on a mat, and the guests squat round in any position they choose. There is no cordiality of manners, but the treatment of the guests shows a keen sense of the duties of hospitality on the part of the host. There is a good deal of formality in the intercourse of these half-wild mamelucos, which, I believe, has been chiefly derived from their Indian forefathers, although a little of it may have been copied from the Portuguese.

A little distance from the house were the open sheds under which the farinha for the use of the establishment was manufactured. In the centre of each shed stood the shallow pans, made of clay and built over ovens, where the meal is roasted. A long flexible cylinder made of the peel of a marantaceous plant, plaited into the proper form, hung suspended from a beam; it is in this that the pulp of the mandioca is pressed, and from it the juice, which is of a highly poisonous nature, although the pulp is wholesome food, runs into pans placed beneath to receive it. A wooden trough, such as is used in all these places for receiving the pulp before the poisonous matter is extracted, stood on the ground, and from the posts hung the long wicker-work baskets, or aturas, in which the women carry the roots from the roca or clearing; a broad ribbon made from the inner bark of the monguba tree is attached to the rims of the baskets, and is passed round the forehead of the carriers, to relieve their backs in supporting the heavy load. Around the shed were planted a number of banana and other fruit trees; amongst them were the never-failing capsicum-pepper bushes, brilliant as holly trees at Christmas time, with their fiery-red fruit, and lemon trees; the one supplying the pungent, the other the acid, for sauce to the perpetual meal of fish. There is never in such places any appearance of careful cultivation, no garden or orchard; the useful trees are surrounded by weeds and bushes, and close behind rises the everlasting forest.

In descending the river we landed frequently, and Mr. Wallace and I lost no chance of adding to our collections; so that before the end of our journey we had got together a very considerable number of birds, insects, and shells, chiefly taken, however, in the low country. Leaving Baiao, we took our last farewell of the limpid waters and varied scenery of the upper river, and found ourselves again in the humid flat region of the Amazons valley. We sailed down this lower part of the river by a different channel from the one we travelled along in ascending, and frequently went ashore on the low islands in mid-river. As already stated, these are covered with water in the wet season; but at this time, there having been three months of fine weather, they were dry throughout, and, by the subsidence of the waters, placed four or five feet above the level of the river. They are covered with a most luxuriant forest, comprising a large number of india-rubber trees. We found several people encamped here, who were engaged in collecting and preparing the rubber, and thus had an opportunity of observing the process.

The tree which yields this valuable sap is the Siphonia elastica, a member of the Euphorbiaceous order; it belongs, therefore, to a group of plants quite different from that which furnishes the caoutchouc of the East Indies and Africa. This latter is the product of different species of Ficus, and is considered, I believe, in commerce an inferior article to the india-rubber of Pará. The Siphonia elastica grows only on the lowlands in the Amazons region; hitherto the rubber has been collected chiefly in the islands and swampy parts of the mainland within a distance of fifty to a hundred miles to the west of Pará; but there are plenty of untapped trees still growing in the wilds of the Tapajos, Madeira, Juruá, and Jauarí, as far as 1800 miles from the Atlantic coast. The tree is not remarkable in appearance; in bark and foliage it is not unlike the European ash; but the trunk, like that of all forest trees, shoots up to an immense height before throwing off branches. The trees seem to be no man's property hereabout. The people we met with told us they came every year to collect rubber on these islands, as soon as the waters had subsided, namely, in

August, and remained till January or February. The process is very simple. Every morning each person, man or woman, to whom is allotted a certain number of trees, goes the round of the whole, and collects in a large vessel the milky sap which trickles from gashes made in the bark on the preceding evening, and which is received in little clay cups, or in ampullaria shells stuck beneath the wounds. The sap, which at first is of the consistence of cream, soon thickens; the collectors are provided with a great number of wooden moulds of the shape in which the rubber is wanted, and when they return to the camp they dip them into the liquid, laying on, in the course of several days, one coat after another. When this is done, the substance is white and hard; the proper colour and consistency are given by passing it repeatedly through a thick black smoke obtained by burning the nuts of certain palm trees, after which process the article is ready for sale. India-rubber is known throughout the province only by the name of seringa, the Portuguese word for syringe; it owes this appellation to the circumstance that it was in this form only that the first Portuguese settlers noticed it to be employed by the aborigines. It is said that the Indians were first taught to make syringes of rubber by seeing natural tubes formed by it, when the spontaneously-flowing sap gathered round projecting twigs. Brazilians of all classes still use it extensively in the form of syringes, for injections form a great feature in the popular system of cures; the rubber for this purpose is made into a pear-shaped bottle, and a quill fixed in the long neck.

September 24th.—Opposite Cametá the islands are all planted with cacao, the tree which yields the chocolate nut. The forest is not cleared for the purpose, but the cacao plants are stuck in here and there almost at random amongst the trees. There are many houses on the banks of the river, all elevated above the swampy soil on wooden piles, and furnished with broad ladders by which to mount to the ground floor. As we passed by in our canoe we could see the people at their occupations in the open verandahs, and in one place saw a ball going on in broad daylight; there were fiddles and guitars hard at work, and a number of lads in white shirts and trousers dancing with brown damsels clad in showy print dresses. The cacao tree produces a curious impression, on account of the flowers and fruit growing directly out of the trunk and branches. There is a whole group of wild fruit trees which have the same habit in this country. In the wildernesses where the cacao is planted, the collecting of the fruit is dangerous from the number of poisonous snakes which inhabit the places. One day, when we were running our montaria to a landing-place, we saw a large serpent on the trees overhead, as we were about to brush past; the boat was stopped just in the nick of time, and Mr. Leavens brought the reptile down with a charge of shot.

September 26th.—At length we got clear of the islands, and saw once more before us the sea-like expanse of waters which forms the mouth of the Tocantins. The river had now sunk to its lowest point, and numbers of fresh-water dolphins were rolling about in shoaly places. There are here two species, one of which was new to science when I sent specimens to England; it is called the Tucuxí (Steno tucuxi of Gray). When it comes to the surface to breathe, it rises horizontally, showing first its back fin; draws an inspiration, and then dives gently down, head foremost. This mode of proceeding distinguishes the Tucuxí at once from the other species, which is called Bouto or porpoise by the natives (Inia Geoffroyi of Desmarest). When this rises, the top of the head is the part first seen; it then blows, and immediately afterwards dips head downwards, its back curving over, exposing successively the whole dorsal ridge with its fin. It seems thus to pitch heels over head, but does not show the tail fin. Besides this peculiar motion, it is distinguished from the Tucuxí by its habit of generally going in pairs. Both species are exceedingly numerous throughout the Amazons and its larger tributaries, but they are nowhere more plentiful than in the shoaly water at the mouth of the Tocantins, especially in the dry season. In the Upper Amazons a third pale flesh-coloured species is also abundant (the Delphinus pallidus of Gervais). With the exception of a species found in the Ganges, all other varieties of dolphin inhabit exclusively the sea. In the broader parts of the Amazons, from its mouth to a distance of fifteen hundred miles in the interior, one or other of the three kinds here mentioned are always heard rolling, blowing, and snorting, especially at night, and these noises contribute much to the impression of sea-wide vastness and desolation which haunts the traveller. Besides dolphins in the water, frigate birds in the air are characteristic of this lower part of the Tocantins. Flocks of them were seen the last two or three days of our journey, hovering above at an immense height. Towards night we were obliged to cast anchor over a shoal in the middle of the river to await the ebb tide. The wind blew very strongly, and this, together with the incoming flow, caused such a heavy sea that it was impossible to sleep. The vessel rolled and pitched until every bone in our bodies ached with the bumps we received, and we were all more or less sea-sick. On the following day we entered the Anapú, and on the 30th of September, after threading again the labyrinth of channels communicating between the Tocantins and the Mojú, arrived at Pará.

I will now give a short account of Cametá, the principal town on the banks of the Tocantins, which I visited for the second time in June, 1849; Mr. Wallace, in the same month, departing from Pará to explore the rivers Guamá and Capim. I embarked as passenger in a Cametá trading vessel, the St. John, a small schooner of thirty tons burthen. I had learnt by this time that the only way to attain the objects for which I had come to this country was to accustom myself to the ways of life of the humbler classes of the inhabitants. A traveller on the Amazons gains little by being furnished with letters of recommendation to persons of note, for in the great interior wildernesses of forest and river the canoe-men have pretty much their own way; the authorities cannot force them to grant passages or to hire themselves to travellers, and therefore a stranger is obliged to ingratiate himself with them in order to get conveyed from place to place. I thoroughly enjoyed the journey to Cametá; the weather was again beautiful in the extreme. We started from Pará at sunrise on the 8th of June, and on the 10th emerged from the narrow channels of the Anapu into the broad Tocantins. The vessel was so full of cargo, that there was no room to sleep in the cabin; so we passed the nights on deck. The captain or supercargo, called in Portuguese *cabo*, was a mameluco, named Manoel, a quiet, good-humoured person, who treated me with the most unaffected civility during the three days' journey. The pilot was also a mameluco, named John Mendez, a handsome young fellow, full of life and spirit. He had on board a wire guitar or viola, as it is here called; and in the bright moonlight nights, as we lay at anchor hour after hour waiting for the tide, he enlivened us all with songs and music. He was on the best of terms with the cabo, both sleeping in the same hammock slung between the masts. I passed the nights wrapped in an old sail outside the roof of the cabin. The crew, five in number, were Indians and half-breeds, all of whom treated their two superiors with the most amusing familiarity, yet I never sailed in a better managed vessel than the St. John.

In crossing to Cametá we had to await the flood-tide in a channel called Entre-as-Ilhas, which lies between two islands in mid-river, and John Mendez, being in good tune, gave us an extempore song, consisting of a great number of verses. The canoe-men of the Amazons have many songs and choruses, with which they are in the habit of relieving the monotony of their slow voyages, and which are known all over the interior. The choruses consist of a simple strain, repeated almost to weariness, and sung generally in unison, but sometimes with an attempt at harmony. There is a wildness and sadness about the tunes which harmonise well with, and in fact are born of, the circumstances of the canoe-man's life: the echoing channels, the endless gloomy forests, the solemn nights, and the desolate scenes of broad and stormy waters and falling banks. Whether they were invented by the Indians or introduced by the Portuguese it is hard to decide, as many of the customs of the lower classes of Portuguese are so similar to those of the Indians, that they have become blended with them. One of the commonest songs is very wild and pretty. It has for refrain the words "Mai, Mai" ("Mother, Mother"), with a long drawl on the second word. The stanzas are very variable. The best wit on board starts the verse, improvising as he goes on, and the others join in the chorus. They all relate to the lonely river life and the events of the voyage; the shoals, the wind; how far they shall go before they stop to sleep, and so forth. The sonorous native names of places, Goajará, Tucumandúba, &c., add greatly to the charm of the wild music. Sometimes they bring in the stars thus:—

A lua está sahindo,
Mai, Mai!
A lua está sahindo,
Mai, Mai!
As sete estrellas estaõ chorando,
Mai, Mai!
Por s'acharem desamparados,
Mai, Mai!

The moon is rising,
Mother, Mother!
The moon is rising,
Mother, Mother!

The seven stars (Pleiades) are weeping,
Mother, Mother!
To find themselves forsaken,
Mother, Mother!

I fell asleep about ten o'clock, but at four in the morning John Mendez woke me to enjoy the sight of the little schooner tearing through the waves before a spanking breeze. The night was transparently clear and almost cold, the moon appeared sharply defined against the dark blue sky, and a ridge of foam marked where the prow of the vessel was cleaving its way through the water. The men had made a fire in the galley to make tea of an acid herb, called *erva cidreira*, a quantity of which they had gathered at the last landing-place, and the flames sparkled cheerily upwards. It is at such times as these that Amazon travelling is enjoyable, and one no longer wonders at the love which many, both natives and strangers. have for this wandering life. The little schooner sped rapidly on with booms bent and sails stretched to the utmost. Just as day dawned, we ran with scarcely slackened speed into the port of Cametá, and cast anchor.

I stayed at Cametá until the 16th of July, and made a considerable collection of the natural productions of the neighbourhood. The town in 1849 was estimated to contain about 5,000 inhabitants, but the municipal district of which Cametá is the capital numbers 20,000; this, however, comprised the whole of the lower part of the Tocantins, which is the most thickly populated part of the province of Pará. The productions of the district are cacao, india-rubber, and Brazil nuts. The most remarkable feature in the social aspect of the place is the hybrid nature of the whole population, the amalgamation of the white and Indian races being here complete. The aborigines were originally very numerous on the western bank of the Tocantins, the principal tribe having been the Camútas, from which the city takes its name. They were a superior nation, settled, and attached to agriculture, and received with open arms the white immigrants who were attracted to the district by its fertility, natural beauty, and the healthfulness of the climate. The Portuguese settlers were nearly all males, the Indian women were good-looking, and made excellent wives; so the natural result has been, in the course of two centuries, a complete blending of the two races. There is now, however, a considerable infusion of negro blood in the mixture, several hundred African slaves having been introduced during the last seventy years. The few whites are chiefly Portuguese, but there are also two or three Brazilian families of pure European descent. The town consists of three long streets, running parallel to the river, with a few shorter ones crossing them at right angles. The houses are very plain, being built, as usual in this country, simply of a strong framework, filled up with mud, and coated with white plaster. A few of them are of two or three stories. There are three churches, and also a small theatre, where a company of native actors at the time of my visit were representing light Portuguese plays with considerable taste and ability. The people have a reputation all over the province for energy and perseverance ; and it is often said, that they are as keen in trade as the Portuguese. The lower classes are as indolent and sensual here as in other parts of the province, a moral condition not to be wondered at in a country where perpetual summer reigns, and where the necessaries of life are so easily obtained. But they are light-hearted. quick-witted, communicative, and hospitable. I found here a native poet, who had written some pretty verses, showing an appreciation of the natural beauties of the country, and was told that the Archbishop of Bahia, the primate of Brazil, was a native of Cametá. It is interesting to find the mamelucos displaying talent and enterprise, for it shows that degeneracy does not necessarily result from the mixture of white and Indian blood. The Cametaenses boast, as they have a right to do, of theirs being the only large town which resisted successfully the anarchists in the great rebellion of 1835-6. Whilst the whites of Pará were submitting to the rule of half-savage revolutionists, the mamelucos of Cametá placed themselves under the leadership of a courageous priest, named Prudencio ; armed themselves, fortified the place, and repulsed the large forces which the insurgents of Pará sent to attack the place. The town not only became the refuge for all loyal subjects, but was a centre whence large parties of volunteers sallied forth repeatedly to attack the anarchists in their various strongholds.

The forest behind Cametá is traversed by several broad roads, which lead over undulating ground many miles into the interior. They pass generally under shade, and part of the way through groves of coffee and orange trees, fragrant plantations of cacao, and tracts of second-growth woods. The narrow brook-watered valleys, with which the land is intersected, alone have remained clothed with primæval forest, at least near the town. The houses along these beautiful roads belong chiefly to mameluco, mulatto, and Indian families, each of which has its own small plantation. There are only a few planters with larger establishments, and these have seldom more than a dozen slaves. Besides the main roads, there are endless bye-paths which thread the forest and communicate with isolated houses. Along these the traveller may wander day after day without leaving the shade, and everywhere meet with cheerful, simple, and hospitable people.

Soon after landing I was introduced to the most distinguished citizen of the place, Dr. Angelo Custodio Correia, whom I have already mentioned. This excellent man was a favourable specimen of the highest class of native Brazilians. He had been educated in Europe, was now a member of the Brazilian Parliament, and had been twice president of his native province. His manners were less formal, and his goodness more thoroughly genuine, perhaps, than is the rule generally with Brazilians. He was admired and loved, as I had ample opportunity of observing, throughout all Amazonia. He sacrificed his life in 1855, for the good of his fellow-townsmen, when Cametà was devastated by the cholera; having stayed behind with a few heroic spirits to succour invalids and direct the burying of the dead, when nearly all the chief citizens had fled from the place. After he had done what he could he embarked for Para, but was himself then attacked with cholera, and died on board the steamer before he reached the capital. Dr. Angelo received me with the usual kindness which he showed to all strangers. He procured me, unsolicited, a charming country house, free of rent, hired a mulatto servant for me, and thus relieved me of the many annoyances and delays attendant on a first arrival in a country town where even the name of an inn is unknown. The rocinha thus given up for my residence belonged to a friend of his, Senhor José Raimundo Furtado, a stout florid-complexioned gentleman, such a one as might be met with any day in a country town in England. To him also I was indebted for many acts of kindness.

The rocinha was situated near a broad grassy road bordered by lofty woods, which leads from Cametá to the Aldeia, a village two miles distant. My first walks were along this road. From it branches another similar but still more picturesque road, which runs to Curimá and Pacajá, two small settlements, several miles distant, in the heart of the forest. The Curimá

BIRD-KILLING SPIDER (MYGALE AVICULARIA) ATTACKING FINCHES.

road is beautiful in the extreme. About half a mile from the house where I lived it crosses a brook flowing through a deep dell, by means of a long rustic wooden bridge. The virgin forest is here left untouched; numerous groups of slender palms, mingled with lofty trees overrun with creepers and parasites, fill the shady glen, and arch over the bridge, forming one of the most picturesque scenes imaginable. A little beyond the bridge there was an extensive grove of orange and other trees, which yielded me a rich harvest. The Aldeia road runs parallel to the river, the land from the border of the road to the indented shore of the Tocantins forming a long slope which was also richly wooded; this slope was threaded by numerous shady paths, and abounded in beautiful insects and birds. At the opposite or southern end of the town there was a broad road called the Estrada da Vacaria; this ran along the banks of the Tocantins at some distance from the river, and continued over hill and dale, through bamboo thickets and palm swamps, for about fifteen miles.

At Cametá I chanced to verify a fact relating to the habits of a large hairy spider of the genus Mygale, in a manner worth recording. The species was M. avicularia, or one very closely allied to it; the individual was nearly two inches in length of body, but the legs expanded seven inches, and the entire body and legs were covered with coarse grey and reddish hairs. I was attracted by a movement of the monster on a tree-trunk; it was close beneath a deep crevice in the tree, across which was stretched a dense white web. The lower part of the web was broken, and two small birds, finches, were entangled in the pieces; they were about the size of the English siskin, and I judged the two to be male and female. One of them was quite dead, the other lay under the body of the spider not quite dead, and was smeared with the filthy liquor or saliva exuded by the monster. I drove away the spider and took the birds, but the second one soon died. The fact of species of Mygale sallying forth at night, mounting trees, and sucking the eggs and young of humming-birds, has been recorded long ago by Madam Merian and Palisot de Beauvois; but, in the absence of any confirmation, it has come to be discredited. From the way the fact has been related it would appear that it had been merely derived from the report of natives, and had not been witnessed by the narrators. Count Langsdorff, in his "Expedition into the Interior of Brazil," states that he totally disbelieved the story. I found the circumstance to be quite a novelty to the residents hereabout. The Mygales are quite common insects: some species make their cells under stones, others form artistical tunnels in the earth, and some build their dens in the thatch of houses. The natives call them Aranhas carangueijeiras, or crab-spiders. The hairs with which they are clothed come off when touched, and cause a peculiar and almost maddening irritation. The first specimen that I killed and prepared was handled incautiously, and I suffered terribly for three days afterwards. I think this is not owing to any poisonous quality residing in the hairs, but to their being short and hard, and thus getting into the fine creases of the skin. Some Mygales are of immense size. One day I saw the children belonging to an Indian family, who collected for me, with one of these monsters secured by a cord round its waist, by which they were leading it about the house as they would a dog.

The only monkeys I observed at Cametá were the Couxio (Pithecia Satanas)—a large species, clothed with long brownish-black hair—and the tiny Midas argentatus. The Couxio has a thick bushy tail, and the hair of the head, which looks as if it had been carefully combed, sits on it like a wig. It inhabits only the most retired parts of the forest, on the terra firma, and I observed nothing of its habits. The little Midas argentatus is one of the rarest of the American monkeys; indeed, I have not heard of its being found anywhere except near Cametá, where I once saw three individuals, looking like so many white kittens, running along a branch in a cacao grove: in their motions they resembled precisely the Midas ursulus already described. I saw afterwards a pet animal of this species, and heard that there were many so kept, and that they were esteemed as great treasures. The one mentioned was full-grown, although it measured only seven inches in length of body. It was covered with long white silky hairs, the tail being blackish, and the face nearly naked and flesh-coloured. It was a most timid and sensitive little thing. The woman who owned it carried it constantly in her bosom, and no money would induce her to part with her pet. She called it Mico. It fed from her mouth and allowed her to fondle it freely, but the nervous little creature would not permit strangers to touch it. If any one attempted to do so, it shrank back, the whole body trembling with fear, and its teeth chattered whilst it uttered its tremulous frightened tones. The expression of its features was like that of its more robust brother Midas ursulus; the eyes, which were black, were full of curiosity and mistrust, and were always kept fixed on the person who attempted to advance towards it.

In the orange groves and other parts humming-birds were plentiful, but I did not notice more than three species. I saw one day a little pigmy belonging to the genus Phaethornis in the act of washing itself in a brook; perched on a thin branch, one end of which was under water. It dipped itself, then fluttered its wings and pruned its feathers, and seemed thoroughly to enjoy itself, alone in the shady nook which it had chosen—a place overshadowed by broad leaves of ferns and Heliconiæ. I thought, as I watched it, that there was no need for poets to invent elves and gnomes, whilst Nature furnishes us with such marvellous little sprites ready to hand.

My return-journey to Pará afforded many incidents characteristic of Amazonian travelling. I left Cametá on the 16th of July. My luggage was embarked in the morning in the Santa Rosa, a vessel of the kind called cuberta, or covered canoe. The cuberta is very much used on these rivers. It is not decked, but the sides forward are raised and arched over, so as to admit of cargo being piled high above the water-line. At the stern is a neat square cabin, also raised, and between the cabin and covered forepart is a narrow piece decked over, on which are placed the cooking arrangements. This is called the tombadilha or quarterdeck, and when the canoe is heavily laden it goes under water as the vessel heels over to the wind. There are two masts, rigged with fore and aft sails. The foremast has often, besides, a main and top sail. The forepart is planked over at the top, and on this raised deck the crew work the vessel, pulling it along, when there is no wind, by means of the long oars already described.

As I have just said, my luggage was embarked in the morning. I was informed that we should start with the ebb-tide in the afternoon; so I thought I should have time to pay my respects to Dr. Angelo and other friends, whose extreme courtesy and goodness had made my residence at Cametá so agreeable. After dinner the guests, according to custom at the house of the Correias, walked into the cool verandah which overlooks the river; and there we saw the Santa Rosa, a mere speck in the offing miles away, tacking down river with a fine breeze. I was now in a fix, for it would be useless attempting to overtake the cuberta, and besides the sea ran too high for any montaria. I was then told, that I ought to have been aboard hours before the time fixed for starting, because when a breeze springs up, vessels start before the tide turns; the last hour of the flood not being very strong. All my precious collections, my clothes, and other necessaries were on board, and it was indispensable that I should be at Pará when the things were disembarked. I tried to hire a montaria and men, but was told that it would be madness to cross the river in a small boat with this breeze. On going to Senhor Laroque, another of my Cametá friends, I was relieved of my embarrassment; for I found there an English gentleman, Mr. Patchett of Pernambuco, who was visiting Pará and its neighbourhood on his way to England, and who, as he was going back to Pará in a small boat with four paddles, which would start at midnight, kindly offered me a passage. The evening from seven to ten o'clock was very stormy. About seven, the night became intensely dark, and a terrific squall of wind burst forth, which made the loose tiles fly over the housetops; to this succeeded lightning and stupendous claps of thunder, both nearly simultaneous. We had had several of these short and sharp storms during the past month. At midnight, when we embarked, all was as calm as though a ruffle had never disturbed air, forest, or river. The boat sped along like an arrow to the rhythmic paddling of the four stout youths we had with us, who enlivened the passage with their wild songs. Mr. Patchett and I tried to get a little sleep, but the cabin was so small and encumbered with boxes placed at all sorts of angles, that we found sleep impossible. I was just dozing when the day dawned, and, on awaking, the first object I saw was the Santa Rosa, at anchor beside a green island in mid-river. I preferred to make the remainder of the voyage in company of my collections, so bade Mr. Patchett good-day. The owner of the Santa Rosa, Senhor Jacinto Machado, whom I had not seen before, received me aboard, and apologised for having started without me. He was a white man, a planter, and was now taking his year's produce of cacao, about twenty tons, to Pará. The canoe was very heavily laden, and I was rather alarmed to see that it was leaking at all points. The crew were all in the water, diving about to feel for the holes, which they stopped with pieces of rag and clay, and an old negro was baling the water out of the hold. This was a pleasant prospect for a three-days' voyage! Senhor Machado treated it as the most ordinary incident possible: "It was always likely to leak, for it was an old vessel that had been left as worthless high and dry on the beach, and he had bought it very cheap."

When the leaks were stopped, we proceeded on our journey, and at night reached the mouth of the Anapú. I wrapped myself in an old sail, and fell asleep on the raised deck. The next day we threaded the Igarapé-mirim, and on the 19th descended the Mojú. Senhor Machado and I by this time had become very good friends. At every interesting spot on the banks of the Mojú, he manned the small boat and took me ashore. There are many large houses on this river, belonging to what were formerly large and flourishing plantations, but which, since the Revolution of 1835-6, had been suffered to go to decay. Two of the largest buildings were constructed by the Jesuits in the early part of the last century. We were told that there were formerly eleven large sugar-mills on the banks of the Mojú, whilst now there are only three. At Burujúba there is a large monastery in a state of ruin ; part of the edifice, however, was still inhabited by a Brazilian family. The walls are four feet in thickness. The long dark corridors and gloomy cloisters struck me as very inappropriate in the midst of this young and radiant nature. They would be better in place on some barren moor in Northern Europe, than here in the midst of perpetual summer. The next turn in the river below Burujúba brought the city of Pará into view. The wind was now against us, and we were obliged to tack about. Towards evening it began to blow stiffly, the vessel heeled over very much, and Senhor Machado, for the first time, trembled for the safety of his cargo ; the leaks burst out afresh, when we were yet two miles from the shore. He ordered another sail to be hoisted, in order to run more quickly into port, but soon afterwards an extra puff of wind came, and the old boat lurched alarmingly, the rigging gave way, and down fell boom and sail with a crash, encumbering us with the wreck. We were then obliged to have recourse to oars ; and as soon as we were near the land, fearing that the crazy vessel would sink before reaching port, I begged Senhor Machado to send me ashore in the boat, with the more precious portion of my collections.

CHAPTER V.

CARIPI AND THE BAY OF MARAJÓ.

River Pará and Bay of Marajó—Journey to Caripí—Negro observance of Christmas—A German Family—Bats—Ant-eaters—Humming-birds—Excursion to the Murucupí—Domestic Life of the Inhabitants—Hunting Excursion with Indians—White Ants.

THAT part of the Pará river which lies in front of the city, as I have already explained, forms a narrow channel; being separated from the main waters of the estuary by a cluster of islands. This channel is about two miles broad, and constitutes part of the minor estuary of Goajará, into which the three rivers Guamá, Mojú, and Acará discharge their waters. The main channel of the Pará lies 10 miles away from the city, directly across the river; at that point, after getting clear of the islands, a great expanse of water is beheld, 10 to 12 miles in width; the opposite shore—the island of Marajó—being visible only in clear weather as a line of tree-tops dotting the horizon. A little further upwards, that is, to the south-west, the mainland on the right or eastern shore appears, this is called Carnapijó; it is rocky, covered with the never-ending forest, and the coast, which is fringed with broad sandy beaches, describes a gentle curve inwards. The broad reach of the Pará in front of this coast is called the Bahia, or Bay of Marajó. The coast and the interior of the land are peopled by civilised Indians and mamelucos, with a mixture of free negroes and mulattos. They are poor, for the waters are not abundant in fish, and they are dependent for a livelihood solely on their small plantations, and the scanty supply of game found in the woods. The district was originally peopled by various tribes of Indians, of whom the principal were the Tupinambás and

Nhengahíbas. Like all the coast tribes, whether inhabiting the banks of the Amazons or the sea-shore between Pará and Bahia, they were far more advanced in civilization than the hordes scattered through the interior of the country, some of which still remain in the wild state, between the Amazons and the Plata. There are three villages on the coast of Carnapijó, and several planters' houses, formerly the centres of flourishing estates, which have now relapsed into forest in consequence of the scarcity of labour and diminished enterprise. One of the largest of these establishments is called Caripí: at the time of which I am speaking it belonged to a Scotch gentleman, Mr. Campbell, who had married the daughter of a large Brazilian proprietor. Most of the occasional English and American visitors to Pará had made some stay at Caripí, and it had obtained quite a reputation for the number and beauty of the birds and insects found there; I therefore applied for and obtained permission to spend two or three months at the place. The distance from Pará was about 23 miles, round by the northern end of the Ilha das onças (Isle of Tigers), which faces the city. I bargained for a passage thither with the cabo of a small trading-vessel, which was going past the place, and started on the 7th of December, 1848.

We were 13 persons aboard: the cabo, his pretty mulatto mistress, the pilot and five Indian canoemen, three young mamelucos (tailor-apprentices who were taking a holiday trip to Cametá), a runaway slave heavily chained, and myself. The young mamelucos were pleasant, gentle fellows: they could read and write, and amused themselves on the voyage with a book containing descriptions and statistics of foreign countries, in which they seemed to take great interest—one reading whilst the others listened. At Uirapiranga, a small island behind the Ilha das onças, we had to stop a short time to embark several pipes of cashaça at a sugar estate. The cabo took the montaria and two men; the pipes were rolled into the water and floated to the canoe, the men passing cables round and towing them through a rough sea. Here we slept, and the following morning, continuing our voyage, entered a narrow channel which intersects the land of Carnapijó. At 2 p.m. we emerged from this channel, which is called the Aititúba, or Arrozal, into the broad Bahia, and then saw, two or three miles away to the left, the red-tiled mansion of Caripí, embosomed in woods on the shores of a charming little bay.

I remained here nine weeks, or until the 12th of February, 1849. The house was very large and most substantially built, but consisted of only one story. I was told it was built by the Jesuits more than a century ago. The front had no verandah, the doors opening on a slightly elevated terrace, about a hundred yards distant from the broad sandy beach. Around the residence the ground had been cleared to the extent of two or three acres, and was planted with fruit-trees. Well-trodden pathways through the forest led to little colonies of the natives, on the banks of retired creeks and rivulets in the interior. I led here a solitary but not unpleasant life; for there was a great charm in the loneliness of the place. The swell of the river beating on the sloping beach caused an unceasing murmur, which lulled me to sleep at night, and seemed appropriate music in those midday hours when all nature was pausing breathless under the rays of a vertical sun. Here I spent my first Christmas-day in a foreign land. The festival was celebrated by the negroes of their own free will, and in a very pleasing manner. The room next to the one I had chosen was the capella, or chapel. It had a little altar which was neatly arranged, and the room was furnished with a magnificent brass chandelier. Men, women, and children were busy in the chapel all day on the 24th of December, decorating the altar with flowers and strewing the floor with orange-leaves. They invited some of their neighbours to the evening prayers; and when the simple ceremony began, an hour before midnight, the chapel was crowded. They were obliged to dispense with the mass, for they had no priest; the service therefore consisted merely of a long litany and a few hymns. There was placed on the altar a small image of the infant Christ, the "Menino Deos" as they called it, or the child-god, which had a long ribbon depending from its waist. An old white-haired negro led off the litany, and the rest of the people joined in the responses. After the service was over they all went up to the altar, one by one, and kissed the end of the ribbon. The gravity and earnestness shown throughout the proceedings were remarkable. Some of the hymns were very simple and beautiful, especially one beginning "Virgem soberana," a trace of whose melody springs to my recollection whenever I think on the dreamy solitude of Caripí.

The first few nights I was much troubled by bats. The room where I slept had not been used for many months, and the roof was open to the tiles and rafters. The first night I slept soundly and did not perceive anything unusual, but on the next I was aroused about midnight by the rushing noise made by vast hosts of bats sweeping about the room. The air was alive with them; they had put out the lamp, and when I relighted it the place appeared blackened with the impish multitudes that were whirling round and round. After I had laid about well with a stick for a few minutes they disappeared amongst the tiles, but when all was still again they returned, and once more extinguished the light. I took no further notice of them, and went to sleep. The next night several got into my hammock; I seized them as they were crawling over me, and dashed them against the wall. The next morning I found a wound, evidently caused by a bat, on my hip. This was rather unpleasant, so I set to work with the negroes, and tried to exterminate them. I shot a great many as they hung from the rafters, and the negroes having mounted with ladders to the roof outside, routed out from beneath the eaves many hundreds of them, including young broods. There were altogether four species — two belonging to the genus Dysopes, one to Phyllostoma, and the fourth to Glossophaga. By far the greater number belonged to the Dysopes perotis, a species having very large ears, and measuring two feet from tip to tip of the wings. The Phyllostoma was a small kind, of a dark-gray colour, streaked with white down the back, and having a leaf-shaped fleshy expansion on the tip of the nose. I was never attacked by bats except on this occasion. The fact of their sucking the blood of persons sleeping, from wounds which they make in the toes, is now well established; but it is only a few persons who are subject to this blood-letting. According to the negroes, the Phyllostoma is the only kind which attacks man. Those which I caught crawling over me were Dysopes, and I am inclined to think many different kinds of bats have this propensity.

One day I was occupied searching for insects in the bark of a fallen tree, when I saw a large cat-like animal advancing towards the spot. It came within a dozen yards before perceiving me. I had no weapon with me but an old chisel, and was getting ready to defend myself if it should make a spring, when it turned round hastily and trotted off. I did not obtain a very distinct view of it, but I could see its colour was that of the Puma, or American Lion, although it was rather too small for that species. The Puma is not a common animal in the Amazons forests. I did not see altogether more than a dozen skins in the possession of the natives. The fur is of a fawn colour. On account of its hue resembling that of a deer common in the forests, the natives call it the Sassú-arána,* or the false deer; that is, an animal which deceives one at first sight by its superficial resemblance to a deer. The hunters are not at all afraid of it, and speak always in disparaging terms of its courage. Of the Jaguar they give a very different account.

The only species of monkey I met with at Caripi was the same dark-coloured little Midas already mentioned as found near Pará. The great Ant-eater, Tamanduá of the natives (Myrmecophaga jubata), was not uncommon here. After the first few weeks of residence I ran short of fresh provisions The people of the neighbourhood had sold me all the fowls they could spare; I had not yet learned to eat the stale and stringy salt-fish which is the staple food in these places, and for several days I had lived on rice-porridge, roasted bananas, and farinha. The housekeeper asked me whether I could eat Tamanduá. I told her almost anything in the shape of flesh would be acceptable; so the same day she went with an old negro named Antonio and the dogs, and in the evening brought one of the animals. The meat was stewed, and turned out very good, something like goose in flavour. The people at Caripí would not touch a morsel, saying it was not considered fit to eat in these parts; I had read, however, that it was an article of food in other countries of South America. During the next two or three weeks, whenever we were short of fresh meat, Antonio was always ready, for a small reward, to get me a Tamanduá. But one day he came to me in great distress, with the news that his favourite dog, Atrevido, had been caught in the grip of an ant-eater, and was killed. We hastened to the place, and found the dog was not dead, but severely torn by the claws of the animal, which itself was mortally wounded, and was now relaxing its grasp.

The habits of the Myrmecophaga jubata are now pretty well known. It is not uncommon in the drier forests of the Amazons valley, but is not found, I believe, in the Ygapó, or flooded lands. The Brazilians call the species the Tamanduá bandeira, or the Banner Ant-eater, the term banner being

* The old zoologist Marcgrave called the Puma the Cuguacuarana, probably (the c's being soft) a misspelling of Sassú-arána; hence the name Cougouar employed by French zoologists, and copied in most works on natural history.

ANT-EATER GRAPPLING WITH DOG.

applied in allusion to the curious coloration of the animal, each side of the body having a broad oblique stripe, half gray and half black, which gives it some resemblance to a heraldic banner. It has an excessively long slender muzzle, and a wormlike extensile tongue. Its jaws are destitute of teeth. The claws are much elongated, and its gait is very awkward. It lives on the ground, and feeds on termites, or white ants; the long claws being employed to pull in pieces the solid hillocks made by the insects, and the long flexible tongue to lick them up from the crevices. All the other species of this singular genus are arboreal. I met with four species altogether. One was the Myrmecophaga tetradactyla; the two others, more curious and less known, were very small kinds, called Tamanduá-i. Both are similar in size—ten inches in length, exclusive of the tail—and in the number of the claws, having two of unequal length to the anterior feet, and four to the hind feet. One species is clothed with grayish-yellow silky hair; this is of rare occurrence. The other has a fur of a dingy brown colour, without silky lustre. One was brought to me alive at Caripí, having been caught by an Indian, clinging motionless inside a hollow tree. I kept it in the house about twenty-four hours. It had a moderately long snout, curved downwards, and extremely small eyes. It remained nearly all the time without motion, except when irritated, in which case it reared itself on its hind legs from the back of a chair to which it clung, and clawed out with its forepaws like a cat. Its manner of clinging with its claws, and the sluggishness of its motions, gave it a great resemblance to a sloth. It uttered no sound, and remained all night on the spot where I had placed it in the morning. The next day I put it on a tree in the open air, and at night it escaped. These small Tamanduás are nocturnal in their habits, and feed on those species of termites which construct earthy nests, that look like ugly excrescences, on the trunks and branches of trees. The different kinds of ant-eaters are thus adapted to various modes of life, terrestrial and arboreal. Those which live on trees are again either diurnal or nocturnal, for Myrmecophaga tetradactyla is seen moving along the main branches in the daytime. The allied group of the Sloths, which are still more exclusively South American forms than ant-eaters are, at the present time furnish arboreal species only, but formerly terrestrial forms of sloths also existed, as the Megatherium, whose mode of life was a puzzle, seeing that it was of too colossal a size to live on trees, until Owen showed how it might have obtained its food from the ground.

In January the orange trees became covered with blossom—at least to a greater extent than usual, for they flower more or less in this country all the year round—and the flowers attracted a great number of humming-birds. Every day, in the cooler hours of the morning, and in the evening from four o'clock till six, they were to be seen whirling about the trees by scores. Their motions are unlike those of all other birds. They dart to and fro so swiftly that the eye can scarcely follow them, and when they stop before a flower it is only for a few moments. They poise themselves in an unsteady manner, their wings moving with inconceivable rapidity; probe the flower, and then shoot off to another part of the tree. They do not proceed in that methodical manner which bees follow, taking the flowers seriatim, but skip about from one part of the tree to another in the most capricious way. Sometimes two males close with each other and fight, mounting upwards in the struggle, as insects are often seen to do when similarly engaged, and then separating hastily and darting back to their work. Now and then they stop to rest, perching on leafless twigs, where they may be sometimes seen probing, from the places where they sit, the flowers within their reach. The brilliant colours with which they are adorned cannot be seen whilst they are fluttering about, nor can the different species be distinguished unless they have a deal of white hue in their plumage, such as Heliothrix auritus, which is wholly white underneath, although of a glittering green colour above, and the white-tailed Florisuga mellivora. There is not a great variety of humming-birds in the Amazons region, the number of species being far smaller in these uniform forest plains than in the diversified valleys of the Andes, under the same parallels of latitude. The family is divisible into two groups, contrasted in form and habits, one containing species which live entirely in the shade of the forest, and the other comprising those which prefer open sunny places. The forest species (Phaethorninæ) are seldom seen at flowers, flowers being, in the shady places where they abide, of rare occurrence; but they search for insects on leaves, threading the bushes and passing above and beneath each leaf with wonderful rapidity. The other group (Trochilinæ) are not quite confined to cleared places, as they come into the forest wherever a tree is in blossom, and descend into sunny openings where flowers are to be found. But it is only where the woods are less dense than usual that this is the case; in the lofty forests and twilight shades of the lowlands and islands they are scarcely ever seen. I searched well at Caripí, expecting to find the Lophornis Gouldii, which I was told had been obtained in the locality. This is one of the most beautiful of all humming-birds, having round the neck a frill of long white feathers tipped with golden green. I was not, however, so fortunate as to meet with it. Several times I shot by mistake a humming-bird hawk-moth instead of a bird. This moth (Macroglossa Titan) is somewhat smaller than humming-birds generally are; but its manner of flight, and the way it poises itself before a

Humming-bird and Humming-bird Hawk-moth.

flower whilst probing it with its proboscis, are precisely like the same actions of humming-birds. It was only after many days' experience that I learnt to distinguish one from the other when on the wing. This resemblance has attracted the notice of the natives, all of whom, even educated whites, firmly believe that one is transmutable into the other. They have observed the metamorphosis of caterpillars into butterflies, and think it not at all more wonderful that a moth should change into a humming-bird. The resemblance between this hawk-moth and a humming-bird is certainly very curious, and strikes one even when both are examined in the hand. Holding them sideways, the shape of the head and position of the eyes in the moth are seen to be nearly the same as in the bird, the extended proboscis representing the long beak. At the tip of the moth's body there is a brush of long hair-scales resembling feathers, which, being expanded, looks very much like a bird's tail. But, of course, all these points of resemblance are merely superficial. The negroes and Indians tried to convince me that the two were of the same species. "Look at their feathers," they said, "their eyes are the same, and so are their tails." This belief is so deeply rooted that it was useless to reason with them on the subject. The Macroglossa moths are found in most countries, and have everywhere the same habits; one well-known species is found in England. Mr. Gould relates that he once had a stormy altercation with an English gentleman, who affirmed that humming-birds were found in England, for he had seen one flying in Devonshire, meaning thereby the moth Macroglossa stellatarum. The analogy between the two creatures has been brought about, probably, by the similarity of their habits, there being no indication of the one having been adapted in outward appearance with reference to the other.

It has been observed that humming-birds are unlike other birds in their mental qualities, resembling in this respect insects rather than warm-blooded vertebrate animals. The want of expression in their eyes, the small degree of versatility in their actions, the quickness and precision of their movements, are all so many points of resemblance between them and insects. In walking along the alleys of the forest a Phaethornis frequently crosses one's path, often stopping suddenly and remaining poised in mid-air, a few feet distant from the face of the intruder. The Phaethorninæ are certainly more numerous in individuals in the Amazons region than the Trochilinæ. They build their nests, which are made of fine vegetable fibres and lichens, densely woven together and thickly lined with silk-cotton from the fruit of the samaüma tree (Eriodendron samaüma), on the inner sides of the tips of palm-fronds. They are long and purse-shaped. The young when first hatched have very much shorter bills than their parents. The only species of Trochilinæ which I found at Caripí were the little brassy-green Polytmus viridissimus, the sapphire and emerald (Thalurania furcata), and the large falcate-winged Campylopterus obscurus.

Snakes were very numerous at Caripí; many harmless species were found near the house, and these sometimes came into the rooms. I was wandering one day amongst the green bushes of Guajará, a tree which yields a grape-like berry (Chrysobalanus

Icaco) and grows along all these sandy shores, when I was startled by what appeared to be the flexuous stem of a creeping plant endowed with life and threading its way amongst the leaves and branches. This animated liana turned out to be a pale-green snake, the Dryophis fulgida. Its whole body is of the same green hue, and it is thus rendered undistinguishable amidst the foliage of the Guajará bushes, where it prowls in search of its prey, tree-frogs and lizards. The forepart of its head is prolonged into a slender pointed beak, and the total length of the reptile was six feet. There was another kind found amongst bushes on the borders of the forest, closely allied to this, but much more slender, viz., the Dryophis acuminata. This grows to a length of 4 feet 8 inches, the tail alone being 22 inches; but the diameter of the thickest part of the body is little more than a quarter of an inch. It is of light-brown colour, with iridescent shades, variegated with obscurer markings, and looks like a piece of whipcord. One individual which I caught of this species had a protuberance near the middle of the body. On opening it I found a half-digested lizard which was much more bulky than the snake itself. Another kind of serpent found here, a species of Helicops, was amphibious in its habits. I saw several of this in wet weather on the beach, which, on being approached, always made straightway for the water, where they swam with much grace and dexterity. Florinda, the housekeeper, one day caught a Helicops whilst angling for fish, it having swallowed the fishhook with the bait. She and others told me these water-snakes lived on small fishes, but I did not meet with any proof of the statement. In the woods, snakes were constantly occurring: it was not often, however, that I saw poisonous species. There were many arboreal kinds besides the two just mentioned; and it was rather alarming, in entomologising about the trunks of trees, to suddenly encounter, on turning round, as sometimes happened, a pair of glittering eyes and a forked tongue within a few inches of one's head. The last kind I shall mention is the Coral-snake, which is a most beautiful object when seen coiled up on black soil in the woods. The one I saw here was banded with black and vermilion, the black bands having each two clear white rings. The state of specimens preserved in spirits can give no idea of the brilliant colours which adorn the Coral-snake in life.

In company with Petzell, a German settler near Caripí, I made many excursions of long extent in the neighbouring forest. We sometimes went to Murucupí, a creek which passes through the forest about four miles behind Caripí, the banks of which are inhabited by Indians and half-breeds, who have lived there for many generations in perfect seclusion from the rest of the world, the place being little known or frequented. A path from Caripí leads to it through a gloomy tract of virgin forest, where the trees are so closely packed together that the ground beneath is thrown into the deepest shade, under which nothing but fetid fungi and rotting vegetable débris is to be seen. On emerging from this unfriendly solitude near the banks of the Murucupí, a charming contrast is presented. A glorious vegetation, piled up to an immense height, clothes the banks of the creek, which traverses a broad tract of semi-cultivated ground, and the varied masses of greenery are lighted up with a sunny glow. Open palm-thatched huts peep forth here and there from amidst groves of banana, mango, cotton, and papaw trees and palms. On our first excursion, we struck the banks of the river in front of a house of somewhat more substantial architecture than the rest, having finished mud walls, plastered and whitewashed, and a covering of red tiles. It seemed to be full of children, and the aspect of the household was improved by a number of good-looking mameluco women, who were busily employed washing, spinning, and making farinha. Two of them, seated on a mat in the open verandah, were engaged sewing dresses; for a festival was going to take place a few days hence at Balcarem, a village eight miles distant from Murucupí, and they intended to be present to hear mass and show their finery. One of the children, a naked boy about seven years of age, crossed over with the montaria to fetch us. We were made welcome at once, and asked to stay for dinner. On our accepting the invitation a couple of fowls were killed, and a wholesome stew of seasoned rice and fowls soon put in preparation. It is not often that the female members of a family in these retired places are familiar with strangers; but these people had lived a long time in the capital, and therefore were more civilised than their neighbours. Their father had been a prosperous tradesman, and had given them the best education the place afforded. After his death the widow with several daughters, married and unmarried, retired to this secluded spot, which had been their sitio, farm or country house, for many years. One of the daughters was married to a handsome young mulatto, who was present and sang us some pretty songs, accompanying himself on the guitar.

After dinner I expressed a wish to see more of the creek; so a lively and polite old man, whom I took to be one of the neighbours, volunteered as guide. We embarked in a little montaria, and paddled some three or four miles up and down the stream. Although I had now become familiarised with beautiful vegetation, all the glow of fresh admiration came again to me in this place. The creek was about 100 yards wide, but narrower in some places. Both banks were masked by lofty walls of green drapery, here and there a break occurring, through which, under overarching trees, glimpses were obtained of the palm-thatched huts of settlers. The projecting boughs of lofty trees, which in some places stretched half-way across the creek, were hung with natural garlands and festoons, and an endless variety of creeping plants clothed the water-frontage, some of which, especially the Bignonias, were ornamented with large gaily-coloured flowers. Art could not have assorted together beautiful vegetable forms so harmoniously as was here done by Nature. Palms, as usual, formed a large proportion of the lower trees; some of them, however, shot up their slim stems to a height of sixty feet or more, and waved their bunches of nodding plumes between us and the sky. One kind of palm, the Pashiúba (Iriartea exorhiza), which grows here in greater abundance than elsewhere, was especially attractive. It is not one of the tallest kinds, for when full-grown its height is not more, perhaps, than forty feet: the leaves are somewhat less drooping, and the leaflets much broader than in other species, so that they have not that feathery appearance which some of those palms have, but still they possess their own peculiar beauty. My guide put me ashore in one place to show me the roots of the Pashiúba. These grow above ground, radiating from the trunk many feet above the surface, so that the tree looks as if supported on stilts; and a person can, in old trees, stand upright amongst the roots with the perpendicular stem wholly above his head. It adds to the singularity of their appearance that these roots, which have the form of straight rods, are studded with stout thorns, whilst the trunk of the tree is quite smooth. The purpose of this curious arrangement is, perhaps, similar to that of the buttress roots already described—namely, to recompense the tree by root growth above the soil for its inability, in consequence of the competition of neighbouring roots, to extend it underground. The great amount of moisture and nutriment contained in the atmosphere may also favour these growths.

On returning to the house, I found Petzell had been well occupied during the hot hours of the day collecting insects in a neighbouring clearing. Our kind hosts gave us a cup of coffee about five o'clock, and we then started for home. The last mile of our walk was performed in the dark. The forest in this part is obscure even in broad daylight, but I was scarcely prepared for the intense opacity of darkness which reigned here on this night, and which prevented us from seeing each other, although walking side by side. Nothing occurred of a nature to alarm us, except that now and then a sudden rush was heard amongst the trees, and once a dismal shriek startled us. Petzell tripped at one place, and fell all his length into the thicket. With this exception, we kept well to the pathway, and in due time arrived safely at Caripí.

One of my neighbours at Murucupí was a hunter of reputation in these parts. He was a civilised Indian, married and settled, named Raimundo, whose habit was to sally forth at intervals to certain productive hunting-grounds, the situation of which he kept secret, and procure fresh provisions for his family. I had found out by this time, that animal food was as much a necessary of life in this exhausting climate as it is in the North of Europe. An attempt which I made to live on vegetable food was quite a failure, and I could not eat the execrable salt-fish which Brazilians use. I had been many days without meat of any kind, and nothing more was to be found near Caripí, so I asked as a favour of Senhor Raimundo permission to accompany him on one of his hunting trips, and shoot a little game for my own use. He consented, and appointed a day on which I was to come over to his house to sleep, so as to be ready for starting with the ebb-tide shortly after midnight.

The locality we were to visit was situated near the extreme point of the land of Carnapijó, where it projects northwardly into the middle of the Pará estuary, and is broken into a number of islands. On the afternoon of January 11th, 1849, I walked through the woods to Raimundo's house, taking nothing with me but a double-barrelled gun, a supply of ammunition, and a box for the reception of any insects I might capture. Raimundo was a carpenter, and seemed to be a very industrious man; he had two apprentices, Indians like himself—one a young lad, and the other apparently about twenty years of age. His wife was of the same race. The Indian women are not always of a taciturn disposition like their husbands. Senhora

Dominga was very talkative; there was another old squaw at the house on a visit, and the tongues of the two were going at a great rate the whole evening, using only the Tupí language. Raimundo and his apprentices were employed building a canoe. Notwithstanding his industry, he seemed to be very poor, and this was the condition of most of the residents on the banks of the Murucupí. They have, nevertheless, considerable plantations of mandioca and Indian corn, besides small plots of cotton, coffee, and sugarcane; the soil is very fertile; they have no rent to pay, and no direct taxes. There is, moreover, always a market in Pará, twenty miles distant, for their surplus produce, and a ready communication with it by water.

In the evening we had more visitors. The sounds of pipe and tabor were heard, and presently a procession of villagers emerged from a pathway through the mandioca fields. They were on a begging expedition for St. Thomé, the patron saint of Indians and Mamelucos. One carried a banner, on which was rudely painted the figure of St. Thomé with a glory round his head. The pipe and tabor were of the simplest description. The pipe was a reed pierced with four holes, by means of which a few unmusical notes were produced, and the tabor was a broad hoop with a skin stretched over each end. A deformed young man played both the instruments. Senhor Raimundo received them with the quiet politeness which comes so natural to the Indian when occupying the position of host. The visitors, who had come from the Villa de Condé, five miles through the forest, were invited to rest. Raimundo then took the image of St. Thomé from one of the party, and placed it by the side of Nossa Senhora in his own oratorio, a little decorated box in which every family keeps its household gods, finally lighting a couple of wax candles before it. Shortly afterwards a cloth was laid on a mat, and all the guests were invited to supper. The fare was very scanty; a boiled fowl with rice, a slice of roasted pirarucú, farinha, and bananas. Each one partook very sparingly, some of the young men contenting themselves with a plateful of rice. One of the apprentices stood behind with a bowl of water and a towel, with which each guest washed his fingers and rinsed his mouth after the meal. They stayed all night: the large open shed was filled with hammocks, which were slung from pole to pole; and on retiring, Raimundo gave orders for their breakfast in the morning.

Raimundo called me at two o'clock, when we embarked (he, his older apprentice Joaquim, and myself) in a shady place where it was so dark that I could see neither canoe nor water, taking with us five dogs. We glided down a winding creek where huge trunks of trees slanted across close overhead, and presently emerged into the Murucupí. A few yards further on we entered the broader channel of the Aititúba. This we crossed, and entered another narrow creek on the opposite side. Here the ebb-tide was against us, and we had great difficulty in making progress. After we had struggled against the powerful current a distance of two miles, we came to a part where the ebb-tide ran in the opposite direction, showing that we had crossed the water-shed. The tide flows into this channel or creek at both ends simultaneously, and meets in the middle, although there is apparently no difference of level, and the breadth of the water is the same. The tides are extremely intricate throughout all the infinite channels and creeks which intersect the lands of the Amazons delta. The moon now broke forth and lighted up the trunks of colossal trees, the leaves of monstrous Jupatí palms which arched over the creek, and revealed groups of arborescent arums standing like rows of spectres on its banks. We had a glimpse now and then into the black depths of the forest, where all was silent except the shrill stridulation of wood-crickets. Now and then a sudden plunge in the water ahead would startle us, caused by heavy fruit or some nocturnal animal dropping from the trees. The two Indians here rested on their paddles, and allowed the canoe to drift with the tide. A pleasant perfume came from the forest, which Raimundo said proceeded from a cane-field. He told me that all this land was owned by large proprietors at Pará, who had received grants from time to time from the Government for political services. Raimundo was quite in a talkative humour; he related to me many incidents of the time of the "Cabanagem," as the revolutionary days of 1835-6 are popularly called. He said he had been much suspected himself of being a rebel, but declared that the suspicion was unfounded. The only complaint he had to make against the white man was, that he monopolised the land without having any intention or prospect of cultivating it. He had been turned out of one place where he had squatted and cleared a large piece of forest. I believe the law of Brazil at this time was that the new lands should become the property of those who cleared and cultivated them, if their right was not disputed within a given term of years by some one who claimed the proprietorship. This land-law has since been repealed, and a new one adopted, founded on that of the United States. Raimundo spoke of his race as the red-skins, "pelle vermelho;" they meant well to the whites, and only begged to be let alone. "God," he said, "had given room enough for us all." It was pleasant to hear the shrewd good-natured fellow talk in this strain. Our companion, Joaquim, had fallen asleep; the night air was cool, and the moonlight lit up the features of Raimundo, revealing a more animated expression than is usually observable in Indian countenances. I always noticed that Indians were more cheerful on a voyage, especially in the cool hours of night and morning, than when ashore. There is something in their constitution of body which makes them feel excessively depressed in the hot hours of the day, especially inside their houses. Their skin is always hot to the touch. They certainly do not endure the heat of their own climate so well as the whites. The negroes are totally different in this respect; the heat of mid-day has very little effect on them, and they dislike the cold nights on the river.

We arrived at our hunting-ground about half-past four. The channel was here broader, and presented several ramifications. It yet wanted an hour and a half to daybreak, so Raimundo recommended me to have a nap. We both stretched ourselves on the benches of the canoe and fell asleep, letting the boat drift with the tide, which was now slack. I slept well, considering the hardness of our bed, and when I awoke, in the middle of a dream about home-scenes, the day was beginning to dawn. My clothes were quite wet with the dew. The birds were astir, the cicadas had begun their music, and the Urania Leilus, a strange and beautiful tailed and gilded moth, whose habits are those of a butterfly, commenced to fly in flocks over the tree-tops. Raimundo exclaimed, "Clareia o dia!"—"The day brightens!" The change was rapid: the sky in the east assumed suddenly the loveliest azure colour, across which streaks of thin white clouds were painted. It is at such moments as this when one feels how beautiful our earth truly is! The channel on whose waters our little boat was floating was about two hundred yards wide; others branched off right and left, surrounding the group of lonely islands which terminate the land of Carnapijó. The forest on all sides formed a lofty hedge without a break: below, it was fringed with mangrove bushes, whose small foliage contrasted with the large glossy leaves of the taller trees, or the feather and fan-shaped fronds of palms.

Being now arrived at our destination, Raimundo turned up his trousers and shirt-sleeves, took his long hunting-knife, and leapt ashore with the dogs. He had to cut a gap in order to enter the forest. We expected to find Pacas and Cutías; and the method adopted to secure them was this: at the present early hour they would be seen feeding on fallen fruits, but would quickly, on hearing a noise, betake themselves to their burrows: Raimundo was then to turn them out by means of the dogs, and Joaquim and I were to remain in the boat with our guns, ready to shoot all that came to the edge of the stream—the habits of both animals, when hard-pressed, being to take to the water. We had not long to wait. The first arrival was a Paca, a reddish, nearly tailless Rodent, spotted with white on the sides, and intermediate in size and appearance between a hog and a hare. My first shot did not take effect; the animal dived into the water and did not reappear. A second was brought down by my companion as it was rambling about under the mangrove bushes. A Cutía next appeared: this is also a Rodent, about one-third the size of the Paca: it swims, but does not dive, and I was fortunate enough to shoot it. We obtained in this way two more Pacas and another Cutía. All the time the dogs were yelping in the forest. Shortly afterwards Raimundo made his appearance, and told us to paddle to the other side of the island. Arrived there, we landed and prepared for breakfast. It was a pretty spot—a clean, white, sandy beach beneath the shade of wide-spreading trees. Joaquim made a fire. He first scraped fine shavings from the midrib of a Bacaba palm-leaf; these he piled into a little heap in a dry place, and then struck a light in his bamboo tinder-box with a piece of an old file and a flint, the tinder being a felt-like soft substance manufactured by an ant (Polyrhachis bispinosus). By gentle blowing the shavings ignited, dry sticks were piled on them, and a good fire soon resulted. He then singed and prepared the cutía, finishing by running a spit through the body, and fixing one end in the ground in a slanting position over the fire. We had brought with us a bag of farinha and a cup containing a lemon, a dozen or two of fiery red peppers, and a few spoonfuls of salt. We breakfasted heartily when our cutía was roasted, and washed the meal down with a calabash full of the pure water of the river.

After breakfast the dogs found another cutía, which was

hidden in its burrow two or three feet beneath the roots of a large tree, and took Raimundo nearly an hour to disinter it. Soon afterwards we left this place, crossed the channel, and, paddling past two islands, obtained a glimpse of the broad river between them, with a long sandy pit, on which stood several scarlet ibises and snow-white egrets. One of the islands was low and sandy, and half of it was covered with gigantic arum trees, the often-mentioned Caladium arborescens, which presented a strange sight. Most people are acquainted with the little British species, Arum maculatum, which grows in hedge-bottoms, and many, doubtless, have admired the larger kinds grown in hothouses; they can therefore form some idea of a forest of arums. On this islet the woody stems of the plants near the bottom were 8 to 10 inches in diameter, and the trees were 12 to 15 feet high; all growing together in such a manner that there was just room for a man to walk freely between them. There was a canoe inshore, with a man and a woman: the man, who was hooting with all his might, told us in passing that his son was lost in the "aningal" (arum-grove). He had strayed whilst walking ashore, and the father had now been an hour waiting for him in vain.

About one o'clock we again stopped at the mouth of a little creek. It was now intensely hot. Raimundo said deer were found here; so he borrowed my gun, as being a more effective weapon than the wretched arms called Lazarinos, which he, in common with all the native hunters, used, and which sell at Pará for seven or eight shillings apiece. Raimundo and Joaquim now stripped themselves quite naked, and started off in different directions through the forest, going naked in order to move with less noise over the carpet of dead leaves, amongst which they stepped so stealthily that not the slightest rustle could be heard. The dogs remained in the canoe, in the neighbourhood of which I employed myself two hours entomologising. At the end of that time my two companions returned, having met with no game whatever.

We now embarked on our return voyage. Raimundo cut two slender poles, one for a mast and the other for a sprit: to these he rigged a sail we had brought in the boat, for we were to return by the open river, and expected a good wind to carry us to Caripí. As soon as we got out of the channel we began to feel the wind—the sea-breeze, which here makes a clean sweep from the Atlantic. Our boat was very small and heavily laden; and when, after rounding a point, I saw the great breadth we had to traverse (seven miles), I thought the attempt to cross in such a slight vessel foolhardy in the extreme. The waves ran very high: there was no rudder; Raimundo steered with a paddle, and all we had to rely upon to save us from falling into the trough of the sea and being instantly swamped were his nerve and skill. There was just room in the boat for our three selves, the dogs, and the game we had killed; and when between the swelling ridges of waves in so frail a shell, our destruction seemed inevitable; as it was, we shipped a little water now and then. Joaquim assisted with his paddle to steady the boat: my time was fully occupied in baling out the water and watching the dogs, which were crowded together in the prow, yelling with fear; one or other of them occasionally falling over the side and causing great commotion in scrambling in again. Off the point was a ridge of rocks, over which the surge raged furiously. Raimundo sat at the stern, rigid and silent; his eye steadily watching the prow of the boat. It was almost worth the risk and discomfort of the passage to witness the seamanlike ability displayed by Indians on the water. The little boat rode beautifully, rising well with each wave, and in the course of an hour and a half we arrived at Caripí, thoroughly tired and wet through to the skin.

On the 16th of January the dry season came abruptly to an end. The sea-breezes, which had been increasing in force for some days, suddenly ceased, and the atmosphere became misty; at length heavy clouds collected where a uniform blue sky had for many weeks prevailed, and down came a succession of heavy showers, the first of which lasted a whole day and night. This seemed to give a new stimulus to animal life. On the first night there was a tremendous uproar—tree-frogs, crickets, goat-suckers, and owls all joining to perform a deafening concert. One kind of goat-sucker kept repeating at intervals throughout the night a phrase similar to the Portuguese words, "Joaõ corta pao,"—"John, cut wood;" a phrase which forms the Brazilian name of the bird. An owl in one of the Genipapa trees muttered now and then a succession of syllables resembling the word "Murucututú." Sometimes the croaking and hooting of frogs and toads were so loud that we could not hear one another's voices within doors. Swarms of dragonflies appeared in the daytime about the pools of water created by the rain, and ants and termites came forth in the winged state in vast numbers. I noticed that the winged termites, or white ants, which came by hundreds to the lamps at night, when alighting on the table, often jerked off their wings by a voluntary movement. On examination I found that the wings were not shed by the roots, for a small portion of the stumps remained attached to the thorax. The edge of the fracture was in all cases straight, not ruptured: there is, in fact, a natural seam crossing the member towards its root, and at this point the long wing naturally drops or is jerked off when the insect has no further use for it. The white ant is endowed with wings simply for the purpose of flying away from the colony, peopled by its wingless companions, to pair with individuals of the same or other

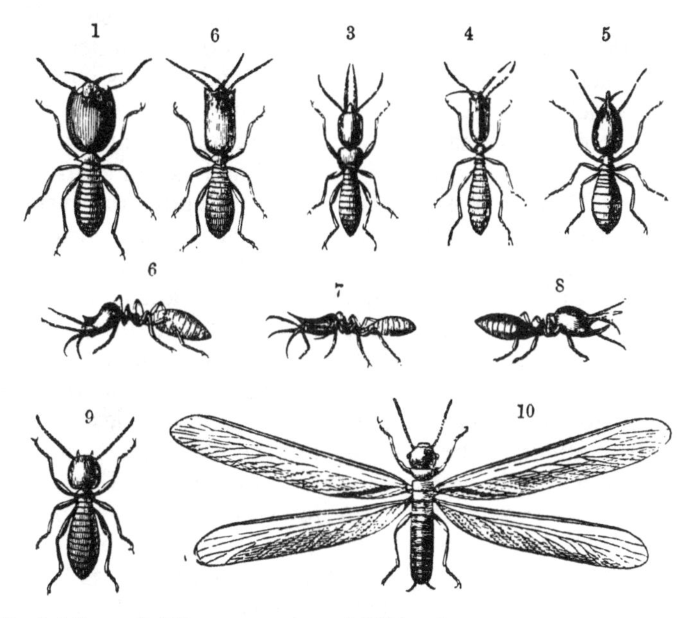

1—8. Soldiers of different species of White Ants.—9. Ordinary shape of worker.—10. Winged class.

colonies, and thus propagate and disseminate its kind. The winged individuals are males and females, whilst the great bulk of their wingless fraternity are of no sex, but are of two castes, soldiers and workers, which are restricted to the functions of building the nests, nursing and defending the young brood. The two sexes mate whilst on the ground after the wings are shed; and then the married couples, if they escape the numerous enemies which lie in wait for them, proceed to the task of founding new colonies. Ants and white ants have much that is analogous in their modes of life: they belong, however, to two widely different orders of insects, strongly contrasted in their structure and manner of growth.

I amassed at Caripí a very large collection of beautiful and curious insects, amounting altogether to about twelve hundred species. The number of Coleoptera was remarkable, seeing that this order is so poorly represented near Pará. I attributed their abundance to the number of new clearings made in the virgin forest by the native settlers. The felled timber attracts lignivorous insects, and these draw in their train the predaceous species of various families. As a general rule the species were smaller and much less brilliant in colours than those of Mexico and South Brazil. The species too, although numerous, were not represented by great numbers of individuals; they were also extremely nimble, and therefore much less easy of capture than insects of the same order in temperate climates. The carnivorous beetles at Caripí were, like those of Pará, chiefly arboreal. Most of them exhibited a beautiful contrivance for enabling them to cling to and run over smooth or flexible surfaces, such as leaves. Their tarsi or feet are broad, and furnished beneath with a brush of short stiff hairs; whilst their claws are toothed in the form of a comb, adapting them for clinging to the smooth edges of leaves, the joint of the foot which precedes the claw being cleft so as to allow free play to the claw in grasping. The common dung-beetles at Caripí, which flew about in the evening like the Geotrupes, the familiar "shardborne beetle with his drowsy hum" of our English lanes, were of colossal size and beautiful colours. One kind had a long spear-shaped horn projecting from the crown of its head (Phanæus lancifer). A blow from this fellow, as he came heavily flying along, was never very pleasant. All the tribes of beetles which feed on vegetable substances, fresh or decayed, were very numerous. The most beautiful of these, but not the most common, were the Longicornes; very graceful insects, having slender bodies and long antennæ, often ornamented with fringes and tufts of hair. They were found on flowers, on trunks of trees, or flying about the new clearings. One small species (Coremia hirtipes) has a tuft of hair on its hind legs, whilst many of its sister species have a similar ornament on the antennæ. It suggests curious reflections when we see an ornament like the feather of a grenadier's cap situated on one part of the body in one species, and in a totally different part in nearly allied ones. I tried in vain to discover the use of these curious brush-like decorations. On the trunk of a living leguminous tree, Petzell found a number of a very rare and handsome species, the Platysternus hebræus, which is of a broad shape, coloured ochreous, but spotted and striped with black, so as to resemble a domino. On the felled trunks of trees, swarms of gilded-green Longicornes occurred, of small size (Chrysoprasis), which looked like miniature musk-beetles, and, indeed, are closely allied to those well-known European insects.

At length, on the 12th of February, I left Caripí, my negro and Indian neighbours bidding me a warm "adeos." I had passed a delightful time, notwithstanding the many privations undergone in the way of food. The wet season had now set in; the lowlands and islands would soon become flooded daily at high water, and the difficulty of obtaining fresh provisions would increase. I intended, therefore, to spend the next three months at Pará, in the neighbourhood of which there was still much to be done in the intervals of fine weather, and then start off on another excursion into the interior.

CHAPTER VI.

THE LOWER AMAZONS—PARA TO OBYDOS.

Modes of Travelling on the Amazons—Preparations for Voyage—Life on board a large Trading-vessel—The narrow Channels joining the Pará to the Amazons—First sight of the Great River—Gurupá—The Great Shoal—Flat-topped Mountains—Santarem—Obydos.

At the time of my first voyage up the Amazons—namely, in 1849—nearly all communication with the interior was by means of small sailing-vessels, owned by traders residing in the remote towns and villages, who seldom came to Pará themselves, but entrusted vessels and cargoes to the care of half-breeds or Portuguese cabos. Sometimes, indeed, they risked all in the hands of the Indian crew, making the pilot, who was also steersman, do duty as supercargo. Now and then, Portuguese and Brazilian merchants at Pará furnished young Portuguese with merchandise, and despatched them to the interior, to exchange the goods for produce among the scattered population. The means of communication, in fact, with the upper parts of the Amazons had been on the decrease for some time, on account of the augmented difficulty of obtaining hands to navigate vessels. Formerly, when the Government wished to send any important functionary, such as a judge or a military commandant, into the interior, they equipped a swift-sailing galliota, manned with ten or a dozen Indians. These could travel, on the average, in one day farther than the ordinary sailing craft could in three. Indian paddlers were now, however, almost impossible to be obtained, and Government officers were obliged to travel as passengers in trading-vessels. The voyage made in this way was tedious in the extreme. When the regular east-wind blew—the "vento geral," or trade-wind of the Amazons—sailing-vessels could get along very well; but when this failed, they were obliged to remain, sometimes many days together, anchored near the shore, or progress loboriously by means of the "espia." The latter mode of travelling was as follows. The montaria, with twenty or thirty fathoms of cable, one end of which was attached to the foremast, was sent ahead with a couple of hands, who secured the other end of the rope to some strong bough or tree-trunk; the crew then hauled the vessel up to the point, after which the men in the boat re-embarked the cable, and paddled forwards to repeat the process. In the dry season, from August to December, when the trade-wind is strong and the currents slack, a schooner could reach the mouth of the Rio Negro, a thousand miles from Pará, in about forty days; but in the wet season, from January to July, when the east-wind no longer blows, and the Amazons pours forth its full volume of water, flooding the banks and producing a tearing current, it took three months to travel the same distance. It was a great blessing to the inhabitants when, in 1853, a line of steamers was established, and this same journey could be accomplished with ease and comfort, at all seasons, in eight days!

Whilst preparing for my voyage it happened, fortunately, that the half-brother of Dr. Angelo Custodio, a young mestizo, named Joaõ da Cunha Correia, was about starting for the Amazons on a trading expedition in his own vessel, a schooner of about forty tons' burthen. A passage for me was soon arranged with him through the intervention of Dr. Angelo, and we started on the 5th of September, 1849. I intended to stop at some village on the northern shore of the Lower Amazons, where it would be interesting to make collections, in order to show the relations of the fauna to those of Pará and the coast region of Guiana. As I should have to hire a house or hut wherever I stayed, I took all the materials for housekeeping—cooking utensils, crockery, and so forth. To these were added a stock of such provisions as it would be difficult to obtain in the interior; also ammunition, chests, store-boxes, a small library of natural history books, and a hundredweight of copper money. I engaged, after some trouble, a Mameluco youth to accompany me as servant—a short, fat, yellow-faced boy named Luco, whom I had already employed at Pará in collecting. We weighed anchor at night, and, on the following day, found ourselves gliding along the dark-brown waters of the Mojú.

Joaõ da Cunha, like most of his fellow-countrymen, took

matters very easily. He was going to be absent in the interior several years, and therefore intended to diverge from his route to visit his native place, Cametá, and spend a few days with his friends. It seemed not to matter to him that he had a cargo of merchandise, vessel, and crew of twelve persons, which required an economical use of time; "pleasure first and business afterwards" appeared to be his maxim. We stayed at Cametá twelve days. The chief motive for prolonging the stay to this extent was a festival at the Aldeia, two miles below Cametá, which was to commence on the 21st, and which my friend wished to take part in. On the day of the festival the schooner was sent down to anchor off the Aldeia, and master and men gave themselves up to revelry. In the evening a strong breeze sprang up, and orders were given to embark. We scrambled down in the dark through the thickets of cacao, orange, and coffee trees which clothed the high bank, and, after running great risk of being swamped by the heavy sea in the crowded montaria, got all aboard by nine o'clock. We made all sail amidst the "adeos" shouted to us by Indian and mulatto sweethearts from the top of the bank, and, tide and wind being favourable, were soon miles away.

Our crew consisted, as already mentioned, of twelve persons. One was a young Portuguese from the province of Traz os Montes, a pretty sample of the kind of emigrants which Portugal sends to Brazil. He was two or three and twenty years of age, and had been about two years in the country, dressing and living like the Indians, to whom he was certainly inferior in manners. He could not read or write, whereas one at least of our Tapuyos had both accomplishments. He had a little wooden image of Nossa Senhora in his rough wooden clothes-chest, and to this he always had recourse when any squall arose, or when we got aground on a shoal. Another of our sailors was a tawny white of Cametá; the rest were Indians, except the cook, who was a Cafuzo, or half-bred between the Indian and negro. It is often said that this class of mestizos is the most evilly disposed of all the numerous crosses between the races inhabiting Brazil; but Luiz was a simple, good-hearted fellow, always ready to do one a service. The pilot was an old Tapuyo of Pará, with regular oval face and well-shaped features. I was astonished at his endurance. He never quitted the helm night or day, except for two or three hours in the morning. The other Indians used to bring him his coffee and meals, and after breakfast one of them relieved him for a time, when he used to lie down on the quarterdeck and get his two hours' nap. The Indians forward had things pretty much their own way. No system of watches was followed; when any one was so disposed, he lay down on the deck and went to sleep; but a feeling of good fellowship seemed always to exist amongst them. One of them was a fine specimen of the Indian race: a man very little short of six feet high, with remarkable breadth of shoulder and full muscular chest. His comrades called him the commandant, on account of his having been one of the rebel leaders when the Indians and others took Santarem in 1835. They related of him that, when the legal authorities arrived with an armed flotilla to recapture the town, he was one of the last to quit, remaining in the little fortress which commands the place to make a show of loading the guns, although the ammunition had given out long ago. Such were our travelling companions. We lived almost the same as on board ship. Our meals were cooked in the galley; but, where practicable, and during our numerous stoppages, the men went in the montaria to fish near the shore, so that our breakfasts and dinners of salt pirarecu were sometimes varied with fresh food.

September 24th.—We passed Entre-as-Ilhas with the morning tide yesterday, and then made across to the eastern shore—the starting-point for all canoes which have to traverse the broad mouth of the Tocantins, going west. Early this morning we commenced the passage. The navigation is attended with danger, on account of the extensive shoals in the middle of the river, which are covered only by a small depth of water at this season of the year. The wind was fresh, and the schooner rolled and pitched like a ship at sea. The distance was about fifteen miles. In the middle, the river-view was very imposing. Towards the north-east there was a long sweep of horizon clear of land, and on the south-west stretched a similar boundless expanse, but varied with islets clothed with fan-leaved palms, which, however, were visible only as isolated groups of columns, tufted at the top, rising here and there amidst the waste of waters. In the afternoon we rounded the westernmost point; the land, which is not terra firma, but simply a group of large islands forming a portion of the Tocantins delta, was then about three miles distant.

On the following day (25th) we sailed towards the west, along the upper portion of the Pará estuary, which extends seventy miles beyond the mouth of the Tocantins. It varies in width from three to five miles, but broadens rapidly near its termination, where it is eight or nine miles wide. The northern shore is formed by the island of Marajó, and is slightly elevated and rocky in some parts. A series of islands conceals the southern shore from view most part of the way. The whole country, mainland and islands, is covered with forest. We had a good wind all day, and about 7 p.m. entered the narrow river of Breves, which commences abruptly the extensive labyrinth of channels that connects the Pará with the Amazons. The sudden termination of the Pará, at a point where it expands to so great a breadth, is remarkable; the water, however, is very shallow over the greater portion of the expanse. I noticed, both on this and on the three subsequent occasions of passing this place, in ascending and descending the river, that the flow of the tide from the east along the estuary, as well as up the Breves, was very strong. This seems sufficient to prove that no considerable volume of water passes by this medium from the Amazons to the Pará, and that the opinion of those geographers is an incorrect one, who believe the Pará to be one of the mouths of the great river. There is, however, another channel connecting the two rivers, which enters the Pará six miles to the south of the Breves. The lower part of its course for eighteen miles is formed by the Uanapú, a large and independent river flowing from the south. The tidal flow is said by the natives to produce little or no current up this river—a fact which seems to afford a little support to the view just stated.

We passed the village of Breves at 3 p.m. on the 26th. It consists of about forty houses, most of which are occupied by Portuguese shopkeepers. A few Indian families reside here, who occupy themselves with the manufacture of ornamental pottery and painted cuyas, which they sell to traders or passing travellers. The cuyas—drinking cups made from gourds—are sometimes very tastefully painted. The rich black ground-colour is produced by a dye made from the bark of a tree called Comateü, the gummy nature of which imparts a fine polish. The yellow tints are made with the Tabatinga clay; the red with the seeds of the Urucú, or anatto plant; and the blue with indigo, which is planted round the huts. The art is indigenous with the Amazonian Indians, but it is only the settled agricultural tribes belonging to the Tupí stock who practise it.

September 27th-30th.—After passing Breves we continued our way slowly along a channel, or series of channels, of variable width. On the morning of the 27th we had a fair wind, the breadth of the stream varying from about 150 to 400 yards. About midday we passed, on the western side, the mouth of the Aturiazal, through which, on account of its swifter current, vessels pass in descending from the Amazons to Pará. Shortly afterwards we entered the narrow channel of the Jaburú, which lies twenty miles above the mouth of the Breves. Here commences the peculiar scenery of this remarkable region. We found ourselves in a narrow and nearly straight canal, not more than eighty to a hundred yards in width, and hemmed in by two walls of forest, which rose quite perpendicularly from the water to a height of seventy or eighty feet. The water was of great and uniform depth, even close to the banks. We seemed to be in a deep gorge, and the strange impression the place produced was augmented by the dull echoes wakened by the voices of our Indians and the splash of their paddles. The forest was excessively varied. Some of the trees, the dome-topped giants of the Leguminous and Bombaceous orders, reared their heads far above the average height of the green walls. The fan-leaved Mirití palm was scattered in some numbers amidst the rest, a few solitary specimens shooting up their smooth columns above the other trees. The graceful Assai palm grew in little groups, forming feathery pictures set in the rounder foliage of the mass. The Ubussú, lower in height, showed only its shuttlecock-shaped crowns of huge undivided fronds, which, being of a vivid pale-green, contrasted forcibly against the sombre hues of the surrounding foliage. The Ubussú grew here in great numbers; the equally remarkable Jupatí palm (Rhaphia tædigera), which, like the Ubussú, is peculiar to this district, occurred more sparsely, throwing its long shaggy leaves, forty to fifty feet in length, in broad arches over the canal. An infinite diversity of smaller-sized palms decorated the water's edge, such as the Marajá-i (Bactris, many species), the Ubim (Geonoma), and a few stately Bacábas (Œnocarpus bacaba). The shape of this last is exceedingly elegant, the size of the crown being in proper proportion to the straight smooth stem. The leaves, down even to the bases of the glossy petioles, are of a rich dark-green colour, and free from spines. "The forest wall"—I am extracting from my journal—"under which we are now moving, consists, besides palms, of a great variety of ordinary forest-trees. From the highest branches of these down to the water sweep ribbons of climbing plants, of the most diverse and ornamental foliage possible. Creeping convolvuli and others have made use of the slender lianas and hanging air-roots as ladders to climb by. Now and then appears a Mimosa or other tree having similar fine pinnate foliage, and thick masses of Inga border the water, from whose branches hang long bean-pods, of different shape and size, according to the species, some of them a yard in length. Flowers there are very few. I see, now and then, a gorgeous crimson blossom on long spikes ornamenting the sombre foliage towards the summits of the forest. I suppose it to belong to a climber of the Combretaceous order. There are also a few yellow and violet Trumpet-flowers (Bignoniæ). The blossoms of the Ingás, although not conspicuous, are delicately beautiful. The forest all along offers so dense a front that one never obtains a glimpse into the interior of the wilderness."

The length of the Jaburú channel is about 35 miles, allowing for the numerous abrupt bends which occur between the middle and the northern end of its course. We were three days and a half accomplishing the passage. The banks on each side seemed to be composed of hard river-mud, with a thick covering of vegetable mould, so that I should imagine this whole district originated in a gradual accumulation of alluvium, through which the endless labyrinths of channels have worked their deep and narrow beds. The flood-tide as we travelled northward became gradually of less assistance to us, as it caused only a feeble current upwards. The pressure of the waters from the Amazons here makes itself felt: as this is not the case lower down, I suppose the currents are diverted through some of the numerous channels which we passed on our right, and which traverse, in their course towards the sea, the north-western part of Marajó. In the evening of the 29th we arrived at a point where another channel joins the Jaburú from the north-east. Up this the tide was flowing; we turned westward, and thus met the flood coming from the Amazons. This point is the object of a strange superstitious observance on the part of the canoemen. It is said to be haunted by a Pajé, or Indian wizard, whom it is necessary to propitiate, by depositing some article on the spot, if the voyager wishes to secure a safe return from the "sertaô," as the interior of the country is called. The trees were all hung with rags, shirts, straw hats, bunches of fruit, and so forth. Although the superstition doubtless originated with the aborigines, yet I observed, in both my voyages, that it was only the Portuguese and uneducated Brazilians who deposited anything. The pure Indians gave nothing, and treated the whole affair as a humbug; but they were all civilised Tapuyos.

On the 30th, at 9 p.m., we reached a broad channel called Macaco, and now left the dark, echoing Jaburú. The Macaco sends off branches towards the north-west coast of Marajó. It is merely a passage amongst a cluster of islands, between which a glimpse is occasionally obtained of the broad waters of the main Amazons. A brisk wind carried us rapidly past its monotonous scenery, and early in the morning of the 1st of October we reached the entrance of the Uituquára, or the Wind-hole, which is 15 miles distant from the end of the Jaburú. This is also a winding channel, 35 miles in length, threading a group of islands, but it is much narrower than the Macaco.

On emerging from the Uituquára on the 2nd, we all went ashore—the men to fish in a small creek; Joaõ da Cunha and I to shoot birds. We saw a flock of scarlet and blue macaws (Macrocercus macao) feeding on the fruits of a bacaba palm, and looking like a cluster of flaunting banners beneath its dark-green crown. We landed about fifty yards from the place, and crept cautiously through the forest, but before we reached them they flew off with loud harsh screams. At a wild-fruit tree we were more successful, as my companion shot an anacá (Derotypus coronatus), one of the most beautiful of the parrot family. It is of a green colour, and has a hood of feathers, red bordered with blue, at the back of its head, which it can elevate or depress at pleasure. The anacá is the only new-world parrot which nearly resembles the cockatoo of Australia. It is found in all the lowlands throughout the Amazons region, but is not a common bird anywhere. Few persons succeed in taming it, and I never saw one that had been taught to speak. The natives are very fond of the bird nevertheless, and keep it in their houses for the sake of seeing the irascible creature expand its beautiful frill of feathers, which it readily does when excited. The men returned with a large quantity of fish. I was surprised at the great variety of species; the prevailing kind was a species of Loricaria, a foot in length, and wholly encased in bony armour. It abounds at certain seasons in shallow water. The flesh is dry, but very palatable. They brought also a small alligator, which they called Jacaré curúa, and said it was a kind found only in shallow creeks. It was not more than two feet in length, although full-grown, according to the statement of the Indians, who said it was a "mai d'ovos," or mother of eggs, as they had pillaged the nest, which they had found near the edge of the water. The eggs were rather larger than a hen's, and regularly oval in shape, presenting a rough hard surface of shell. Unfortunately, the alligator was cut up ready for cooking when we returned to the schooner, and I could not therefore make a note of its peculiarities. The pieces were skewered and roasted over the fire, each man being his own cook. I never saw this species of alligator afterwards.

Acari Fish (Loricaria duodecimalis).

October 3rd.—About midnight the wind, for which we had long been waiting, sprang up, the men weighed anchor, and we were soon fairly embarked on the Amazons. I rose long before sunrise, to see the great river by moonlight. There was a spanking breeze, and the vessel was bounding gaily over the waters. The channel along which we were sailing was only a narrow arm of the river, about two miles in width : the total breadth at this point is more than 20 miles, but the stream is divided into three parts by a series of large islands. The river, notwithstanding this limitation of its breadth, had a most majestic appearance. It did not present that lake-like aspect which the waters of the Pará and Tocantins affect, but had all the swing, so to speak, of a vast flowing stream. The ochre-coloured turbid waters offered also a great contrast to the rivers belonging to the Pará system. The channel formed a splendid reach, sweeping from south-west to north-east, with an horizon of water and sky both up stream and down. At 11

FLAT-TOPPED MOUNTAINS OF PARAUA-QUARA, LOWER AMAZONS.

a.m. we arrived at Gurupá, a small village situated on a rocky bank 30 or 40 feet high. Here we landed, and I had an opportunity of rambling in the neighbouring woods, which are intersected by numerous pathways, carpeted with Lycopodia growing to a height of 8 or 10 inches, and enlivened by numbers of glossy blue butterflies of the Theclidæ or hair-streak family. At 5 p.m. we were again under weigh. Soon after sunset, as we were crossing the mouth of the Xingú, the first of the great tributaries of the Amazons, 1200 miles in length, a black cloud arose suddenly in the north-east. Joaõ da Cunha ordered all sails to be taken in, and immediately afterwards a furious squall burst forth, tearing the waters into foam, and producing a frightful uproar in the neighbouring forests. A drenching rain followed: but in half an hour all was again calm, and the full moon appeared sailing in a cloudless sky.

From the mouth of the Xingú the route followed by vessels leads straight across the river, here 10 miles broad. Towards midnight the wind failed us, when we were close to a large shoal called the Baixo Grande. We lay here becalmed in the sickening heat for two days, and when the trade-wind recommenced with the rising moon at 10 p.m. on the 6th, we found ourselves on a lee-shore. Notwithstanding all the efforts of our pilot to avoid it, we ran aground. Fortunately the bottom consisted only of soft mud, so that by casting anchor to windward, and hauling in with the whole strength of crew and passengers, we got off after spending an uncomfortable night. We rounded the point of the shoal in two fathoms' water; the head of the vessel was then put westward, and by sunrise we were bounding forward before a steady breeze, all sail set and everybody in good humour.

The weather was now delightful for several days in succession, the air transparently clear, and the breeze cool and invigorating. At daylight, on the 6th, a chain of blue hills, the Serra de Almeyrim, appeared in the distance, on the north bank of the river. The sight was most exhilarating after so long a sojourn in a flat country. We kept to the southern shore, passing in the course of the day the mouths of the Urucuricáya and the Aquiquí, two channels which communicate with the Xingú. The whole of this southern coast hence to near Santarem, a distance of 130 miles, is lowland and quite uninhabited. It is intersected by short arms or back waters of the Amazons, which are called in the Tupí language Paranámirims, or little rivers. By keeping to these, small canoes can travel great part of the distance without being much exposed to the heavy seas of the main river. The coast throughout has a most desolate aspect: the forest is not so varied as on the higher land; and the water-frontage, which is destitute of the green mantle of climbing plants that form so rich a decoration in other parts, is encumbered at every step with piles of fallen trees, peopled by white egrets, ghostly storks, and solitary herons. In the evening we passed Almeyrim. The hills, according to Von Martius, who landed here, are about 800 feet above the level of the river, and are thickly wooded to the summit. They commence on the east by a few low, isolated, and rounded elevations; but towards the west of the village they assume the appearance of elongated ridges, which seem as if they had been planed down to a uniform height by some external force. The next day we passed in succession a series of similar flat-topped hills, some isolated and of a truncated-pyramidal shape, others prolonged to a length of several miles. There is an interval of low country between these and the Almeyrim range, which has a total length of about 25 miles: then commences abruptly the Serra de Marauquá, which is succeeded in a similar way by the Velha Pobre range, the Serras de Tapaiuna-quára, and Parauá-quára. All these form a striking contrast to the Serra de Almeyrim, in being quite destitute of trees. They have steep rugged sides, apparently clothed with short herbage, but here and there exposing bare white patches. Their total length is about 40 miles. In the rear, towards the interior, they are succeeded by other ranges of hills, communicating with the central mountain-chain of Guiana, which divides Brazil from Cayenne.

As we sailed along the southern shore, during the 6th and two following days, the table-topped hills on the opposite side occupied most of our attention. The river is from four to five miles broad, and in some places long, low, wooded islands intervene in mid-stream, whose light-green vivid verdure formed a strangely beautiful foreground to the glorious landscape of broad stream and grey mountain. Ninety miles beyond Almeyrim stands the village of Monte Alegre, which is built near the summit of the last hill visible of this chain. At this point the river bends a little towards the south, and the hilly country recedes from its shores to re-appear at Obydos, greatly decreased in height, about a hundred miles further west.

We crossed the river three times between Monte Alegre and

the next town, Santarem. In the middle the waves ran very high, and the vessel lurched fearfully, hurling everything that was not well secured from one side of the deck to the other. On the morning of the 9th of October, a gentle wind carried us along a "remanso," or still water, under the southern shore. These tracts of quiet water are frequent on the irregular sides of the stream, and are the effect of counter movements caused by the rapid current of its central parts. At 9 a.m. we passed the mouth of a Paraná-mirim, called Mahicá, and then found a sudden change in the colour of the water and aspect of the banks. Instead of the low and swampy water-frontage which had prevailed from the mouth of the Xingú, we saw before us a broad sloping beach of white sand. The forest, instead of being an entangled mass of irregular and rank vegetation as hitherto, presented a rounded outline, and created an impression of repose that was very pleasing. We now approached, in fact, the mouth of the Tapajos, whose clear olive-green waters here replaced the muddy current against which we had so long been sailing. Although this is a river of great extent—1,000 miles in length, and, for the last eighty miles of its course, four to ten in breadth—its contribution to the Amazons is not perceptible in the middle of the stream. The white turbid current of the main river flows disdainfully by, occupying nearly the whole breadth of the channel, whilst the darker water of its tributary seems to creep along the shore, and is no longer distinguishable four or five miles from its mouth.

We reached Santarem at 11 a.m. The town has a clean and cheerful appearance from the river. It consists of three long streets, with a few short ones crossing them at right angles, and contains about 2,500 inhabitants. It lies just within the mouth of the Tapajos, and is divided into two parts, the town and the aldeia or village. The houses of the white and trading classes are substantially built, many being of two and three stories, and all white-washed and tiled. The aldeia, which contains the Indian portion of the population, or did so formerly, consists mostly of mud huts, thatched with palm-leaves. The situation of the town is very beautiful. The land, although but slightly elevated, does not form, strictly speaking, a portion of the alluvial river plains of the Amazons, but is rather a northern prolongation of the Brazilian continental land. It is scantily wooded, and toward the interior consists of undulating campos, which are connected with a series of hills extending southward as far as the eye can reach. I subsequently made this place my head-quarters for three years; an account of its neighbourhood is therefore reserved for another chapter. At the first sight of Santarem, one cannot help being struck with the advantages of its situation. Although 400 miles from the sea, it is accessible to vessels of heavy tonnage coming straight from the Atlantic. The river has only two slight bends between this port and the sea, and for five or six months in the year the Amazonian trade wind blows with very little interruption, so that sailing ships coming from foreign countries could reach the place with little difficulty. We ourselves had accomplished 200 miles, or about half the distance from the sea, in an ill-rigged vessel, in three days and a half. Although the land in the immediate neighbourhood is perhaps ill adapted for agriculture, an immense tract of rich soil, with forest and meadow land, lies on the opposite banks of the river, and the Tapajos leads into the heart of the mining provinces of interior Brazil. But where is the population to come from to develop the resources of this fine country? At present the district within a radius of twenty-five miles contains barely 6,500 inhabitants; behind the town, towards the interior, the country is uninhabited, and jaguars roam nightly, at least in the rainy season, close up to the ends of the suburban streets.

From information obtained here, I fixed upon the next town, Obydos, as the best place to stay at a few weeks, in order to investigate the natural productions of the north side of the Lower Amazons. We started at sunrise on the 10th, and being still favoured by wind and weather, made a pleasant passage, reaching Obydos, which is nearly fifty miles distant from Santarem, by midnight. We sailed all day close to the southern shore, and found the banks here and there dotted with houses of settlers, each surrounded by its plantation of cacao, which is the staple product of the district. This coast has an evil reputation for storms and mosquitoes, but we fortunately escaped both. It was remarkable that we had been troubled by mosquitoes only on one night, and then to a small degree, during the whole of our voyage.

I landed at Obydos the next morning, and then bid adieu to my kind friend Joaõ da Cunha, who, after landing my baggage, got up his anchor and continued on his way. The town contains about 1,200 inhabitants, and is airily situated on a high bluff, 90 or 100 feet above the level of the river. The coast is precipitous for two or three miles hence to the west. The cliffs consist of the parti-coloured clay, or Tabatinga, which occurs so frequently throughout the Amazons region ; the strong current of the river sets full against them in the season of high water, and annually carries away large portions. The clay in places is stratified alternately pink and yellow, the pink beds being the thickest, and of much harder texture than the others. When I descended the river in 1859, a German Major of Engineers, in the employ of the Government, told me that he had found calcareous layers, thickly studded with marine shells interstratified with the clay. On the top of the Tabatinga lies a bed of sand, in some places several feet thick, and the whole formation rests on strata of sandstone, which are exposed only when the river reaches its lowest level. Behind the town rises a fine rounded hill, and a range of similar elevations extends six miles westward, terminating at the mouth of the Trombetas, a large river flowing through the interior of Guiana. Hills and lowlands alike are covered with a sombre rolling forest. The river here is contracted to a breadth of rather less than a mile (1738 yards), and the entire volume of its waters, the collective product of a score of mighty streams, is poured through the strait with tremendous velocity. It must be remarked, however, that the river valley itself is not contracted to this breadth, the opposite shore not being continental land, but a low alluvial tract, subject to inundation more or less in the rainy season. Behind it lies an extensive lake, called the Lago Grande da Ville Franca, which communicates with the Amazons, both above and below Obydos, and has therefore the appearance of a by-water or an old channel of the river. This lake is about thirty-five miles in length, and from four to ten in width ; but its waters are of little depth, and in the dry season its dimensions are much lessened. It has no perceptible current, and does not therefore now divert any portion of the waters of the Amazons from their main course past Obydos.

I remaided at Obydos from the 11th of October to the 19th of November. I spent three weeks here, also, in 1859, when the place was much changed, through the influx of Portuguese immigrants and the building of a fortress on the top of the bluff. It is one of the pleasantest towns on the river. The houses are all roofed with tiles, and are mostly of substantial architecture. Most of the Obydos townsfolk are owners of cacao plantations, which are situated on the lowlands in the vicinity. Some are large cattle proprietors, and possess estates of many square leagues' extent in the campo, or grass-land districts, which border the Lago Grande and other similar inland lakes, near the villages of Faro and Alemquer. These campos bear a crop of nutritious grass; but in certain seasons, when the rising of the Amazons exceeds the average, they are apt to be flooded, and then the large herds of half-wild cattle suffer great mortality from drowning, hunger, and the aligators. Neither in cattle-keeping nor cacao-growing are any but the laziest and most primitive methods followed, and the consequence is, that the proprietors are generally poor.

The forest at Obydos seemed to abound in monkeys, for I rarely passed a day without seeing several. I noticed four species: the Coaitá (Ateles paniscus), the Chrysothrix sciureus, the Callithrix torquatus, and our old Pará friend, Midas ursulus. The Coaitá is a large black monkey, covered with coarse hair, and having the prominent parts of the face of a tawny flesh-coloured hue. It is the largest of the Amazonian monkeys in stature, but is excelled in bulk by the "Barrigudo" (Lagothrix Humboldtii) of the Upper Amazons. It occurs throughout the lowlands of the Lower and Upper Amazons; but does not range to the south beyond the limits of the river plains. At that point an allied species, the White-whiskered Coaitá (Ateles marginatus) takes its place. The Coaitás are called by zoologists spider-monkeys, on account of the length and slenderness of their body and limbs. In these apes the tail, as a prehensile organ, reaches its highest degree of perfection; and on this account it would, perhaps, be correct to consider the Coaitás as the extreme development of the American type of apes. As far as we know, from living and fossil species, the New World has progressed no farther than the Coaitá, towards the production of a higher form of the Quadrumanous order. The tendency of Nature here has been, to all appearance, simply to perfect those organs which adapt the species more and more completely to a purely arboreal life: and no nearer approach has been made towards the more advanced forms of anthropoid apes, which are the products of the Old World solely. The flesh of this monkey is much esteemed by the natives in this part of the country, and the Military Commandant of Obydos, Major Gama, every week sent a negro hunter to shoot one for his table. One day I went on a Coaitá hunt, borrowing a negro slave of a friend to show me the way. When in the deepest part of a ravine, we heard a rustling sound in the trees overhead, and Manoel soon pointed out a Coaitá to me. There was something human-like in its appearance, as the lean, dark, shaggy creature moved deliberately amongst the branches at a great height. I fired, but unfortunately only wounded it in the belly. It fell with a crash, headlong, about twenty or thirty feet, and then caught a bough with its tail, which grasped it instantaneously, so that the animal remained suspended in mid-air. Before I could reload it recovered itself, and mounted nimbly to the topmost branches, out of the reach of a fowling-piece, where we could perceive the poor thing apparently probing the wound with its fingers. Coaitás are more frequently kept in a tame state than any other kind of monkey. The Indians are very fond of them as pets, and the women often suckle them when young at their breasts. They become attached to their masters, and will sometimes follow them on the ground to considerable distances. I once saw a most ridiculously tame Coaitá. It was an old female, which accompanied its owner, a trader on the river, in all his voyages. By way of giving me a specimen of its intelligence and feeling, its master set to and rated it soundly, calling it scamp, heathen, thief, and so forth, all through the copious Portuguese vocabulary of vituperation. The poor monkey, quietly seated on the ground, seemed to be in sore trouble at this display of anger. It began by looking earnestly at him, then it whined, and lastly rocked its body to and fro with emotion, crying piteously, and passing its long gaunt arms continually over its forehead; for this was its habit when excited, and the front of the head was worn quite bald in consequence. At length its master altered his tone. "It's all a lie, my old woman; you're an angel, a flower, a good affectionate old creature," and so forth. Immediately the poor monkey ceased its wailing, and soon after came over to where the man sat. The disposition of the Coaitá is mild in the extreme: it has none of the painful, restless vivacity of its kindred, the Cebi, and no trace of the surly, untameable temper of its still nearer relatives, the Mycetes, or howling monkeys. It is, however, an arrant thief, and shows considerable cunning in pilfering small articles of clothing, which it conceals in its sleeping place. The natives of the Upper Amazons procure the Coaitá, when full grown, by shooting it with the blowpipe and poisoned darts, and restoring life by putting a little salt (the antidote to the Urarí poison with which the darts are tipped) in its mouth. The animals thus caught become tame forthwith. Two females were once kept at the Jardin des Plantes of Paris, and Geoffroy St. Hilaire relates of them that they rarely quitted each other,

9

remaining most part of the time in close embrace, folding their tails round one another's bodies. They took their meals together; and it was remarked on such occasions, when the friendship of animals is put to a hard test, that they never quarrelled or disputed the possession of a favourite fruit with each other.

The neighbourhood of Obydos was rich also in insects. In the broad alleys of the forest a magnificent butterfly of the genus Morpho, six to eight inches in expanse, the Morpho Hecuba,

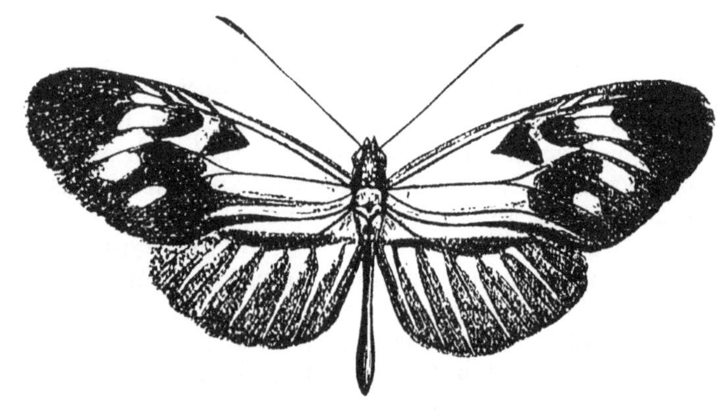

Heliconius Thelxiope.

Heliconius Melpomene.

was seen daily gliding along at a height of twenty feet or more from the ground. Amongst the lower trees and bushes numerous kinds of Heliconii, a group of butterflies peculiar to tropical America, having long narrow wings, were very abundant. The prevailing ground colour of the wings of these insects is a deep black, and on this are depicted spots and streaks of crimson, white, and bright yellow, in different patterns according to the species. Their elegant shape, showy colours, and slow, sailing mode of flight, make them very attractive objects, and their numbers are so great that they form quite a feature in the physiognomy of the forest, compensating for the scarcity of flowers. Next to the Heliconii, the Catagrammas (C. astarte and C. peristera) were the most conspicuous. These have a very rapid and short flight, settling frequently and remaining stationary for a long time on the trunks of trees. The colours of their wings are vermilion and black, the surface having a rich velvety appearance. The genus owes its Greek name Catagramma (signifying "a letter beneath") to the curious markings of the underside of the wings, resembling Arabic numerals. The species and varieties are of almost endless diversity, but the majority inhabit the hot valleys of the eastern parts of the Andes. Another butterfly nearly allied to these, Callithea Leprieurii, was also very abundant here, at the marshy head of the pool before mentioned. The wings are of a rich dark-blue colour, with a broad border of silvery-green. These two groups of Callithea and Catagramma are found only in tropical America, chiefly near the equator, and are certainly amongst the most beautiful productions of a region where the animals and plants seem to have been fashioned in nature's choicest moulds. A great variety of other beautiful and curious insects adorned these pleasant woods. Others were seen only in the sunshine in open places. As the waters retreated from the beach, vast numbers of sulphur-yellow and orange-coloured butterflies congregated on the moist sand. The greater portion of them belonged to the genus Callidryas. They assembled in densely-packed masses, sometimes two or three yards in circumference, their wings all held in an upright position, so that the beach looked as though variegated with beds of crocuses. These Callidryades seem to be migratory insects, and have large powers of dissemination. During the last two days of our voyage the great numbers constantly passing over the river attracted the attention of every one on board. They all crossed in one direction, namely, from north to south, and the processions were uninterrupted, from an early hour in the morning until sunset. All the individuals which resort to the margins of sandy beaches are of the male sex. The females are much more rare, and are seen only on the borders of the forest, wandering from tree to tree, and depositing their eggs on low mimosas which grow in the shade. The migrating hordes, as far as I could ascertain, are composed only of males, and on this account I believe their wanderings do not extend very far.

A strange kind of wood-cricket is found in this neighbourhood, the males of which produce a very loud and not unmusical noise by rubbing together the overlapping edges of their wing-cases. The notes are certainly the loudest and most extraordinary that I ever heard produced by an orthopterous insect. The natives call it the Tananá, in allusion to its music, which is a

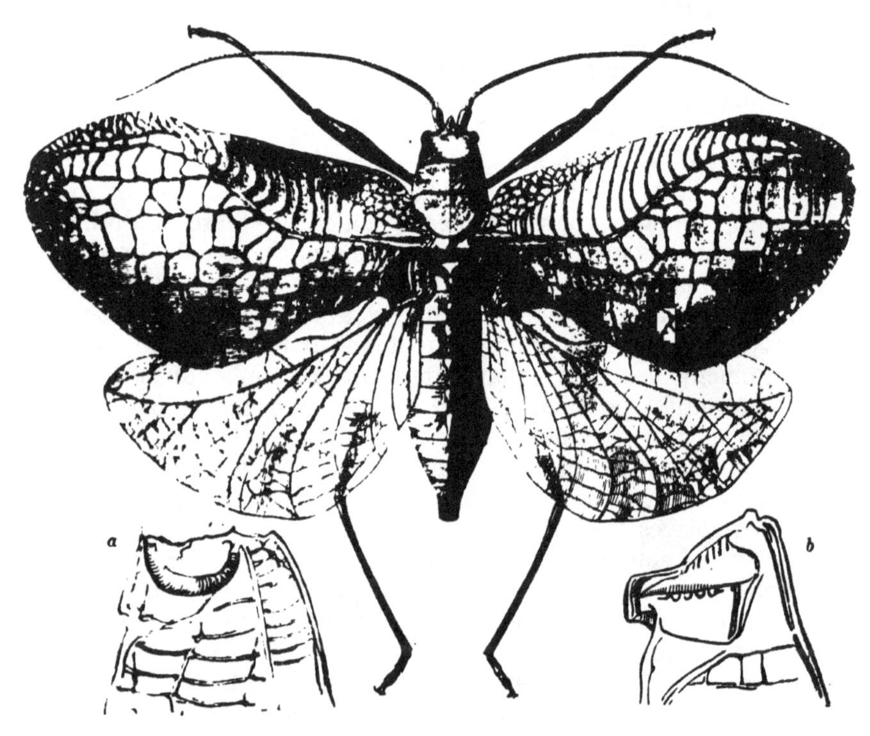

Musical Cricket (Chlorocœlus Tananá).

a. b. Lobes of wing-cases transformed into a musical instrument.

sharp, resonant stridulation resembling the syllables ta-na-ná, ta-na-ná, succeeding each other with little intermission. It seems to be rare in the neighbourhood. When the natives capture one, they keep it in a wicker-work cage for the sake of hearing it sing. A friend of mind kept one six days. It was lively only for two or three, and then its loud note could be heard from one end of the village to the other. When it died, he gave me the specimen, the only one I was able to procure. It is a member of the family Locustidæ, a group intermediate between the Crickets (Achetidæ) and the Grasshoppers (Acridiidæ). The total length of the body is two inches and a quarter; when the wings are closed, the insect has an inflated vesicular or bladder-like shape, owing to the great convexity of the thin, but firm, parchmenty wing-cases, and the colour is wholly pale-green. The instrument by which the Tananá produces its music is curiously contrived out of the ordinary nervures of the wing-cases. In each wing-case the inner edge, near its origin, has a horny expansion or lobe; on one wing (*b*) this lobe has sharp raised margins; on the other (*a*), the strong nervure which traverses the lobe on the other side is crossed by a number of fine sharp furrows like those of a file. When the insect rapidly moves its wings, the file of the one lobe is scraped sharply across the horny margin of the other, thus producing the sounds; the parchmenty wing-cases and the hollow drum-like space which they enclose assisting to give resonance to the tones. The projecting portions of both wing-cases are traversed by a similar strong nervure, but this is scored like a file only in one of them, in the other remaining perfectly smooth. Other species of the family to which the Tananá belongs have similar stridulating organs, but in none are these so highly developed as in this insect; they exist always in the males only, the other sex having the edges of the wing-cases quite straight and simple. The mode of producing the sounds, and their object, have been investigated by several authors with regard to certain European species. They are the call-notes of the males. In the common field-cricket of Europe the male has been observed to place itself, in the evening, at the entrance of its burrow, and stridulate until a female approaches, when the louder notes are succeeded by a more subdued tone, whilst the successful musician caresses with his antennæ the mate he has won. Any one, who will take the trouble, may observe a similar proceeding in the common house-cricket. The nature and object of this insect music are more uniform than the structure and situation of the instrument by which it is produced. This differs in each of the three allied families above mentioned. In the crickets the wing-cases are symmetrical; both have straight edges and sharply-scored nervures adapted to produce the stridulation. A distinct portion of their edges is not, therefore, set apart for the elaboration of a sound-producing instrument. In this family the wing-cases lie flat on the back of the insect, and overlap each other for a considerable portion of their extent. In the Locustidæ the same members have a sloping position on each side of the body, and do not overlap, except to a small extent near their bases; it is out of this small portion that the stridulating organ is contrived. Greater resonance is given in most species by a thin transparent plate, covered by a membrane, in the centre of the overlapping lobes. In the Grasshoppers (Acridiidæ) the wing-cases meet in a straight suture, and the friction of portions of their edges is no longer possible. But Nature exhibits the same fertility of resource here as elsewhere; and, in contriving other methods of supplying the males with an instrument for the production of call-notes, indicates the great importance which she attaches to this function. The music in the males of the Acridiidæ is produced by the scraping of the long hind thighs against the horny nervures of the outer edges of the wing-cases; a drum-shaped organ placed in a cavity near the insertion of the thighs being adapted to give resonance to the tones.

I obtained very few birds at Obydos. There was no scarcity of birds, but they were mostly common Cayenne species. In early morning the woods near my house were quite animated with their songs—an unusual thing in this country. I heard here for the first time the pleasing wild notes of the Carashué, a species of thrush, probably the Mimus lividus of ornithologists, I found it afterwards to be a common bird in the scattered woods of the campo district near Santarem. It is a much smaller and plainer-coloured bird than our thrush, and its song is not so loud, varied, or so long sustained; but the tone is of a sweet and plaintive quality, which harmonizes well with the wild and silent woodlands, where alone it is heard, in the mornings and evenings of sultry tropical days. In course of time the song of this humble thrush stirred up pleasing associations in my mind, in the same way as those of its more highly endowed congeners formerly did at home. There are several allied species in Brazil; in the southern provinces they are called Sabiahs. The Brazilians are not insensible to the charms of this their best songster, for I often heard some pretty verses in praise of the Sabiah, sung by young people to the accompaniment of the guitar. I found several times the nest of the Carashué, which is built of dried grass and slender twigs, and lined with mud; the eggs are coloured and spotted like those of our blackbird, but they are considerably smaller. I was much pleased with a brilliant little red-headed mannikin which I shot here (Pipra cornuta). There were three males seated on a low branch, and hopping slowly backwards and forwards, near to one another, as though engaged in a kind of dance. In the pleasant airy woods surrounding the sandy shores of the pool behind the town, the yellow-bellied Trogon (T. viridis) was very common. Its back is of a brilliant metallic-green colour, and the breast steel blue. The natives call it the Suruquá do Ygapó, or Trogon of the flooded lands, in contradistinction to the various red-breasted species, which are named Suruquás da terra firma. I often saw small companies of half a dozen individuals, quietly seated on the lower branches of trees. They remained almost motionless, for an hour or two at a time, simply moving their heads, on the watch for passing insects; or, as seemed more generally to be the case, scanning the neighbouring trees for fruit; which they darted off now and then, at long intervals, to secure, returning always to the same perch.

CHAPTER VII.

THE LOWER AMAZONS—OBYDOS TO MANAOS, OR THE BARRA OF THE RIO NEGRO.

Departure from Obydos—River banks and by-channels—Cacao planters—Daily life on board our vessel—Great storm—Sand island and its birds—Hill of Parentins—Negro trader and Mauhés Indians—Villa Nova, its inhabitants, forest, and animal productions—Cararaucú—A rustic festival—Lake of Cararaucú—Motúca flies—Serpa—Christmas holidays—River Madeira—A mameluco farmer—Mura Indians—Rio Negro—Description of Barra—Descent to Pará—Yellow fever.

A TRADER of Obydos, named Penna, was about proceeding in a cuberta laden with merchandise to the Rio Negro, intending to stop frequently on the road; so I bargained with him for a passage. He gave up a part of the toldo, or fore-cabin as it may be called, and here I slung my hammock and arranged my boxes, so as to be able to work as we went along. The stoppages I thought would be an advantage, as I could collect in the woods whilst he traded, and thus acquire a knowledge of the productions of many places on the river which, in a direct voyage, it would be impossible to do. I provided a stock of groceries for two months' consumption; and, after the usual amount of unnecessary fuss and delay on the part of the owner, we started on the 19th of November. Penna took his family with him: this comprised a smart, lively mameluco woman, named Catarina, whom we called Senhora Katita, and two children. The crew consisted of three men, one a sturdy Indian, another a Cafuzo, godson of Penna, and the third, our best hand, a steady, good-natured mulatto, named Joaquim. My boy Luco was to assist in rowing and so forth. Penna was a timid middle-aged man, a white with a slight cross of Indian; when he was surly and obstinate, he used to ask me to excuse him on account of the Tapuyo blood in his veins. He tried to make me as comfortable as the circumstances admitted, and provided a large stock of eatables and drinkables; so that altogether the voyage promised to be a pleasant one.

On leaving the port of Obydos we crossed over to the right bank, and sailed with a light wind all day, passing numerous houses, each surrounded by its grove of cacao trees. On the 20th we made slow progress. After passing the high land at the mouth of the Trombetas, the banks were low, clayey, or earthy on both sides. The breadth of the river varies hereabout from two and a half to three miles, but neither coast is the true terra firma. On the northern side a by-channel runs for a long distance inland, communicating with the extensive lake of Faro; on the south, three channels lead to the similar fresh-water sea of Villa Franca; these are in part arms of the river, so that the land they surround consists, properly speaking, of islands. When this description of land is not formed wholly of river deposit, as sometimes happens, or is raised above the level of the highest floods, it is called *Ygapó alto*, and is distinguished by the natives from the true islands of mid-river, as well as from the terra firma. We landed at one of the cacao plantations. The house was substantially built; the walls formed of strong upright posts, lathed across, plastered with mud and whitewashed, and the roof tiled. The family were mamelucos, and seemed to be an average sample of the poorer class of cacao growers. All were loosely dressed and barefooted. A broad verandah extended along one side of the house, the floor of which was simply the well-trodden earth; and here hammocks were slung between the bare upright supports, a large rush mat being spread on the ground, upon which the stout matron-like mistress, with a tame parrot perched upon her shoulder, sat sewing with two pretty little mulatto girls. The master, coolly clad in shirt and drawers, the former loose about the neck, lay in his hammock smoking a long gaudily-painted wooden pipe. The household utensils, earthenware jars, water-pots and saucepans, lay at one end, near which was a wood fire, with the ever-ready coffee-pot simmering on the top of a clay tripod. A large shed stood a short distance off, embowered in a grove of banana, papaw, and mango trees; and under it were the ovens, troughs, sieves, and all other apparatus for the preparation of mandioca. The cleared space around the house was only a few yards in extent; beyond it lay the cacao plantations, which stretched on each side parallel to the banks of the river. There was a path through the forest which led to the mandioca fields, and several miles beyond to other houses on the banks of an interior channel. We were kindly received, as is always the case when a stranger visits these out-of-the-way habitations; the people being invariably civil and hospitable. We had a long chat, took coffee, and on departing one of the daughters sent a basketful of oranges for our use down to the canoe.

The cost of a cacao plantation in the Obydos district is after the rate of 240 reis or sixpence per tree, which is much higher than at Cametá, where I believe the yield is not so great. The forest here is cleared before planting, and the trees are grown in rows. The smaller cultivators are all very poor. Labour is scarce; one family generally manages its own small plantation of 10,000 to 15,000 trees, but at the harvest time neighbours assist each other. It appeared to me to be an easy, pleasant life; the work is all done under shade, and occupies only a few weeks in the year. The incorrigible nonchalance and laziness of the people alone prevent them from surrounding themselves with all the luxuries of a tropical country. They might plant orchards of the choicest fruit-trees around their houses, grow Indian corn, and rear cattle and hogs, as intelligent settlers from Europe would certainly do, instead of indolently relying solely on the produce of their small plantations, and living on a meagre diet of fish and farinha. In preparing the cacao they have not devised any means of separating the seeds well from the pulp, or drying it in a systematic way; the consequence is that, although naturally of good quality, it moulds before reaching the merchants' stores, and does not fetch more than half the price of the same article grown in other parts of tropical America. The Amazons region is the original home of the principal species of chocolate tree, the Theobroma cacao; and it grows in abundance in the forests of the upper river. The cultivated crop appears to be a precarious one; little or no care, however, is bestowed on the trees, and even weeding is done very inefficiently. The plantations are generally old, and have been made on the low ground near the river, which renders them liable to inundation when this rises a few inches more than the average. There is plenty of higher land quite suitable to the tree, but it is uncleared, and the want of labour and enterprise prevents the establishment of new plantations.

We passed the last houses in the Obydos district on the 20th, and the river scenery then resumed its usual wild and solitary character, which the scattered human habitations relieved, although in a small degree. We soon fell into a regular mode of life on board our little ark. Penna would not travel by night; indeed, our small crew, wearied by the day's labour, required rest, and we very rarely had wind in the night. We used to moor the vessel to a tree, giving out plenty of cable, so as to sleep at a distance from the banks and free of mosquitoes, which although swarming in the forest, rarely came many yards out into the river at this season of the year. The strong current, at a distance of thirty or forty yards from the coast, steadied the cuberta head to stream, and kept us from drifting ashore. We all slept in the open air, as the heat of the cabins was stifling in the early part of the night. Penna, Senhora Katita, and I, slung our hammocks in triangle between the mainmast and two stout poles fixed in the raised deck. A sheet was the only covering required, besides our regular clothing; for the decrease of temperature at night on the Amazons is never so great as to be felt otherwise than as a delightful coolness, after the sweltering heat of the afternoons. We used to rise when the first gleam of dawn showed itself above the long dark line of forest. Our clothes and hammocks were then generally soaked with dew, but this was not felt to be an inconvenience. The Indian Manoel used to revive himself by a plunge in the river, under the bows of the vessel. It is the habit of all Indians, male and female, to bathe early in the morning; they do it sometimes for warmth's sake, the temperature of the water being often considerably higher than that of the air. Penna and I lolled in our hammocks, whilst Katita prepared the indispensable cup of strong coffee, which she did with wonderful celerity, smoking meanwhile her early morning pipe of tobacco. Liberal owners of river craft allow a cup of coffee sweetened with molasses, or a ration of cashaça, to each man of their crews; Penna gave them coffee. When all were served, the day's work began. There was seldom any wind at this early hour; so if there was still water along the shore the men rowed, if not there was no way of progressing but by espia. In some places the currents ran with great force close to the banks, especially where these receded to form long bays or *enseadas*, as they are called, and then we made very little headway. In such places the banks consist of loose earth, a rich crumbly vegetable mould, supporting a growth of most luxuriant forest, of which the currents almost daily carry away large portions, so that the stream for several yards out is encumbered with fallen trees, whose branches quiver in the current. When projecting points of land were encountered, it was impossible, with our weak crew, to pull the cuberta against the whirling torrents which set round them; and in such cases we had to cross the river, drifting often with the current, a mile or two lower down on the opposite shore. There generally sprung up a light wind as the day advanced, and then we took down our hammocks, hoisted all sail, and bowled away merrily. Penna generally preferred to cook the dinner ashore, when there was little or no wind. About midday on these calm days we used to look out for a nice shady nook in the forest, with cleared space sufficient to make a fire upon. I then had an hour's hunting in the neighbouring wilderness, and was always rewarded by the discovery of some new species. During the greater part of our voyage, however, we stopped at the house of some settler, and made our fire in the port. Just before dinner it was our habit to take a bath in the river, and then, according to the universal custom on the Amazons, where it seems to be suitable on account of the weak fish diet, we each took half a tea-cupful of neat cashaça, the "abre" or "opening," as it is called, and set to on our mess of stewed pirarecú, beans, and bacon. Once or twice a week we had fowls and rice; at supper, after sunset, we often had fresh fish caught by our men in the evening. The mornings were cool and pleasant until towards midday; but in the afternoons the heat became almost intolerable, especially in gleamy, squally weather, such as generally prevailed. We then crouched in the shade of the sails, or went down to our hammocks in the cabin, choosing to be half stifled rather than expose ourselves on deck to the sickening heat of the sun. We generally ceased travelling about nine o'clock, fixing upon a safe spot wherein to secure the vessel for the night. The cool evening hours were delicious; flocks of whistling ducks (Anas autumnalis), parrots, and hoarsely-screaming macaws, pair by pair, flew over from their feeding to their resting places, as the glowing sun plunged abruptly beneath the horizon. The brief evening chorus of animals then began, the chief performers being the howling monkeys, whose frightful unearthly roar deepened the feeling of solitude which crept on as darkness closed around us. Soon after, the fireflies in great diversity of species came forth and flitted about the trees. As night advanced, all became silent in the forest, save the occasional hooting of tree-frogs, or the monotonous chirping of wood-crickets and grasshoppers.

We made but little progress on the 20th and two following

days, on account of the unsteadiness of the wind. The dry season had been of very brief duration this year; it generally lasts in this part of the Amazons from July to January, with a short interval of showery weather in November. The river ought to sink thirty or thirty-five feet below its highest point; this year it had declined only about twenty-five feet, and the November rains threatened to be continuous. The drier the weather, the stronger blows the east wind; it now failed us altogether, or blew gently for a few hours merely in the afternoons. I had hitherto seen the great river only in its sunniest aspect; I was now about to witness what it could furnish in the way of storms.

On the night of the 22nd the moon appeared with a misty halo. As we went to rest, a fresh watery wind was blowing, and a dark pile of clouds gathering up river in a direction opposite to that of the wind. I thought this betokened nothing more than a heavy rain, which would send us all in a hurry to our cabins. The men moored the vessel to a tree alongside a hard clayey bank, and after supper all were soon fast asleep, scattered about the raised deck. About eleven o'clock I was awakened by a horrible uproar, as a hurricane of wind suddenly swept over from the opposite shore. The cuberta was hurled with force against the clayey bank; Penna shouted out, as he started to his legs, that a trovoada de cima, or a squall from up river, was upon us. We took down our hammocks, and then all hands were required to save the vessel from being dashed to pieces. The moon set, and a black pall of clouds spread itself over the dark forests and river; a frightful crack of thunder now burst over our heads, and down fell the drenching rain. Joaquim leapt ashore through the drowning spray with a strong pole, and tried to pass the cuberta round a small projecting point, whilst we on deck aided in keeping her off and lengthened the cable. We succeeded in getting free, and the stout-built boat fell off into the strong current farther away from the shore, Joaquim swinging himself dexterously aboard by the bowsprit as it passed the point. It was fortunate for us that we happened to be on a sloping clayey bank, where there was no fear of falling trees; a few yards farther on, where the shore was perpendicular and formed of crumbly earth, large portions of loose soil, with all their superincumbent mass of forest, were being washed away; the uproar thus occasioned adding to the horrors of the storm.

The violence of the wind abated in the course of an hour,

but the deluge of rain continued until about three o'clock in the morning; the sky being lighted up by almost incessant flashes of pallid lightning, and the thunder pealing from side to side without interruption. Our clothing, hammocks, and goods were thoroughly soaked by the streams of water which trickled through between the planks. In the morning all was quiet; but an opaque, leaden mass of clouds overspread the sky, throwing a gloom over the wild landscape that had a most dispiriting effect. These squalls from the west are always expected about the time of the breaking up of the dry season, in these central parts of the Lower Amazons. They generally take place about the beginning of February, so that this year they had commenced much earlier than usual. The soil and climate are much drier in this part of the country than in the region lying farther to the west, where the denser forests and more clayey, humid soil produce a considerably cooler atmosphere. The storms may be therefore attributed to the rush of cold moist air from up river, when the regular trade-wind coming from the sea has slackened or ceased to blow.

On the 26th we arrived at a large sandbank connected with an island in mid-river, in front of an inlet called Maracá-uassú. Here we anchored and spent half a day ashore. Penna's object in stopping was simply to enjoy a ramble on the sands with the children, and give Senhora Katita an opportunity to wash the linen. The sandbank was now fast going under water with the rise of the river; in the middle of the dry season it is about a mile long and half a mile in width. The canoe-men delight in these open spaces, which are a great relief to the monotony of the forest that clothes the land in every other part of the river. Farther westward they are much more frequent, and of larger extent. They lie generally at the upper end of islands; in fact, the latter originate in accretions of vegetable matter, formed by plants and trees growing on a shoal. The island was wooded chiefly with the trumpet tree (Cecropia peltata), which has a hollow stem and smooth pale bark. The leaves are similar in shape to those of the horse-chestnut, but immensely larger; beneath they are white, and when the welcome trade-wind blows they show their silvery undersides,—a pleasant signal to the weary canoe traveller. The mode of growth of this tree is curious: the branches are emitted at nearly right angles with the stem, the branchlets in minor whorls around these, and so forth, the leaves growing at their extremities; so that the total appearance is that of a huge candelabrum. Cecropiæ of different species are characteristic of Brazilian forest scenery; the kind of which I am speaking grows in great numbers everywhere on the banks of the Amazons where the land is low. In the same places the curious Monguba tree (Bombax ceiba) is also plentiful; the dark-green bark of its huge tapering trunk, scored with gray, forming a conspicuous object. The principal palm-tree on the lowlands is the Jauarí (Astryocaryum Jauarí), whose stem, surrounded by whorls of spines, shoots up to a great height. On the borders of the island were large tracts of arrow-grass (Gynerium saccharoides), which bears elegant plumes of feathers, like those of the reed, and grows to a height of twenty feet, the leaves arranged in a fan-shaped figure near the middle of the stem. I was surprised to find on the higher parts of the sandbank the familiar foliage of a willow (Salix Humboldtiana). It is a dwarf species, and grows in patches resembling beds of osiers; as in the English willows, the leaves were peopled by small chrysomelideous beetles. In wandering about, many features reminded me of the seashore. Flocks of white gulls were flying overhead, uttering their well-known cry, and sandpipers coursed along the edge of the water. Here and there lonely wading-birds were stalking about; one of these, the Curicáca (Ibis melanopis), flew up with a low cackling noise, and was soon joined by an unicorn bird (Palamedea cornuta), which I startled up from amidst the bushes, whose harsh screams, resembling the bray of a jackass, but shriller, disturbed unpleasantly the solitude of the place. Amongst the willow bushes were flocks of a handsome bird belonging to the Icteridæ or troupial family, adorned with a rich plumage of black and saffron-yellow. I spent some time watching an assemblage of a species of bird called by the natives Tamburí-pará, on the Cecropia trees. It is the Monasa nigrifrons of ornithologists, and has a plain slate-coloured plumage, with the beak of an orange hue. It belongs to the family of Barbets, most of whose members are remarkable for their dull inactive temperament. Those species which are arranged by ornithologists under the genus Bucco are called by the Indians, in the Tupí language, Tai-assú uirá, or pig-birds. They remain seated sometimes for hours together on low branches in the shade, and are stimulated to exertion only when attracted by passing insects. This flock of Tamburí-pará were the reverse of dull; they were gambolling and chasing each other amongst the branches. As they sported about, each emitted a few short tuneful notes, which altogether produced a ringing, musical chorus that quite surprised me.

On the 27th we reached an elevated wooden promontory, called Parentins, which now forms the boundary between the provinces of Pará and the Amazons. Here we met a small canoe descending to Santarem. The owner was a free negro named Lima, who, with his wife, was going down the river to exchange his year's crop of tobacco for European merchandize. The long shallow canoe was laden nearly to the water level. He resided on the banks of the Abacaxí, a river which discharges its waters into the Canomá, a broad interior channel which extends from the river Madeira to the Parentins, a distance of 180 miles. Penna offered him advantageous terms, so a bargain was struck, and the man saved his long journey. The negro seemed a frank, straightforward fellow; he was a native of Pernambuco, but had settled many years ago in this part of the country. He had with him a little Indian girl belonging to the Mauhés tribe, whose native seat is the district of country lying in the rear of the Canomá, between the Madeira and the Tapajos. The Mauhés are considered, I think with truth, to be a branch of the great Mundurucú nation, having segregated from them at a remote period, and by long isolation acquired different customs and a totally different language, in a manner which seems to have been general with the Brazilian aborigines. The Mundurucús seem to have retained more of the general characteristics of the original Tupí stock than the Mauhés. Senhor Lima told me, what I afterwards found to be correct, that there were scarcely two words alike in the languages of the two peoples, although there are words closely allied to Tupí in both. The little girl had not the slightest trace of the savage in her appearance. Her features were finely shaped, the cheek-bones not at all prominent, the lips thin, and the expression of her countenance frank and smiling. She had been brought only a few weeks previously from a remote settlement of her tribe on the banks of the Abacaxí, and did not yet know five words of Portuguese. The Indians, as a general rule, are very manageable when they are young, but it is a frequent complaint that when they reach the age of puberty they become restless and discontented The rooted impatience of all restraint then shows itself, and the kindest treatment will not prevent them running away from their masters; they do not return to the malocas of their tribes, but join parties who go out to collect the produce of the forests and rivers, and lead a wandering semi-savage kind of life.

We remained under the Serra dos Parentins all night. Early the next morning a light mist hung about the tree-tops, and the forest resounded with the yelping of Whaiápu-sai monkeys. I went ashore with my gun and got a glimpse of the flock, but did not succeed in obtaining a specimen. They were of small size and covered with long fur of a uniform gray colour. I think the species was the Callithrix donacophilus. The rock composing the elevated ridge of the Parentins is the same coarse iron-cemented conglomerate which I have often spoken of as occurring near Pará and in several other places. Many loose blocks were scattered about. The forest was extremely varied, and inextricable coils of woody climbers stretched from tree to tree. Thongs of cacti were spread over the rocks and tree-trunks. The variety of small, beautifully-shaped ferns, lichens, and boleti, made the place quite a museum of cryptogamic plants. I found here two exquisite species of Longicorn beetles, and a large kind of grasshopper (Pterochroza), whose broad fore-wings resembled the leaf of a plant, providing the insect with a perfect disguise when they were closed; whilst the hind wings were decorated with gaily-coloured eye-like spots.

The negro left us and turned up a narrow channel, the Paraná-mirim dos Ramos (the little river of the branches, *i.e.*, having many ramifications), on the road to his home, 130 miles distant. We then continued our voyage, and in the evening arrived at Villa Nova, a straggling village containing about seventy houses, many of which scarcely deserve the name, being mere mud-huts roofed with palm leaves. We stayed here four days. The village is built on a rocky bank, composed of the same coarse conglomerate as that already so often mentioned. In some places a bed of Tabatinga clay rests on the conglomerate. The soil in the neighbourhood is sandy, and the forest, most of which appears to be of second growth, is traversed by broad alleys which terminate to the south and east on the banks of pools and lakes, a chain of which extends through the interior of the land. As soon as we anchored I set off with Luco to explore the district. We walked about a mile along the marly shore, on which was a thick carpet of flowering shrubs, enlivened by a great variety of lovely little butterflies, and then entered the forest by a dry watercourse. About a furlong inland this opened on a broad placid pool, whose banks, clothed with grass of the softest green hue, sloped gently from the water's edge to the compact wall of forest which encompassed the whole. The pool swarmed with water-fowl; snowy egrets, dark-coloured striped herons, and storks of various species standing in rows around its margins. Small flocks of macaws were stirring about the topmost branches of the trees. Long-legged piosócas (Parra Jacana) stalked over the water-plants on the surface of the pool, and in the bushes on its margin were great numbers of a kind of canary (Sycalis brasiliensis) of a greenish-yellow colour, which has a short and not very melodious song. We had advanced but a few steps when we startled a pair of the Jaburú-moleque (Mycteria americana), a powerful bird of the stork family, four and a half feet in height, which flew up and alarmed the rest, so that I got only one bird out of the tumultuous flocks which passed over our heads. Passing towards the farther end of the pool, I saw, resting on the surface of the water, a number of large round leaves, turned up at their edges; they belonged to the Victoria water-lily. The leaves were just beginning to expand (December 3rd) some were still under water, and the largest of those which had reached the surface measured not quite three feet in diameter. We found a montaria with a paddle in it, drawn up on the bank, which I took leave to borrow of the unknown owner, and Luco paddled me amongst the noble plants to search for flowers, meeting, however, with no success. I learnt afterwards that the plant is common in nearly all the lakes of this neighbourhood. The natives call it the furno do Piosoca, or oven of the Jacana, the shape of the leaves being like that of the ovens on which mandioca meal is roasted. We saw many kinds of hawks and eagles, one of which, a black species, the Caracára-í (Milvago nudicollis), sat on the top of a tall naked stump, uttering its hypocritical whining notes. This eagle is considered a bird of ill omen by the Indians; it often perches on the tops of trees in the neighbourhood of their huts, and is then said to bring a warning of death to some member of the household. Others say that its whining cry is intended to attract other defenceless birds within its reach. The little courageous flycatcher Bem-ti-vi (Saurophagus sulphuratus) assembles in companies of four or five, and attacks it boldly, driving it from the perch where it would otherwise sit for hours. I shot three hawks of as many different species; and these, with a Magoary stork, two beautiful gilded-green jacamars (Galbula chalcocephala), and half-a-dozen leaves of the water-lily, made a heavy load, with which we trudged off back to the canoe.

A few years after this visit, namely, in 1854-5, I passed eight months at Villa Nova. The district of which it is the chief town is very extensive, for it has about forty miles of linear extent along the banks of the river: but the whole does not contain more than 4,000 inhabitants. More than half of these are pure-blood Indians, who live in a semi-civilized condition on the banks of the numerous channels and lakes. The trade of the place is chiefly in india-rubber, balsam of copaiba (which are collected on the banks of the Madeira and the numerous rivers that enter the Canomá channel), and salt fish prepared in the dry season, nearer home. These articles are sent to Pará in exchange for European goods. The few Indian and half-breed families who reside in the town, are many shades inferior in personal qualities and social condition to those I lived amongst near Pará and Cametá. They live in wretched dilapidated mud-hovels; the women cultivate small patches of mandioca; the men spend most of their time in fishing, selling what they do not require themselves, and getting drunk with the most exemplary regularity on cashaça, purchased with the proceeds.

I made, in this second visit to Villa Nova, an extensive collection of the natural productions of the neighbourhood. A few remarks on some of the more interesting of these must suffice. The forests are very different in their general character from those of Pará, and in fact those of humid districts generally throughout the Amazons. The same scarcity of large-leaved Musaceous and Marantaceous plants was noticeable here as at Obydos. The low-lying areas of forest or Ygapós, which alternate everywhere with the more elevated districts, did not furnish the same luxuriant vegetation as they do in the Delta region of the Amazons. They are flooded during three or four months in the year, and when the waters retire, the soil—to which the very thin coating of alluvial deposit imparts little fertility—remains bare, or covered with a matted bed of dead leaves, until the next flood season. These tracts have then a barren appearance; the trunks and lower branches of the trees are coated with dried slime, and disfigured by rounded masses of fresh-water sponges, whose long horny spiculæ and dingy colours give them the appearance of hedgehogs. Dense bushes of a harsh, cutting grass, called

Tirirfca, form almost the only fresh vegetation in the dry season. Perhaps the dense shade, the long period during which the land remains under water, and the excessively rapid desiccation when the waters retire; all contribute to the barrenness of these Ygapós. The higher and drier land is everywhere sandy, and tall coarse grasses line the borders of the broad alleys which have been cut through the second-growth woods. These places swarm with carapátos, ugly ticks belonging to the genus Ixodes, which mount to the tips of blades of grass, and attach themselves to the clothes of passers-by. They are a great annoyance. It occupied me a full hour daily to pick them off my flesh after my diurnal ramble. There are two species; both are much flattened in shape, have four pairs of legs, a thick short proboscis and a horny integument. Their habit is to attach themselves to the skin by plunging their proboscides into it, and then suck the blood until their flat bodies are distended into a globular form. The whole proceeding, however, is very slow, and it takes them several days to pump their fill. No pain or itching is felt, but serious sores are caused if care is not taken in removing them, as the proboscis is liable to break off and remain in the wound. A little tobacco juice is generally applied to make them loosen their hold. They do not cling firmly to the skin by their legs, although each of these has a pair of sharp and fine claws, connected with the tips of the member by means of a flexible pedicle. When they mount to the summits of slender blades of grass, or the tips of leaves, they hold on by their fore-legs only, the other three pairs being stretched out so as to fasten on any animal which comes in their way. The smaller of the two species is of a yellowish colour; it is much the most abundant, and sometimes falls upon one by scores. When distended, it is about the size of a No. 8 shot; the larger kind, which fortunately comes only singly to the work, swells to the size of a pea.

In some parts of the interior the soil is composed of very coarse sand and small fragments of quartz; in these places no trees grow. I visited, in company with the priest, Padre Torquato, one of these treeless spaces or campos, as they are called, situated five miles from the village. The road thither led through a varied and beautiful forest, containing many gigantic trees. I missed the Assai, Mirití, Paxiúba, and other palms which are all found only on rich moist soils, but the noble Bacába was not uncommon, and there was a great diversity of dwarf species of Marajá palms (Bactris), one of which, called the Peuriríma, was very elegant, growing to a height of twelve or fifteen feet, with a stem no thicker than a man's finger. On arriving at the campo all this beautiful forest abruptly ceased, and we saw before us an oval tract of land, three or four miles in circumference, destitute even of the smallest bush. The only vegetation was a crop of coarse hairy grass growing in patches. The forest formed a hedge all round the isolated field, and its borders were composed in great part of trees which do not grow in the dense virgin forest, such as a great variety of bushy Melastomas, low Byrsomina trees, myrtles, and Lacre trees, whose berries exude globules of wax resembling gamboge. On the margins of the campo wild pine-apples also grew in great quantity. The fruit was of the same shape as our cultivated kind, but much smaller, the size being that of a moderately large appple. We gathered several quite ripe; they were pleasant to the taste, of the true pine-apple flavour, but had an abundance of fully developed seeds, and only a small quantity of eatable pulp. There was no path beyond this campo; in fact, all beyond is terra incognita to the inhabitants of Villa Nova.

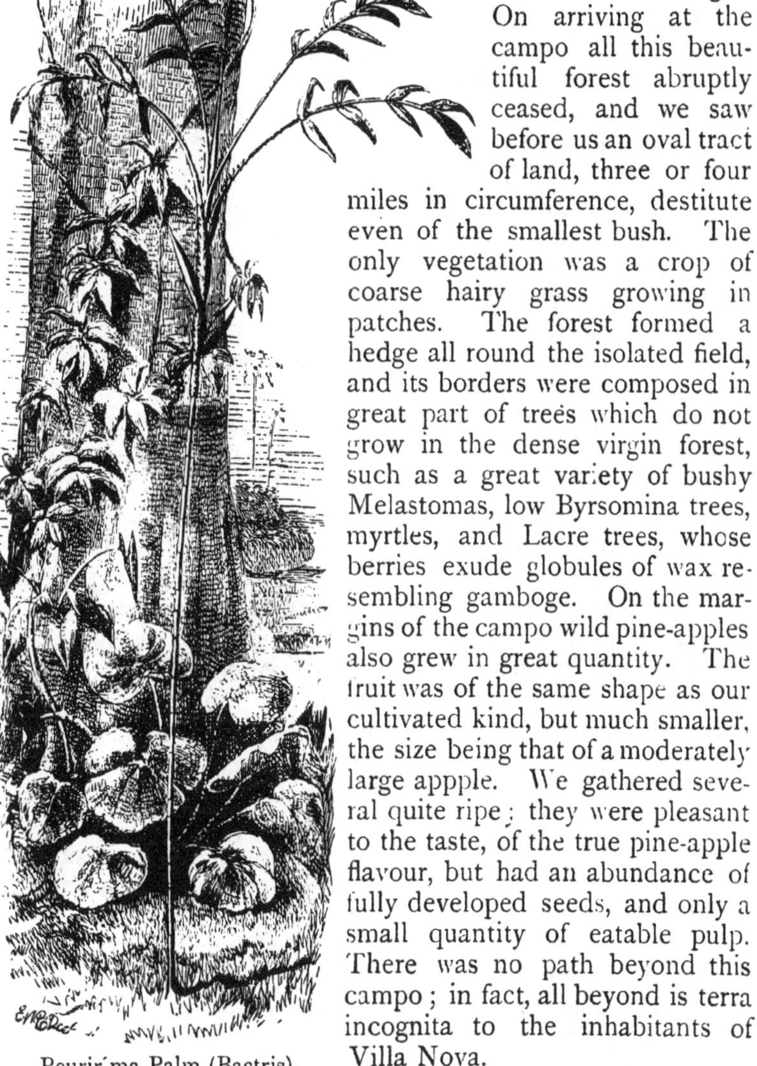

Peuririma Palm (Bactris).

The only interesting Mammalian animal which I saw at Villa Nova was a monkey of a species new to me: it was not, however, a native of the district, having been brought by a trader from the river Madeira, a few miles above Borba. It was a howler, probably the Mycetes stramineus of Geoffroy St. Hilaire. The howlers are the only kinds of monkey which the natives have not succeeded in taming. They are often caught, but they do not survive captivity many weeks. The one of which I am speaking was not quite full grown. It measured sixteen inches in length, exclusive of the tail; the whole body was covered with rather long and shining dingy-white hair, the whiskers and beard only being of a tawny hue. It was kept in a house, together with a Coaita and a Caiarára monkey (Cebus albifrons). Both these lively members of the monkey order seemed rather to court attention, but the Mycetes slunk away when any one approached it. When it first arrived, it occasionally made a gruff subdued howling noise early in the morning. The deep volume of sound in the voice of the howling monkeys, as is well known, is produced by a drum-shaped expansion of the larynx. It was curious to watch the animal whilst venting its hollow cavernous roar, and observe how small was the muscular exertion employed. When howlers are seen in the forest, there are generally three or four of them mounted on the topmost branches of a tree. It does not appear that their harrowing roar is emitted from sudden alarm; at least, it was not so in captive individuals. It is probable, however, that the noise serves to intimidate their enemies. I did not meet with the Mycetes stramineus in any other part of the Amazons region; in the neighbourhood of Pará a reddish-coloured species prevails (M. Belzebuth); in the narrow channels near Breves I shot a large, entirely black kind; another yellow-handed species, according to the report of the natives, inhabits the island of Macajó, which is probably the M. flavimanus of Kuhl; some distance up the Tapajos the only howler found is a brownish-black species; and on the Upper Amazons the sole species seen was the Mycetes ursinus, whose fur is of a shining yellowish-red colour.

In the dry forests of Villa Nova I saw a rattlesnake for the first time. I was returning home one day through a narrow alley, when I heard a pattering noise close to me. Hard by was a tall palm tree, whose head was heavily weighted with parasitic plants, and I thought the noise was a warning that it was about to fall. The wind lulled for a few moments, and then there was no doubt that the noise proceeded from the ground. On turning my head in that direction, a sudden plunge startled me, and a heavy gliding motion betrayed a large serpent making off almost from beneath my feet. The ground is always so encumbered with rotting leaves and branches, that one only discovers snakes when they are in the act of moving away. The residents of Villa Nova would not believe that I had seen a rattlesnake in their neighbourhood; in fact, it is not known to occur in the forests at all, its place being the open campos, where, near Santarem, I killed several. On my second visit to Villa Nova I saw another. I had then a favourite little dog, named Diamante, who used to accompany me in my rambles. One day he rushed into the thicket, and made a dead set at a large snake, whose head I saw raised above the herbage. The foolish little brute approached quite close, and then the serpent reared its tail slightly in a horizontal position and shook its terrible rattle. It was many minutes before I could get the dog away; and this incident, as well as the one already related, shows how slow the reptile is to make the fatal spring.

I was much annoyed, and at the same time amused, with the Urubú vultures. The Portuguese call them corvos or crows: in colour and general appearance they somewhat resemble rooks, but they are much larger, and have naked, black, wrinkled skin about their face and throat. They assemble in great numbers in the villages about the end of the wet season, and are then ravenous with hunger. My cook could not leave the open kitchen at the back of the house for a moment, whilst the dinner was cooking, on account of their thievish propensities. Some of them were always loitering about, watching their opportunity, and the instant the kitchen was left unguarded, the bold marauders marched in and lifted the lids of the saucepans with their beaks to rob them of their contents. The boys of the village lie in wait, and shoot them with bow and arrow; and vultures have consequently acquired such a dread of these weapons, that they may be often kept off by hanging a bow from the rafters of the kitchen. As the dry season advances, the hosts of Urubús follow the fishermen to the lakes, where they gorge themselves with the offal of the fisheries. Towards February they return to the villages, and are then not nearly so ravenous as before their summer trips.

The insects of Villa Nova are, to a great extent, the same as those of Santarem and the Tapajos. A few species of all orders, however, are found here, which occurred nowhere else on the Amazons, besides several others which are properly considered local varieties or races of others found at Pará, on the northern shore of the Amazons, or in other parts of Tropical America. The Hymenoptera were especially numerous, as they always are in districts which possess a sandy soil: but the many interesting facts which I gleaned relative to their habits will be more conveniently introduced when I treat of the same or similar species found in the localities above named. In the broad alleys of the forest several species of Morpho were common. One of these is a sister form to the Morpho Hecuba, which I have mentioned as occurring at Obydos. The Villa Nova kind differs from Hecuba sufficiently to be considered a distinct species, and has been described under the name of M. Cisseis; but it is clearly only a local variety of it, the range of the two being limited by the barrier of the broad Amazons. It is a grand sight to see these colossal butterflies by twos and threes floating at a great height in the still air of a tropical morning. They flap their wings only at long intervals, for I have noticed them to sail a very considerable distance without a stroke. Their wing-muscles, and the thorax to which they are attached, are very feeble in comparison with the wide extent and weight of the wings: but the large expanse of these members doubtless assists the insects in maintaining their aërial course. Morphos are amongst the most conspicuous of the insect denizens of Tropical American forests, and the broad glades of the Villa Nova woods seemed especially suited to them, for I noticed here six species. The largest specimens of Morpho Cisseis measure seven inches and a half in expanse. Another smaller kind, which I could not capture, was of a pale silvery-blue colour, and the polished surface of its wings flashed like a silver speculum, as the insect flapped its wings at a great elevation in the sunlight.

To resume our voyage. We left Villa Nova on the 4th of December. A light wind on the 5th carried us across to the opposite shore and past the mouth of the Paraná-mirím do arco, or the little river of the bow, so called on account of its being a short arm of the main river, of a curved shape, rejoining the Amazons a little below Villa Nova. On the 6th, after passing a large island in mid-river, we arrived at a place where a line of perpendicular clay cliffs, called the Barreiros de Cararaucú, diverts slightly the course of the main stream, as at Obydos. A little below these cliffs were a few settlers' houses: here Penna remained ten days to trade, a delay which I turned to good account in augmenting very considerably my collections.

At the first house a festival was going forward. We anchored at some distance from the shore, on account of the water being shoaly, and early in the morning three canoes put off, laden with salt fish, oil of manatee, fowls and bananas, wares which the owners wished to exchange for different articles required for the festa. Soon after I went ashore. The head man was a tall, well-made, civilised Tapuyo, named Marcellino, who, with his wife, a thin, active, wiry old squaw, did the honours of their house, I thought, admirably. The company consisted of 50 or 60 Indians and Mamelucos; some of them knew Portuguese, but the Tupí language was the only one used amongst themselves. The festival was in honour of our Lady of Conception; and when the people learnt that Penna had on board an image of the saint handsomer than their own, they put off in their canoes to borrow it; Marcellino taking charge of the doll, covering it carefully with a neatly-bordered white towel. On landing with the image, a procession was formed from the port to the house, and salutes fired from a couple of lazarino guns, the saint being afterwards carefully deposited in the family oratorio. After a litany and hymn were sung in the evening, all assembled to supper around a large mat spread on a smooth terrace-like space in front of the house. The meal consisted of a large boiled Pirarecú, which had been harpooned for the purpose in the morning, stewed and roasted turtle, piles of mandioca-meal and bananas. The old lady, with two young girls, showed the greatest activity in waiting on the guests, Marcellino standing gravely by, observing what was wanted and giving the necessary orders to his wife. When all was done, hard drinking began, and soon after there was a dance, to which Penna and I were invited. The liquor served was chiefly a spirit distilled by the people themselves from mandioca cakes. The dances were all of the same class, namely, different varieties of the "Landum," an erotic dance similar to the fandango, originally learnt from the Portuguese. The music was supplied by a couple of wire-stringed guitars, played alternately by the young men. All passed off very quietly considering the amount of strong liquor drunk, and the ball was kept up until sunrise the next morning.

We visited all the houses one after the other. One of them was situated in a charming spot, with a broad sandy beach before it, at the entrance to the Paraná-mirím do Mucambo, a channel leading to an interior lake, peopled by savages of the

Múra tribe. This seemed to be the abode of an industrious family, but all the men were absent, salting Pirarecú on the lakes. The house, like its neighbours, was simply a framework of poles thatched with palm leaves, the walls roughly latticed and plastered with mud: but it was larger and much cleaner inside than the others. It was full of women and children, who were busy all day with their various employments; some weaving hammocks in a large clumsy frame, which held the warp whilst the shuttle was passed by the hand slowly across the six feet breadth of web; others spinning cotton, and others again scraping, pressing, and roasting mandioca. The family had cleared and cultivated a large piece of ground: the soil was of extraordinary richness, the perpendicular banks of the river, near the house, revealing a depth of many feet of crumbling vegetable mould. There was a large plantation of tobacco, besides the usual patches of Indian-corn, sugar-cane, and mandioca; and a grove of cotton, cacao, coffee and fruit trees, surrounded the house. We passed two nights at anchor in shoaly water off the beach. The weather was most beautiful, and scores of Dolphins rolled and snorted about the canoe all night.

We crossed the river at this point, and entered a narrow channel which penetrates the interior of the island of Tupinambarána, and leads to a chain of lakes called the Lagos de Cararaucú. A furious current swept along the coast, eating into the crumbling earthy banks, and strewing the river with débris of the forest. The mouth of the channel lies about twenty-five miles from Villa Nova; the entrance is only about forty yards broad, but it expands, a short distance inland, into a large sheet of water. We suffered terribly from insect pests during the twenty-four hours we remained here. At night it was quite impossible to sleep for mosquitoes; they fell upon us by myriads, and without much piping came straight at our faces as thick as raindrops in a shower. The men crowded into the cabins, and then tried to expel the pests by the smoke from burnt rags, but it was of little avail, although we were half suffocated during the operation. In the daytime the Motúca, a much larger and more formidable fly than the mosquito, insisted upon levying his tax of blood. We had been tormented by it for many days past, but this place seemed to be its metropolis. The species has been described by Perty, the author of the Entomological portion of Spix and Martius' travels, under the name of Hadrus lepidotus. It is a member of the Tabanidæ family, and indeed is closely related to the Hæmatopota pluvialis, a brown fly which haunts the borders of woods in summer time in England. The Motúca is of a bronzed-black colour; its proboscis is formed of a bundle of horny lancets, which are shorter and broader than is usually the case in the family to which it belongs. Its puncture does not produce much pain, but it makes such a large gash in the flesh that the blood trickles forth in little streams. Many scores of them were flying about the canoe all day, and sometimes eight or ten would settle on one's ankles at the same time. It is sluggish in its motions, and may be easily killed with the fingers when it settles. Penna went forward in the montaria to the Pirarecú fishing stations, on a lake lying further inland; but he did not succeed in reaching them on account of the length and intricacy of the channels; so after wasting a day, during which, however, I had a profitable ramble in the forest, we again crossed the river, and on the 16th continued our voyage along the northern shore.

The clay cliffs of Cararaucú are several miles in length. The hard pink-and-red-coloured beds are here extremely thick, and in some places present a compact stony texture. The total height of the cliff is from thirty to sixty feet above the mean level of the river, and the clay rests on strata of the same coarse iron-cemented conglomerate which has already been so often mentioned. Large blocks of this latter have been detached and rolled by the force of currents up parts of the cliff, where they are seen resting on terraces of the clay. On the top of all lies a bed of sand and vegetable mould, which supports a lofty forest, growing up to the very brink of the precipice. After passing these barreiros we continued our way along a low uninhabited coast, clothed, wherever it was elevated above high-water mark, with the usual vividly-coloured forests of the higher Ygapó lands, to which the broad and regular fronds of the Murumurú palm, here extremely abundant, served as a great decoration. Wherever the land was lower than the flood height of the Amazons, Cecropia trees prevailed, sometimes scattered over meadows of tall broad-leaved grasses, which surrounded shallow pools swarming with water-fowl. Alligators were common on most parts of the coast; in some places we saw also small herds of Capybaras (a large Rodent animal, like a colossal Guinea-pig) amongst the rank herbage on muddy banks, and now and then flocks of the graceful squirrel monkey (Chrysothrix sciureus), and the vivacious Caiarára (Cebus albifrons) were seen taking flying leaps from tree to tree. On the 22nd we passed the mouth of the most easterly of the numerous channels which lead to the large interior lake of Saracá, and on the 23rd threaded a series of passages between islands, where we again saw human habitations, ninety miles distant from the last house at Cararaucú. On the 24th we arrived at Serpa.

Serpa is a small village, consisting of about eighty houses, built on a bank elevated twenty-five feet above the level of the river. The beds of Tabatinga clay, which are here intermingled with scoria-looking conglomerate, are in some parts of the declivity prettily variegated in colour; the name of the town in the Tupí language, Ita-coatiára, takes its origin from this circumstance, signifying striped or painted rock. It is an old settlement, and was once the seat of the district government, which had authority over the Barra of the Rio Negro. It was in 1849 a wretched-looking village, but it has since revived, on account of having been chosen by the Steamboat Company of the Amazons as a station for steam saw-mills and tile manufactories. We arrived on Christmas-eve, when the village presented an animated appearance from the number of people congregated for the holidays. The port was full of canoes, large and small—from the montaria, with its arched awning of woven lianas and Maranta leaves, to the two-masted cuberta of the peddling trader, who had resorted to the place in the hope of trafficking with settlers coming from remote sitios to attend the festival. We anchored close to an igarité, whose owner was an old Jurí Indian, disfigured by a large black tatooed patch in the middle of his face, and by his hair being close cropped, except a fringe in front of the head. In the afternoon we went ashore. The population seemed to consist chiefly of semi-civilised Indians, living as usual in half-finished mud hovels. The streets were irregularly laid out, and overrun with weeds and bushes swarming with "mocuim," a very minute scarlet acarus, which sweeps off to one's clothes in passing, and attaching itself in great numbers to the skin causes a most disagreeable itching. The few whites and better class of mameluco residents live in more substantial dwellings, white-washed and tiled. All, both men and women, seemed to me much more cordial, and at the same time more brusque in their manners than any Brazilians I had yet met with. One of them, Captain Manoel Joaquim, I knew for a long time afterwards; a lively, intelligent, and thoroughly good-hearted man, who had quite a reputation throughout the interior of the country for generosity, and for being a firm friend of foreign residents and stray travellers. Some of these excellent people were men of substance, being owners of trading vessels, slaves, and extensive plantations of cacao and tobacco.

We stayed at Serpa five days. Some of the ceremonies observed at Christmas were interesting, inasmuch as they were the same, with little modification, as those taught by the Jesuit missionaries, more than a century ago, to the aboriginal tribes whom they had induced to settle on this spot. In the morning all the women and girls, dressed in white gauze chemises and showy calico print petticoats, went in procession to church, first going the round of the town to take up the different "mordomos," or stewards, whose office is to assist the Juiz of the festa. These stewards carried each a long white reed, decorated with coloured ribbons; several children also accompanied, grotesquely decked with finery. Three old squaws went in front, holding the "sairé," a large semicircular frame, clothed with cotton and studded with ornaments, bits of looking-glass, and so forth. This they danced up and down, singing all the time a monotonous whining hymn in the Tupí language, and at frequent intervals turning round to face the followers, who then all stopped for a few moments. I was told that this sairé was a device adopted by the Jesuits to attract the savages to church, for these everywhere followed the mirrors, in which they saw as it were magically reflected their own persons. In the evening, good-humoured revelry prevailed on all sides. The negroes, who had a saint of their own colour—St. Benedito—had their holiday apart from the rest; and spent the whole night singing and dancing, to the music of a long drum (gambá) and the caracashá. The drum was a hollow log, having one end covered with skin, and was played by the performer sitting astride upon it and drumming with his knuckles. The caracashá is a notched bamboo tube, which produces a harsh rattling noise by passing a hard stick over the notches. Nothing could exceed in dreary monotony this music and the singing and dancing, which were kept up with unflagging vigour all night long. The Indians did not get up a dance; for the whites and mamelucos had monopolised all the pretty coloured girls for their own ball, and the older squaws preferred looking on to taking a part themselves.

Some of their husbands joined the negroes, and got drunk very quickly. It was amusing to notice how voluble the usually taciturn red-skins became under the influence of liquor. The negroes and Indians excused their own intemperance by saying the whites were getting drunk at the other end of the town, which was quite true.

We left Serpa on the 29th of December, in company of an old planter named Senhor Joao (John) Trinidade; at whose sitio, situated opposite the mouth of the Madeira, Penna intended to spend a few days. Our course on the 29th and 30th lay through narrow channels between islands. On the 31st we passed the last of these, and then beheld to the south a sea-like expanse of water, where the Madeira, the greatest tributary of the Amazons, after 2000 miles of course, blends its waters with those of the king of rivers. I was hardly prepared for a junction of waters on so vast a scale as this, now nearly 900 miles from the sea. While travelling week after week along the somewhat monotonous stream, often hemmed in between islands, and becoming thoroughly familiar with it, my sense of the magnitude of this vast water system had become gradually deadened; but this noble sight renewed the first feelings of wonder. One is inclined, in such places as these, to think the Paraenses do not exaggerate much when they call the Amazons the Mediterranean of South America. Beyond the mouth of the Madeira, the Amazons sweeps down in a majestic reach, to all appearance not a whit less in breadth before, than after, this enormous addition to its waters. The Madeira does not ebb and flow simultaneously with the Amazons; it rises and sinks about two months earlier, so that it was now fuller than the main river. Its current therefore poured forth freely from its mouth, carrying with it a long line of floating trees and patches of grass, which had been torn from its crumbly banks in the lower part of its course. The current, however, did not reach the middle of the main stream, but swept along nearer to the southern shore.

A few items of information which I gleaned relative to this river may find a place here. The Madeira is navigable for about 480 miles from its mouth; a series of cataracts and rapids then commences, which extends, with some intervals of quiet water, about 160 miles, beyond which is another long stretch of navigable stream. Canoes sometimes descend from Villa Bella, in the interior province of Matto Grosso, but not so frequently as formerly, and I could hear of very few persons who had attempted of late years to ascend the river to that point. It was explored by the Portuguese in the early part of the eighteenth century; the chief and now the only town on its banks, Borba, 150 miles from its mouth, being founded in 1756. Up to the year 1853, the lower part of the river, as far as about 100 miles beyond Borba, was regularly visited by traders from Villa Nova, Serpa, and Barra, to collect salsaparilla, copaüba balsam, turtle-oil, and to trade with the Indians, with whom their relations were generally on a friendly footing. In that year many india-rubber collectors resorted to this region, stimulated by the high price (2*s*. 6*d*. per pound) which the article was at that time fetching at Pará; and then the Araras, a fierce and intractable tribe of Indians, began to be troublesome. They attacked several canoes and massacred every one on board, the Indian crews as well as the white traders. Their plan was to lurk in ambush near the sandy beaches where canoes stop for the night, and then fall upon the people whilst asleep. Sometimes they came under pretence of wishing to trade, and then as soon as they could get the trader at a disadvantage shot him and his crew from behind trees. Their arms were clubs, bows, and Taquára arrows, the latter a formidable weapon tipped with a piece of flinty bamboo shaped like a spear-head; they could propel it with such force as to pierce a man completely through the body. The whites of Borba made reprisals, inducing the warlike Mundurucús, who had an old feud with the Aráras, to assist them. This state of things lasted two or three years, and made a journey up the Madeira a risky undertaking, as the savages attacked all comers. Besides the Aráras and the Mundurucús, the latter a tribe friendly to the whites, attached to agriculture, and inhabiting the interior of the country from the Madeira to beyond the Tapajos, two other tribes of Indians now inhabit the lower Madeira, namely, the Parentintins and the Muras. Of the former I did not hear much; the Muras lead a lazy quiet life on the banks of the labyrinths of lakes and channels which intersect the low country on both sides of the river below Borba. The Araras are one of those tribes which do not plant mandioca; and indeed have no settled habitations. They are very similar in stature and other physical features to the Mundurucús, although differing from them so widely in habits and social condition. They paint their chins red with Urucú (Anatto), and have usually a black tattooed streak on each side of the face, running from the corner of the mouth to the temple. They have not yet learnt the use of firearms, have no canoes, and spend their lives roaming over the interior of the country, living on game and wild fruits. When they wish to cross a river, they make a temporary canoe with the thick bark of trees, which they secure in the required shape of a boat by means of lianas. I heard it stated by a trader of Santarem, who narrowly escaped being butchered by them in 1854, that the Aráras numbered two thousand fighting men. The number I think must be exaggerated, as it generally is with regard to Brazilian tribes. When the Indians show a hostile disposition to the whites, I believe it is most frequently owing to some provocation they have received at their hands; for the first impulse of the Brazilian red-man is to respect Europeans; they have a strong dislike to be forced into their service, but if strangers visit them with a friendly intention they are well treated. It is related, however, that the Indians of the Madeira were hostile to the Portuguese from the first; it was then the tribes of Muras and Torazes who attacked travellers. In 1855 I met with an American, an odd character, named Kemp, who had lived for many years amongst the Indians on the Madeira, near the abandoned settlement of Crato. He told me his neighbours were a kindly-disposed and cheerful people, and that the onslaught of the Aráras was provoked by a trader from Barra, who wantonly fired into a family of them, killing the parents. and carrying off their children to be employed as domestic servants.

We remained nine days at the sitio of Senhor John Trinidade. It is situated on a tract of high Ygapo land, which is raised, however, only a few inches above high-water mark. This skirts the northern shore for a long distance; the soil consisting of alluvium and rich vegetable mould, and exhibiting the most exuberant fertility. Such districts are the first to be settled on in this country, and the whole coast for many miles was dotted with pleasant-looking sitios like that of our friend. The establishment was a large one, the house and out-buildings covering a large space of ground. The industrious proprietor seemed to be Jack-of-all-trades; he was planter, trader, fisherman, and canoe-builder, and a large igarité was now on the stocks under a large shed. There was great pleasure in contemplating this prosperous farm, from its being worked almost entirely by free labour; in fact, by one family, and its dependants. John Trinidade had only one female slave; his other workpeople were a brother and sister-in-law, two godsons, a free negro, one or two Indians and a family of Muras. Both he and his wife were mamelucos; the negro children called them always father and mother. The order, abundance, and comfort about the place showed what industry and good management could effect in this country without slave-labour. But the surplus produce of such small plantations is very trifling. All we saw had been done since the disorders of 1835-6, during which John Trinidade was a great sufferer; he was obliged to fly, and the Mura Indians destroyed his house and plantations. There was a large, well-weeded grove of cacao along the banks of the river, comprising about 8,000 trees, and further inland considerable plantations of tobacco, mandioca, Indian corn, fields of rice, melons, and water-melons. Near the house was a kitchen garden, in which grew cabbages and onions, introduced from Europe, besides a wonderful variety of tropical vegetables. It must not be supposed that these plantations and gardens were enclosed or neatly kept, such is never the case in this country where labour is so scarce; but it was an unusual thing to see vegetables grown at all, and the ground tolerably well weeded. The space around the house was plentifully planted with fruit-trees, some, belonging to the Anonaceous order, yielding delicious fruits large as a child's head, and full of custardy pulp which it is necessary to eat with a spoon; besides oranges, lemons, guavas, alligator pears, Abíus (Achras cainito), Genipapas, and bananas. In the shade of these, coffee trees grew in great luxuriance. The table was always well supplied with fish, which the Mura, who was attached to the household as fisherman, caught every morning a few hundred yards from the port. The chief kinds were the Surubim, Pira-peëua, and Piramutába, three species of Siluridæ, belonging to the genus Pimelodus. To these we used a sauce in the form of a yellow paste, quite new to me, called Arubé, which is made of the poisonous juice of the mandioca root, boiled down before the starch or tapioca is precipitated, and seasoned with capsicum peppers. It is kept in stone bottles several weeks before using, and is a most appetising relish to fish. Tucupí, another sauce made also from mandioca juice, is much more common in the interior of the country than Arubé. This is made by boiling or heating the pure liquid, after the tapioca has been separated, daily for several days in succession, and seasoning it with peppers and small fishes; when old it has the taste of essence of anchovies. It is generally made as a liquid, but the Jurí and Miranha tribes on the Japura, make it up in the form of a black paste, by a mode of preparation I could not learn; it is then called Tucupí-pixúna, or black Tucupí. I have seen the Indians on the Tapajos, where fish is scarce, season Tucupí with Saüba ants. It is there used chiefly as a sauce to Tacaca, another preparation from mandioca, consisting of the starch beaten up in boiling water.

I thoroughly enjoyed the nine days we spent at this place. Our host and hostess took an interest in my pursuit; one of the best chambers in the house was given up to me, and the young men took me long rambles in the neighbouring forests. I saw very little hard work going forward. Every one rose with the dawn, and went down to the river to bathe; then came the never-failing cup of rich and strong coffee, after which all proceeded to their avocations. At this time nothing was being done at the plantations; the cacao and tobacco crops were not ripe; weeding time was over, and the only work on foot was the preparation of a little farinha by the women. The men dawdled about; went shooting and fishing, or did trifling jobs about the house. The only laborious work done during the year in these establishments is the felling of timber for new clearings; this happens at the beginning of the dry season, namely, from July to September. Whatever employment the people were engaged in, they did not intermit it during the hot hours of the day. Those who went into the woods took their dinners with them—a small bag of farinha and a slice of salt fish. About sunset all returned to the house; they then had their frugal suppers, and towards eight o'clock, after coming to ask a blessing of the patriarchal head of the household, went off to their hammocks to sleep.

John Trinidade was famous for his tobacco and cigarettes, as he took great pains in preparing the Tauarí, or envelope, which is formed of the inner bark of a tree, separated into thin papery layers. Many trees yield it, amongst them the Courataria Guianensis and the Sapucaya nut-tree, both belonging to the same natural order. The bark is cut in long strips, of a breadth suitable for folding the tobacco; the inner portion is then separated, boiled, hammered with a wooden mallet, and exposed to the air for a few hours. Some kinds have a reddish colour and an astringent taste, but the sort prepared by our host was of a beautiful satiny-white hue, and perfectly tasteless. He obtained sixty, eighty, and sometimes a hundred layers from the same strip of bark. The best tobacco in Brazil is grown in the neighbourhood of Borba, on the Madeira, where the soil is a rich black loam; but tobacco of very good quality was grown by John Trinidade and his neighbours along this coast, on similar soil. It is made up into slender rolls, an inch and a half in diameter and six feet in length, tapering at each end. When the leaves are gathered and partially dried, layers of them, after the mid-ribs are plucked out, are placed on a mat and rolled up into the required shape. This is done by the women and children, who also manage the planting, weeding, and gathering of the tobacco. The process of tightening the rolls is a long and heavy task, and can be done only by men. The cords used for this purpose are of very great strength. They are made of the inner bark of a peculiar light-wooded and slender tree, called Uaissíma, which yields, when beaten out, a great quantity of most beautiful silky fibre, many feet in length. I think this might be turned to some use by English manufacturers, if they could obtain it in large quantity. The tree is abundant on light soils on the southern side of the Lower Amazons, and grows very rapidly. When the rolls are sufficiently well pressed, they are bound round with narrow thongs of remarkable toughness, cut from the bark of the climbing Jacitára palm-tree (Desmoncus macracanthus), and are then ready for sale or use.

It was very pleasant to roam in our host's cacaoal. The ground was clear of underwood, the trees were about thirty feet in height, and formed a dense shade. Two species of monkey frequented the trees, and I was told committed great depredations when the fruit was ripe. One of these, the macaco prego (Cebus cirrhifer?), is a most impudent thief; it destroys more than it eats, by its random, hasty way of plucking and breaking the fruits, and when about to return to the forest carries away all it can in its hands or under its arms. The other species, the pretty little Chrysothrix sciureus, contents itself with devouring what it can on the spot. A variety of beautiful insects basked on the foliage, where stray gleams of sunlight glanced through the canopy of broad soft-green leaves, and numbers of an elegant long-legged tiger-beetle (Odontocheila egregia) ran and flew about over the herbage.

We left this place on the 8th of January, and on the afternoon of the 9th arrived at Matarí, a miserable little settlement of Múra Indians. Here we again anchored and went ashore. The place consisted of about twenty slightly-built mud hovels, and had a most forlorn appearance, notwithstanding the luxuriant forest in its rear. A horde of these Indians settled here many years ago, on the site of an abandoned missionary station, and the government had lately placed a resident director over them, with the intention of bringing the hitherto intractable savages under authority. This, however, seemed to promise no other result than that of driving them to their old solitary haunts, on the banks of the interior waters, for many families had already withdrawn themselves. The absence of the usual cultivated trees and plants gave the place a naked and poverty-stricken aspect. I entered one of the hovels, where several women were employed cooking a meal. Portions of a large fish were roasting over a fire made in the middle of the low chamber and the entrails were scattered about the floor, on which the women with their children were squatted. These had a timid, distrustful expression of countenance, and their bodies were begrimed with black mud, which is smeared over the skin as a protection against mosquitoes. The children were naked, the women wore petticoats of coarse cloth, ragged round the edges, and stained in blotches with murixí, a dye made from the bark of a tree. One of them wore a necklace of monkey's teeth. There were scarcely any household utensils; the place was bare with the exception of two dirty grass hammocks hung in the corners. I missed the usual mandioca sheds behind the house, with their surrounding cotton, cacao, coffee, and lemon trees. Two or three young men of the tribe were lounging about the low open doorway. They were stoutly-built fellows, but less well-proportioned than the semi-civilized Indians of the Lower Amazons generally are. Their breadth of chest was remarkable, and their arms were wonderfully thick and muscular. The legs appeared short in proportion to the trunk; the expression of their countenances was unmistakably more sullen and brutal, and the skin of a darker hue, than is common in the Brazilian red man. Before we left the hut an old couple came in; the husband carrying his paddle, bow, arrows, and harpoon, the woman bent beneath the weight of a large basket filled with palm fruits. The man was of low stature and had a wild appearance from the long coarse hair which hung over his forehead. Both his lips were pierced with holes, as is usual with the older Múras seen on the river. They used formerly to wear tusks of the wild hog in these holes whenever they went out to encounter strangers or their enemies in war. The gloomy savagery, filth, and poverty of the people in this place made me feel quite melancholy, and I was glad to return to the canoe. They offered us no civilities; they did not even pass the ordinary salutes, which all the semi-civilized and many savage Indians proffer on a first meeting. The men persecuted Penna for cashaça, which they seemed to consider the only good thing the white man brings with him. As they had nothing whatever to give in exchange, Penna declined to supply them. They followed us as we descended to the port, becoming very troublesome when about a dozen had collected together. They brought their empty bottles with them and promised fish and turtle, if we would only trust them first with the coveted aguardente, or cau-im, as they called it. Penna was inexorable: he ordered the crew to weigh anchor, and the disappointed savages remained hooting after us with all their might, from the top of the bank, as we glided away.

After leaving Matarí we continued our voyage along the northern shore. The banks of the river were of moderate elevation during several days' journey; the terra firma lying far in the interior, and the coast being either low land, or masked with islands of alluvial formation. On the 14th we passed the upper mouth of the Parana-mirim de Eva, an arm of the river of small breadth, formed by a straggling island, some ten miles in length, lying parallel to the northern bank. On passing the western end of this the mainland again appeared, a rather high rocky coast, clothed with a magnificent forest of rounded outline, which continues hence for twenty miles to the mouth of the Rio Negro, and forms the eastern shore of that river. Many houses of settlers, built at a considerable elevation on the wooded heights, now enlivened the river banks. One of the first objects which here greeted us was a beautiful bird we had not hitherto met with, namely, the scarlet and black tanager (Ramphocœlus nigrogularis), flocks of which were seen sporting about the trees on the edge of the water, their flame-coloured liveries lighting up the masses of dark-green foliage.

The weather, from the 14th to the 18th, was wretched; it rained sometimes for twelve hours in succession, not heavily, but in a steady drizzle, such as we are familiar with in our English climate. We landed at several places on the coast, Penna to trade as usual, and I to ramble in the forest in search of birds and insects. In one spot the wooded slope enclosed a very picturesque scene: a brook, flowing through a ravine in the high bank, fell in many little cascades to the broad river beneath, its margins decked out with an infinite variety of beautiful plants. Wild bananas arched over the watercourse, and the trunks of the trees in its vicinity were clothed with ferns, large-leaved species belonging to the genus Lygodium, which, like Osmunda, have their spore-cases collected together on contracted leaves. On the 18th we arrived at a large fazenda (plantation and cattle farm), called Jatuarána. A rocky point here projects into the stream, and as we found it impossible to stem the strong current which whirled round it, we crossed over to the southern shore. Canoes in approaching the Rio Negro generally prefer the southern side on account of the slackness of the current near the banks. Our progress, however, was most tediously slow, for the regular east wind had now entirely ceased, and the vento de cima or wind from up river, having taken its place, blew daily for a few hours dead against us. The weather was oppressively close, and every afternoon a squall arose, which, however, as it came from the right quarter and blew for an hour or two, was very welcome. We made acquaintance on this coast with a new insect pest, the Pium, a minute fly, two-thirds of a line in length, which here commences its reign, and continues henceforward as a terrible scourge along the upper river, or Solimoens, to the end of the navigation on the Amazons. It comes forth only by day, relieving the mosquito at sunrise with the greatest punctuality, and occurs only near the muddy shores of the stream, not one ever being found in the shade of the forest. In places where it is abundant, it accompanies canoes in such dense swarms as to resemble thin clouds of smoke. It made its appearance in this way the first day after we crossed the river. Before I was aware of the presence of flies I felt a slight itching on my neck, wrist, and ankles, and on looking for the cause saw a number of tiny objects having a disgusting resemblance to lice, adhering to the skin. This was my introduction to the much-talked-of Piúm. On close examination they are seen to be minute two-winged insects, with dark-coloured body and pale legs and wings, the latter closed lengthwise over the back. They alight imperceptibly, and squatting close, fall at once to work, stretching forward their long front legs, which are in constant motion and seem to act as feelers, and then applying their short, broad snouts to the skin. Their abdomens soon become distended and red with blood, and then, their thirst satisfied, they slowly move off, sometimes so stupefied with their potations that they can scarcely fly. No pain is felt whilst they are at work, but they each leave a small circular raised spot on the skin and a disagreeable irritation. The latter may be avoided in great measure by pressing out the blood which remains in the spot; but this is a troublesome task, when one has several hundred punctures in the course of a day. I took the trouble to dissect specimens to ascertain the way in which the little pests operate. The mouth consists of a pair of thick fleshy lips, and two triangular horny lancets, answering to the upper lip and tongue of other insects. This is applied closely to the skin, a puncture is made with the lancets, and the blood then sucked through between these into the œsophagus, the circular spot which results coinciding with the shape of the lips. In the course of a few days the red spots dry up, and the skin in time becomes blackened with the endless number of discoloured punctures that are crowded together. The irritation they produce is more acutely felt by some persons than others. I once travelled with a middle-aged Portuguese, who was laid up for three weeks from the attacks of Piúm, his legs being swollen to an enormous size, and the punctures aggravated into spreading sores.

A brisk wind from the east sprang up early in the morning of the 22nd, we then hoisted all sail, and made for the mouth of the Rio Negro. This noble stream at its junction with the Amazons seems, from its position, to be a direct continuation of the main river, whilst the Solimoens which joins at an angle and is somewhat narrower than its tributary, appears to be a branch instead of the main trunk of the vast water-system. One sees therefore at once how the early explorers came to give a separate name to this upper part of the Amazons. The Brazilians have lately taken to applying the convenient term Alto Amazonas (High or Upper Amazons) to the Solimoens, and it is probable that this will gradually prevail over the old name. The Rio Negro broadens considerably from its mouth upwards, and presents the appearance of a great lake; its black-dyed waters having no current, and seeming to be dammed up by the impetuous flow of the yellow, turbid Solimoens, which here belches forth a continuous line of uprooted trees and patches of grass, and forms a striking contrast with its tributary. In crossing we passed the line, a little more than half-way over, where the waters of the two rivers meet and are sharply demarcated from each other. On reaching the opposite shore we found a remarkable change. All our insect pests had disappeared, as if by magic, even from the hold of the canoe: the turmoil of an agitated, swiftly flowing river, and its torn, perpendicular, earthy banks, had given place to tranquil water and a coast indented with snug little bays, fringed with sloping sandy beaches. The low shore and vivid light green endlessly-varied foliage, which prevailed on the south side of the Amazons, were exchanged for a hilly country, clothed with a sombre, rounded, and monotonous forest. Our tedious voyage now approached its termination; a light wind carried us gently along the coast to the city of Barra, which lies about seven or eight miles within the mouth of the river. We stopped for an hour in a clean little bay, to bathe and dress, before showing ourselves again among civilised people. The bottom was visible at a depth of six feet, the white sand taking a brownish tinge from the stained but clear water. In the evening I went ashore, and was kindly received by Senhor Henriques Antony, a warm-hearted Italian, established here in a high position as merchant, who was the never-failing friend of stray travellers. He placed a couple of rooms at my disposal, and in a few hours I was comfortably settled in my new quarters, sixty-four days after leaving Obydos.

I found at Barra my companion, Mr. Wallace, who, since our joint Tocantins expedition, had been exploring, partly with his brother, lately arrived from England, the north-eastern coast of Marajó, the river Capim (a branch of the Guamá, near Pará), Monte Alegre, and Santarem. He had passed us by night below Serpa, on his way to Barra, and so had arrived about three weeks before me. Besides ourselves there were half-a-dozen other foreigners here congregated—Englishmen, Germans, and Americans; one of them a natural-history collector, the rest traders on the rivers. In the pleasant society of these, and of the family of Senhor Henriques, we passed a delightful time; the miseries of our long river voyages were soon forgotten, and in two or three weeks we began to talk of further explorations. Meantime we had almost daily rambles in the neighbouring forest. The whole surface of the land, down to the water's edge, is covered by the uniform dark green rolling forest, the *caá-apoam* (convex woods) of the Indians, characteristic of the Rio Negro. This clothes also the extensive areas of low land, which are flooded by the river in the rainy season. The olive-brown tinge of the water seems to be derived from the saturation in it of the dark green foliage during these annual inundations. The great contrast in form and colour between the forests of the Rio Negro and those of the Amazons arises from the predominance in each of different families of plants. On the main river palms of twenty or thirty different species form a great proportion of the mass of trees, whilst on the Rio Negro they play a very subordinate part. The characteristic kind in the latter region is the Jará (Leopoldinia pulchra), a species not found on the margins of the Amazons, which has a scanty head of fronds, with narrow leaflets of the same dark green hue as the rest of the forest. The stem is smooth, and about two inches in diameter; its height is not more than twelve to fifteen feet; it does not, therefore, rise amongst the masses of foliage of the exogenous trees, so as to form a feature in the landscape, like the broad-leaved Murumurú and Urucurí, the slender Assaí, the tall Jauarí, and the fan-leaved Muritî of the banks of the Amazons. On the shores of the main river the mass of the forest is composed, besides palms, of Leguminosæ, or trees of the bean family, in endless variety as to height, shape of foliage, flowers, and fruit; of silk-cotton trees, colossal nut-trees (Lecythideæ), and Cecropiæ; the underwood and water-frontage consisting in great part of broad-leaved Musaceæ, Marantaceæ, and succulent grasses: all of which are of light shades of green. The forests of the Rio Negro are almost destitute of these large-leaved plants and grasses, which give so rich an appearance to the vegetation wherever they grow; the margins of the stream being clothed with bushes or low trees, having the same gloomy monotonous aspect as the mangroves of the shores of creeks near the Atlantic. The uniformly small but elegantly leaved exogenous trees, which constitute the mass of the forest, consist in great part of members of the Laurel, Myrtle, Bignoniaceous, and Rubiaceous orders. The soil is generally a stiff loam, whose chief component part is the Tabatinga clay, which also forms low cliffs on the coast in some places, where it overlies strata of coarse sandstone. This kind of soil and the same geological formation prevail, as we have seen, in many places on the banks of the Amazons, so that the great contrast in the forest clothing of the two rivers cannot arise from this cause.

The forest was very pleasant for rambling. In some directions broad pathways led down gentle slopes, through what one might fancy were interminable shrubberies of evergreens, to moist hollows where springs of water bubbled up, or shallow brooks ran over their beds of clean white sand. But the most beautiful road was one that ran through the heart of the forest to a waterfall, which the citizens of Barra consider as the chief natural curiosity of their neighbourhood. The waters of one of the larger rivulets which traverse the gloomy wilderness here fall over a ledge of rock about ten feet high. It is not the cascade itself, but the noiseless solitude, and the marvellous diversity and richness of trees, foliage, and flowers, encircling the water basin, that form the attraction of the place. Families make pic-nic excursions to this spot; and the gentlemen—it is said the ladies also—spend the sultry hours of midday bathing in the cold and bracing waters. The place is classic ground to the naturalist, from having been a favourite spot with the celebrated travellers Spix and Martius, during their stay at Barra in 1820. Von Martius was so much impressed by its magical beauty, that he commemorated the visit by making a sketch of the scenery serve as background in one of the plates of his great work on the palms.

Birds and insects, however, were scarce amidst these charming sylvan scenes. I have often traversed the whole distance from Barra to the waterfall, about two miles by the forest road, without seeing or hearing a bird, or meeting with so many as a score of Lepidopterous and Coleopterous insects. In the thinner woods near the borders of the forest many pretty little blue and green creepers of the Dacnidæ group were daily seen feeding on berries, and a few very handsome birds occurred in the forest. But the latter were so rare that we could obtain them only by employing a native hunter, who used to spend a whole day and go a great distance to obtain two or three specimens. In this way I obtained, amongst others, specimens of the Trogon pavoninus (the Suruquá grande of the natives), a most beautiful creature, having soft golden green plumage, red breast, and an orange-coloured beak; also the Ampelis Pompadoura, a rich glossy-purple chatterer with wings of a snowy-white hue.

After we had rested some weeks in Barra, we arranged our plans for further explorations in the interior of the country. Mr. Wallace chose the Rio Negro for his next trip, and I agreed to take the Solimoens. My colleague has already given to the world an account of his journey on the Rio Negro, and his adventurous ascent of its great tributary the Uapés. I left Barra for Ega, the first town of any importance on the Solimoens, on the 26th of March, 1850. The distance is nearly 400 miles, which we accomplished in a small cuberta, manned by ten stout Cucama Indians, in thirty-five days. On this occasion, I spent twelve months in the upper region of the Amazons; circumstances then compelled me to return to Pará. I revisited the same country in 1855, and devoted three years and a half to a fuller exploration of its natural productions. The results of both journeys will be given together in subse quent chapters of this work; in the meantime, I will proceed to give an account of Santarem and the river Tapajos, whose neighbourhoods I investigated in the years 1851-4.

A few words on my visit to Pará in 1851 may be here introduced. I descended the river from Ega to the capital, a distance of 1400 miles, in a heavily-laden schooner belonging to a trader of the former place. The voyage occupied no less than twenty-nine days, although we were favoured by the powerful currents of the rainy season. The hold of the vessel was filled with turtle oil contained in large jars, the cabin was crammed with Brazil nuts, and a great pile of salsaparilla, covered with a thatch of palm leaves, occupied the middle of the deck. We had therefore (the master and two passengers) but rough accommodation, having to sleep on deck, exposed to the wet and stormy weather, under little toldos or arched shelters, arranged with mats of woven lianas and maranta leaves. I awoke many a morning with clothes and bedding soaked through with the rain. With the exception, however, of a slight cold at the commencement, I never enjoyed better health than during this journey. When the wind blew from up river or off the land, we sped away at a great rate; but it was often squally from those quarters, and then it was not safe to hoist the sails. The weather was generally calm, a motionless mass of leaden clouds covering the sky, and the broad expanse of waters flowing smoothly down with no other motion than the ripple of the current. When the wind came from below, we tacked down the stream; sometimes it blew very strong, and then the schooner, having the wind abeam, laboured through the waves, shipping often heavy seas which washed everything that was loose from one side of the deck to the other.

On arriving at Pará, I found the once cheerful and healthful city desolated by two terrible epidemics. The yellow fever, which visited the place the previous year (1850) for the first time since the discovery of the country, still lingered, after having carried off nearly 5 per cent. of the population. The number of persons who were attacked, namely, three-fourths of the entire population, showed how general is the onslaught of an epidemic on its first appearance in a place. At the heels of this plague came the smallpox. The yellow fever had fallen most severely on the whites and mamelucos, the negroes wholly escaping; but the smallpox attacked more especially the Indians, negroes, and people of mixed colour, sparing the whites almost entirely, and taking off about a twentieth part of the population in the course of the four months of its stay. I heard many strange accounts of the yellow fever. I believe Pará was the second port in Brazil attacked by it. The news of its ravages in Bahia, where the epidemic first appeared, arrived some few days before the disease broke out. The government took all the sanitary precautions that could be thought of: amongst the rest was the singular one of firing cannon at the street corners, to purify the air. Mr. Norris, the American consul, told me the first cases of fever occurred near the port, and that it spread rapidly and regularly from house to house, along the streets which run from the waterside to the suburbs, taking about twenty-four hours to reach the end. Some persons related that for several successive evenings before the fever broke out the atmosphere was thick, and that a body of murky vapour, accompanied by a strong stench, travelled from street to street. This moving vapour was called the "Maî da peste" ("the mother or spirit of the plague"); and it was useless to attempt to reason them out of the belief that this was the forerunner of the pestilence. The progress of the disease was very rapid. It commenced in April, in the middle of the wet season. In a few days, thousands of persons lay sick, dying, or dead. The state of the city during the time the fever lasted, may be easily imagined. Towards the end of June it abated, and very few cases occurred during the dry season from July to December.

As I said before, the yellow fever still lingered in the place when I arrived from the interior in April. I was in hopes I should escape it, but was not so fortunate; it seemed to spare no new-comer. At the time I fell ill, every medical man in the place was worked to the utmost in attending the victims of the other epidemic; it was quite useless to think of obtaining their aid, so I was obliged to be my own doctor, as I had been in many former smart attacks of fever. I was seized with shivering and vomit at 9 o'clock in the morning. Whilst the people of the house went down to the town for the medicines I ordered, I wrapped myself in a blanket and walked sharply to and fro along the verandah, drinking at intervals a cup of warm tea, made of a bitter herb in use amongst the natives, called Pajémarióba, a leguminous plant growing in all waste places. About an hour afterwards I took a good draught of a decoction of elder blossoms as a sudorific, and soon after fell insensible into my hammock. Mr. Philipps, an English resident with whom I was then lodging, came home in the afternoon and found me sound asleep and perspiring famously. I did not wake till towards midnight, when I felt very weak and aching in every bone of my body. I then took as a purgative a small dose of Epsom salts and manna. In forty-eight hours the fever left me, and in eight days from the first attack I was able to get about my work. Little else happened during my stay, which need be recorded here. I shipped off all my collections to England, and received thence a fresh supply of funds. It took me several weeks to prepare for my second and longest journey into the interior. My plan now was first to make Santarem head-quarters for some time, and ascend from that place the river Tapajos, as far as practicable. Afterwards I intended to revisit the marvellous country of the Upper Amazons, and work well its natural history at various stations I had fixed upon, from Ega to the foot of the Andes.

CHAPTER VIII.

SANTAREM.

Situation of Santarem—Manners and customs of the inhabitants—Climate—Grassy campos and woods—Excursions to Mapirí, Mahicá, and Irurá, with sketches of their Natural History; Palms, wild fruit-trees, Mining Wasps, Masen Wasps, Bees, and Sloths.

I HAVE already given a short account of the size, situation, and general appearance of Santarem. Although containing not more than 2,500 inhabitants, it is the most civilised and important settlement on the banks of the main river from Peru to the Atlantic. The pretty little town, or city as it is called, with its rows of tolerably uniform white-washed and red-tiled houses, surrounded by green gardens and woods, stands on gently sloping ground on the eastern side of the Tapajos, close to its point of junction with the Amazons. A small eminence on which a fort has been erected, but which is now in a dilapidated condition, overlooks the streets, and forms the eastern limit of the mouth of the tributary. The Tapajos at Santarem is contracted to a breadth of about a mile and a half by an accretion of low alluvial land, which forms a kind of delta on the western side; fifteen miles further up, the river is seen at its full width of ten or a dozen miles, and the magnificent hilly country, through which it flows from the south, is then visible on both shores. This high land, which appears to be a continuation of the central table-lands of Brazil, stretches almost without interruption on the eastern side of the river down to its mouth at Santarem. The scenery, as well as the soil, vegetation, and animal tenants of this region, are widely different from those of the flat and uniform country which borders the Amazons along most part of its course. After travelling week after week on the main river, the aspect of Santarem, with its broad white sandy beach, limpid dark green waters, and line of picturesque hills rising behind over the fringe of green forest, affords an agreeable surprise. On the main Amazons, the prospect is monotonous unless the vessel runs near the shore, when the wonderful diversity and beauty of the vegetation afford constant entertainment. Otherwise, the unvaried, broad yellow stream, and the long low line of forest, which dwindles away in a broken line of trees on the sea-like horizon, and is renewed reach after reach, as the voyager advances, weary by their uniformity.

I arrived at Santarem on my second journey into the interior, in November, 1851, and made it my head-quarters for a period, as it turned out, of three years and a half. During this time I made, in pursuance of the plan I had framed, many excursions up the Tapajos, and to other places of interest in the surrounding region. On landing, I found no difficulty in hiring a suitable house on the outskirts of the place. It was pleasantly situated near the beach, going towards the aldeia or Indian part of the town. The ground sloped from the back premises down to the waterside, and my little raised verandah overlooked a beautiful flower-garden, a great rarity in this country, which belonged to the neighbours. The house contained only three rooms, one with brick and two with boarded floors. It was substantially built, like all the better sort of houses in Santarem, and had a stuccoed front. The kitchen, as is usual, formed an outhouse placed a few yards distant from the other rooms. The rent was 12,000 reis, or about twenty-seven shillings a month. In this country, a tenant has no extra payments to make; the owners of house property pay a dizimo or tithe, to the "collectoria geral," or general treasury, but with this the occupier of course has nothing to do. In engaging servants, I had the good fortune to meet with a free mulatto, an industrious and trustworthy young fellow, named José, willing to arrange with me; the people of his family cooking for us, whilst he assisted me in collecting; he proved of the greatest service in the different excursions we subsequently made. Servants of any kind were almost impossible to be obtained at Santarem, free people being too proud to hire themselves, and slaves too few and valuable to their masters, to be let out to others. These matters arranged, the house put in order, and a rude table, with a few chairs, bought or borrowed to furnish the house with, I was ready in three or four days to commence my natural-history explorations in the neighbourhood.

Santarem is a pleasant place to live in, irrespective of its society. There are no insect pests, mosquito, pium, sand-fly, or motuca. The climate is glorious; during six months of the year, from August to February, very little rain falls, and the sky is cloudless for weeks together, the fresh breezes from the sea, nearly 400 miles distant, moderating the great heat of the sun. The wind is sometimes so strong for days together that it is difficult to make way against it in walking along the streets, and it enters the open windows and doors of houses, scattering loose clothing and papers in all directions. The place is considered healthy, but at the changes of season severe colds and ophthalmia are prevalent. I found three Englishmen living here, who had resided many years in the town or its neighbourhood, and who still retained their florid complexions; the plump and fresh appearance of many of the middle-aged Santarem ladies also bore testimony to the healthfulness of the climate. The streets are always clean and dry, even in the height of the wet season; good order is always kept, and the place pretty well supplied with provisions. Very good bread was hawked round the town every morning, with milk, and a great variety of fruits and vegetables. Amongst the fruits there was a kind called atta, which I did not see in any other part of the country. It belongs to the Anonaceous order, and the tree which produces it grows apparently wild in the neighbourhood of Santarem. It is a little larger than a good sized orange, and the rind, which encloses a mass of rich custardy pulp, is scaled like the pineapple, but green when ripe, and encrusted on the inside with sugar. To finish this account of the advantages of Santarem, the delicious bathing in the clear waters of the Tapajos may be mentioned. There is here no fear of alligators. When the east wind blows, a long swell rolls in on the clean sandy beach, and the bath is most exhilirating.

The country around Santarem is not clothed with dense and lofty forest, like the rest of the great humid river plain of the Amazons. It is a campo region; a slightly elevated and undulating tract of land, wooded only in patches, or with single scattered trees. A good deal of the country on the borders of the Tapajos, which flows from the great campo area of Interior Brazil, is of this description. It is on this account that I consider the eastern side of the river, towards its mouth, to be a northern prolongation of the continental land, and not a portion of the alluvial flats of the Amazons. The soil is a coarse gritty sand; the substratum, which is visible in some places, consisting of sandstone conglomerate probably of the same formation as that which underlies the Tabatinga clay in other parts of the river valley. The surface is carpeted with slender hairy grasses, unfit for pasture, growing to a uniform height of about a foot. The patches of wood look like copses in the middle of green meadows; they are called by the natives "ilhas de mato," or islands of jungle; the name being, no doubt, suggested by their compactness of outline, neatly demarcated in insular form from the smooth carpet of grass around them. They are composed of a great variety of trees, loaded with succulent parasites, and lashed together by woody climbers like the forest in other parts. A narrow belt of dense wood, similar in character to these ilhas, and like them sharply limited along its borders, runs everywhere parallel and close to the river. In crossing the campo, the path from the town ascends a little for a mile or two, passing through this marginal strip of wood; the grassy land then slopes gradually to a broad valley, watered by rivulets, whose banks are clothed with lofty and luxuriant forest. Beyond this, a range of hills extends as far as the eye can reach towards the yet untrodden interior. Some of these hills are long ridges, wooded or bare; others are isolated conical peaks, rising abruptly from the valley. The highest are probably not more than a thousand feet above the level of the river. One remarkable hill, the Serra de Muruarú, about fifteen miles from Santarem, which terminates the prospect to the south, is of the same truncated pyramidal form as the range of hills near Almeyrim. Complete solitude reigns over the whole of this stretch of beautiful country. The inhabitants of Santarem know nothing of the interior, and seem to feel little curiosity concerning it. A few tracks from the town across the campo lead to some small clearings four or five miles off, belonging to the poorer inhabitants of the place; but, excepting these, there are no roads, or signs of the proximity of a civilised settlement.

The appearance of the campos changes very much according to the season. There is not that grand uniformity of aspect throughout the year which is observed in the virgin forest, and which makes a deeper impression on the naturalist the longer he remains in this country. The seasons in this part of the Amazons region are sharply contrasted, but the difference is not so great as in some tropical countries, where, during the dry monsoon, insects and reptiles go into a summer sleep, and the trees simultaneously shed their leaves. As the dry season advances (August, September), the grass on the campos withers, and the shrubby vegetation near the town becomes a mass of parched yellow stubble. The period, however, is not one of general torpidity or repose for animal or vegetable life. Birds certainly are not so numerous as in the wet season, but some kinds remain and lay their eggs at this time—for instance, the ground doves (Chamæpelia). The trees retain their verdure throughout, and many of them flower in the dry months. Lizards do not become torpid, and insects are seen both in the larvæ and the perfect states, showing that the aridity of the climate has not a general influence on the development of the species. Some kinds of butterflies, especially the little hair-streaks (Theclæ), whose caterpillars feed on the trees, make their appearance only when the dry season is at its height. The land molluscs of the district are the only animals which æstivate; they are found in clusters, Bulimi and Helices, concealed in hollow trees, the mouths of their shells closed by a film of mucus. The fine weather breaks up often with great suddenness about the beginning of February. Violent squalls from the west, or the opposite direction to the trade-wind, then occur. They give very little warning, and the first generally catches the people unprepared. They fall in the night, and blowing directly into the harbour, with the first gust sweep all vessels from their anchorage; in a few minutes a mass of canoes, large and small, including schooners of fifty tons burthen, are clashing together, pell-mell, on the beach. I have reason to remember these storms, for I was once caught in one myself whilst crossing the river in an undecked boat, about a day's journey from Santarem. They are accompanied with terrific electric explosions, the sharp claps of thunder falling almost simultaneously with the blinding flashes of lightning. Torrents of rain follow the first outbreak; the wind then gradually abates, and the rain subsides into a steady drizzle, which continues often for the greater part of the succeeding day. After a week or two of showery weather the aspect of the country is completely changed. The parched ground in the neighbourhood of Santarem breaks out, so to speak, in a rash of greenery: the dusty, languishing trees gain, without having shed their old leaves, a new clothing of tender green foliage; a wonderful variety of quick-growing leguminous plants spring up, and leafy creepers overrun the ground, the bushes, and the trunks of trees. One is reminded of the sudden advent of spring after a few warm showers in northern climates; I was the more struck by it as nothing similar is witnessed in the virgin forests amongst which I had passed the four years previous to my stay in this part. The grass on the campos is renewed, and many of the campo trees, especially the myrtles, which grow abundantly in one portion of the district, begin to flower, attracting by the fragrance of their blossoms a great number and variety of insects, more particularly Coleoptera. Many kinds of birds, parrots, toucans, and barbets, which live habitually in the forest, then visit the open places. A few weeks of comparatively dry weather generally intervene in March, after a month or two of rain. The heaviest rains fall in April, May, and June; they come in a succession of showers, with sunny gleamy weather in the intervals. June and July are the months when the leafy luxuriance of the campos, and the activity of life, are at their highest. Most birds have then completed their moulting, which extends over the period from February to May. The flowering shrubs are then mostly in bloom, and numberless kinds of Dipterous and Hymenopterous insects appear simultaneously with the flowers. This season might be considered the equivalent of summer in temperate climates, as the bursting forth of the foliage in February represents the spring; but under the equator there is not that simultaneous march in the annual life of animals and plants which we see in high latitudes; some species, it is true, are dependent upon others in their periodical acts of life, and go hand-in-hand with them, but they are not all simultaneously and similarly affected by the physical changes of the seasons.

I will now give an account of some of my favourite collecting places in the neighbourhood of Santarem, incorporating with the description a few of the more interesting observations made on the Natural History of the localities. To the west of the town there was a pleasant path along the beach to a little bay, called Mapirí, about five miles within the mouth of the Tapajos. The road was practicable only in the dry season. The river at Santarem rises on the average about thirty feet, varying in different years about ten feet; so that in the four months, from April to July, the water comes up to the edge of the marginal belt of wood already spoken of. This Mapirí excursion was most pleasant and profitable in the months from January to March, before the rains became too continuous. The sandy beach beyond the town is very irregular; in some places forming long spits on which, when the east wind is blowing, the waves break in a line of foam; at others receding to shape out quiet little bays and pools. On the outskirts of the town a few scattered huts of Indians and coloured people are passed, prettily situated on the margin of the white beach, with a background of glorious foliage; the cabin of the pure-blood Indian being distinguished from the mud hovels of the free negroes and mulattoes by its light construction, half of it being an open shed, where the dusky tenants are seen at all hours of the day lounging in their open-meshed grass hammocks. About two miles on the road we come to a series of shallow pools, called the Laguinhos, which are connected with the river in the wet season, but separated from it at other times by a high bank of sand topped with bushes. There is a break here in the fringe of wood, and a glimpse is obtained of the grassy campo. When the waters have risen to the level of the pools, this place is frequented by many kinds of wading birds. Snow-white egrets of two species stand about the margins of the water, and dusky-striped herons may be seen half hidden under the shade of the bushes. The pools are covered with a small kind of water-lily, and surrounded by a dense thicket. Amongst the birds which inhabit this spot is the rosy-breasted Troupial (Trupialis Guianensis), a bird resembling our starling in size and habits, and not unlike it in colour, with the exception of the rich rosy vest. The water at this time of the year overflows a large level tract of campo bordering the pools, and the Troupials come to feed on the larvæ of insects which then abound in the moist soil.

Beyond the Laguinhos there succeeds a tract of level beach, covered with trees which form a beautiful grove. About the month of April, when the water rises to this level, the trees are covered with blossom, and a handsome orchid, an Epidendron with large white flowers, which clothes thickly the trunks, is profusely in bloom. Several kinds of kingfisher resort to the place: four species may be seen within a small space: the largest as big as a crow, of a mottled-grey hue, and with an enormous beak; the smallest not larger than a sparrow. The large one makes its nest in clay cliffs, three or four miles distant from this place. None of the kingfishers are so brilliant in colour as our English species. The blossoms on the trees attract two or three species of humming-birds, the most conspicuous of which is a large swallow-tailed kind (Eupetomena macroura), with a brilliant livery of emerald green and steel blue. I noticed that it did not remain so long poised in the air before the flowers as the other smaller species; it perched more frequently, and sometimes darted after small insects on the wing. Emerging from the grove there is a long stretch of sandy beach ; the land is high and rocky, and the belt of wood which skirts the river banks is much broader than it is elsewhere. At length, after rounding a projecting bluff, the bay of Mapirí is reached. The river view is characteristic of the Tapajos: the shores are wooded, and on the opposite side is a line of clay cliffs, with hills in the background clothed with a rolling forest. A long spit of sand extends into mid-river, beyond which is an immense expanse of dark water, the further shore of the Tapajos being barely visible as a thin grey line of trees on the horizon. The transparency of air and water in the dry season when the brisk east wind is blowing, and the sharpness of outline of hills, woods, and sandy beaches, give a great charm to this spot.

Whilst resting in the shade during the great heat of the early hours of afternoon, I used to find amusement in watching the proceedings of the sand wasps. A small pale green kind of Bembex (Bembex ciliata) was plentiful near the bay of Mapirí. When they are at work, a number of little jets of sand are seen shooting over the surface of the sloping bank. The little miners excavate with their fore feet, which are strongly built and furnished with a fringe of stiff bristles ; they work with wonderful rapidity, and the sand thrown out beneath their bodies issues in continuous streams. They are solitary wasps, each female working on her own account. After making a gallery two or three inches in length, in a slanting direction from the surface, the owner backs out and takes a few turns round the orifice, apparently to see whether it is well made, but in reality, I believe, to take note of the locality, that she may find it again. This done the busy workwoman flies away ; but returns, after an absence varying in different cases from a few minutes to an hour or more, with a fly in her grasp, with which she re-enters her mine. On again emerging, the entrance is carefully closed with sand. During this interval she has laid an egg on the body of the fly, which she had previously benumbed with her sting, and which is to serve as food for the soft footless grub soon to be hatched from the egg. From what I could make out, the Bembex makes a fresh excavation for every egg to be deposited ; at least, in two or three of the galleries which I opened there was only one fly enclosed.

I have said that the Bembex on leaving her mine took note

of the locality: this seemed to be the explanation of the short delay previous to her taking flight; on rising in the air, also, the insects generally flew round over the place before making straight off. Another nearly allied but much larger species, the Monedula signata, whose habits I observed on the banks of the Upper Amazons, sometimes excavates its mine solitarily on sand-banks recently laid bare in the middle of the river, and closes the orifice before going in search of prey. In these cases the insect has to make a journey of at least half a mile to procure the kind of fly, the Motúca (Hadrus lepidotus), with which it provisions its cell. I often noticed it to take a few turns in the air round the place before starting; on its return it made without hesitation straight for the closed mouth of the mine. I was convinced that the insects noted the bearings of their nests, and the direction they took in flying from them. The proceeding in this and similar cases (I have read of something analogous having been noticed in hive bees) seems to be a mental act of the same nature as that which takes place in ourselves when recognising a locality. The senses, however, must be immeasurably more keen, and the mental operation much more certain, in them than they are in man; for to my eye there was absolutely no land-mark on the even surface of sand which could serve as guide, and the borders of the forest were not nearer than half a mile. The action of the wasp would be said to be instinctive; but it seems plain that the instinct is no mysterious and unintelligible agent, but a mental process in each individual, differing from the same in man only by its unerring certainty. The mind of the insect appears to be so constituted that the impression of external objects, or the want felt, causes it to act with a precision which seems to us like that of a machine constructed to move in a certain given way. I have noticed in Indian boys a sense of locality almost as keen as that possessed by the sand-wasp. An old Portuguese and myself, accompanied by a young lad about ten years of age, were once lost in the forest in a most solitary place on the banks of the main river. Our case seemed hopeless, and it did not for some time occur to us to consult our little companion, who had been playing with his bow and arrow all the way whilst we were hunting, apparently taking no note of the route. When asked, however, he pointed out, in a moment, the right direction of our canoe. He could not explain how he knew; I believe he had noted the course we had taken almost unconsciously. The sense of locality in his case seemed instinctive

The Monedula signata is a good friend to travellers in those parts of the Amazons which are infested by the blood-thirsty Motúca. I first noticed its habit of preying on this fly one day when we landed to make our fire and dine on the borders of the forest adjoining a sand-bank. The insect is as large as a hornet, and has a most waspish appearance. I was rather startled when one out of the flock which was hovering about us flew straight at my face: it had espied a Motúca on my neck, and was thus pouncing upon it. It seizes the fly not with its jaws, but with its fore and middle feet, and carries it off tightly held to its breast. Wherever the traveller lands on the Upper Amazons in the neighbourhood of a sand-bank he is sure to be attended by one or more of these useful vermin-killers.

The bay of Mapirí was the limit of my day excursions by the river-side, to the west of Santarem. A person may travel, however, on foot, as Indians frequently do, in the dry season for fifty or sixty miles along the broad clean sandy beaches of the Tapajos. The only obstacles are the rivulets, most of which are fordable when the waters are low. To the east my rambles extended to the banks of the Mahicá inlet. This enters the Amazons about three miles below Santarem, where the clear stream of the Tapajos begins to be discoloured by the turbid waters of the main river. The Mahicá has a broad margin of rich level pasture, limited on each side by the straight, tall hedge of forest. On the Santarem side it is skirted by high wooded ridges. A landscape of this description always produced in me an impression of sadness and loneliness, which the riant virgin forests that closely hedge in most of the by-waters of the Amazons never created. The pastures are destitute of flowers, and also of animal life, with the exception of a few small plain-coloured birds and solitary Caracára eagles, whining from the topmost branches of dead trees on the forest borders. A few settlers have built their palm-thatched and mud-walled huts on the banks of the Mahicá, and occupy themselves chiefly in tending small herds of cattle. They seemed to be all wretchedly poor. The oxen however, though small, were sleek and fat, and the district was most promising for agricultural and pastoral employments. In the wet season the waters gradually rise and cover the meadows, but there is plenty of room for the removal of the cattle to higher ground. The lazy and ignorant people seem totally unable to profit by these advantages. The houses have no gardens or plantations near them. I was told it was useless to plant anything, because the cattle devoured the young shoots. In this country grazing and planting are very rarely carried on together, for the people seem to have no notion of enclosing patches of ground for cultivation. They say it is too much trouble to make enclosures. The construction of a durable fence is certainly a difficult matter, for it is only two or three kinds of tree which will serve the purpose in being free from the attacks of insects, and these are scattered far and wide through the woods.

Although the meadows were unproductive ground to a Naturalist, the woods on their borders teemed with life; the number and variety of curious insects of all orders which occurred here was quite wonderful. The belt of forest was intersected by numerous pathways leading from one settler's house to another. The ground was moist, but the trees were not so lofty or their crowns so densely packed together as in other parts; the sun's light and heat therefore had freer access to the soil, and the underwood was much more diversified than in the virgin forest. I never saw so many kinds of dwarf palms together as here; pretty miniature species; some not more than five feet high, and bearing little clusters of round fruit not larger than a good bunch of currants. A few of the forest trees had the size and strongly-branched figures of our oaks, and a similar bark. One noble palm grew here in great abundance, and gave a distinctive character to the district. This was the Œnocarpus distichus, one of the kinds called Bacába by the natives. It grows to a height of forty to fifty feet. The crown is of a lustrous dark-green colour, and of a singularly flattened or compressed shape; the leaves being arranged on each side in nearly the same plane. When I first saw this tree on the campos, where the east wind blows with great force night and day for several months, I thought the shape of the crown was due to the leaves being prevented from radiating equally by the constant action of the breezes. But the plane of growth is not always in the direction of the wind, and the crown has the same shape when the tree grows in the sheltered woods. The fruit of this fine palm ripens towards the end of the year, and is much esteemed by the natives, who manufacture a pleasant drink from it similar to the assai described in a former chapter, by rubbing off the coat of pulp from the nuts, and mixing it with water. A bunch of fruit weighs thirty or forty pounds. The beverage has a milky appearance, and an agreeable nutty flavour. The tree is very difficult to climb, on account of the smoothness of its stem; consequently the natives, whenever they want a bunch of fruit for a bowl of Bacába, cut down and thus destroy a tree which has taken a score or two of years to grow, in order to get at it.

In the lower part of the Mahicá woods, towards the river, there is a bed of stiff white clay, which supplies the people of Santarem with material for the manufacture of coarse pottery and cooking utensils; all the kettles, saucepans, mandioca ovens, coffee-pots, washing-vessels, and so forth, of the poorer classes throughout the country, are made of this same plastic clay, which occurs at short intervals over the whole surface of the Amazons valley, from the neighbourhood of Pará to within the Peruvian borders, and forms part of the great Tabatinga marl deposit. To enable the vessels to stand the fire, the bark of a certain tree, called Caraipé, is burnt and mixed with the clay, which gives tenacity to the ware. Caraipé is an article of commerce, being sold, packed in baskets, at the shops in most of the towns. The shallow pits, excavated in the marly soil at Mahicá, were very attractive to many kinds of mason bees and wasps, who made use of the clay to build their nests with. So that we have here another example of the curious analogy that exists between the arts of insects and those of man. I spent many an hour, watching their proceedings: a short account of the habits of some of these busy creatures may be interesting.

The most conspicuous was a large yellow and black wasp, with a remarkably long and narrow waist, the Pelopæus fistularis. This species collected the clay in little round pellets, which it carried off, after rolling them into a convenient shape, in its mouth. It came straight to the pit with a loud hum, and on alighting, lost not a moment in beginning to work; finishing the kneading of its little load in two or three minutes. The nest of this wasp is shaped like a pouch, two inches in length, and is attached to a branch or other projecting object. One of these restless artificers once began to build on the handle of a chest in the cabin of my canoe, when we were stationary at a place for several days. It was so intent on its work that it allowed me to inspect the movements of its mouth with a lens whilst it was laying on the mortar. Every fresh pellet was brought in with a triumphant song, which changed to a cheerful busy hum when it alighted and began to work. The little ball of moist clay was laid on the edge of the cell, and then spread out around the circular rim, by means of the lower lip guided by the mandibles. The insect placed itself astride over the rim to work, and, on finishing each addition to the structure, took a turn round, patting the sides with its feet inside and out before flying off to gather a fresh pellet. It worked only in sunny weather, and the previous layer was sometimes not quite dry when the new coating was added. The whole structure takes about a week to complete. I left the place before the gay little builder had quite finished her task: she did not accompany the canoe, although we moved along the bank of the river very slowly. On opening closed nests of this species, which are common in the neighbourhood of Mahicá, I always found them to be stocked with small spiders of the genus Gastracantha, in the usual half-dead state to which the mother wasps reduce the insects which are to serve as food for their progeny.

Pelopæus Wasp building nest.

Besides the Pelopæus there were three or four kinds of Trypoxylon, a genus also found in Europe, and which some Naturalists have supposed to be parasitic, because the legs are not furnished with the usual row of strong bristles for digging, characteristic of the family to which it belongs. The species of Trypoxylon, however, are all building wasps; two of them which I observed (T. albitarse and an undescribed species) provision their nests with spiders, a third (T. aurifrons) with small caterpillars. Their habits are similar to those of the Pelopæus: namely, they carry off the clay in their mandibles, and have a different song when they hasten away with the burden, from that which they sing whilst at work. Trypoxylon albitarse, which is a large black kind, three-quarters of an inch in length, makes a tremendous fuss whilst building

Cells of Trypoxylon aurifrons.

its cell. It often chooses the walls or doors of chambers for this purpose, and when two or three are at work in the same place their loud humming keeps the house in an uproar. The cell is a tubular structure about three inches in length. T. aurifrons, a much smaller species, makes a neat little nest shaped like a carafe; building rows of them together in the corners of verandahs.

But the most numerous and interesting of the clay artificers are the workers of a species of social bee, the Melipona fasciculata. The Meliponæ in tropical America take the place of the true Apides, to which the European hive-bee belongs, and which are here unknown; they are generally much smaller insects than the hive-bees and have no sting. The M. fasciculata is about a third shorter than the Apis mellifica; its colonies are composed of an immense number of individuals. The workers are generally seen collecting pollen in the same

Melipona Bees gathering clay.

way as other bees, but great numbers are employed gathering clay. The rapidity and precision of their movements whilst thus engaged are wonderful. They first scrape the clay with their jaws: the small portions gathered are then cleared by the anterior paws and passed to the second pair of feet, which in their turn, convey them to the large foliated expansions of the hind shanks, which are adapted normally in bees, as every one knows, for the collection of pollen. The middle feet pat the growing pellets of mortar on the hind legs to keep them in a compact shape as the particles are successively added. The little hodsmen soon have as much as they can carry, and they then fly off. I was for some time puzzled to know what the bees did with the clay; but I had afterwards plenty of opportunity for ascertaining. They construct their combs in any suitable crevice in trunks of trees or perpendicular banks, and the clay is required to build up a wall so as to close the gap, with the exception of a small orifice for their own entrance and exit. Most kinds of Meliponæ are in this way masons as well as workers in wax and pollen-gatherers. One little species (undescribed), not more than two lines long, builds a neat tubular gallery of clay, kneaded with some viscid substance, outside the entrance to its hive, besides blocking up the crevice in the tree within which it is situated. The mouth of the tube is trumpet-shaped, and at the entrance a number of the pigmy bees are always stationed, apparently acting as sentinels.

A hive of the Melipona fasciculata, which I saw opened, contained about two quarts of pleasantly-tasted liquid honey. The bees, as already remarked, have no sting, but they bite furiously when their colonies are disturbed. The Indian who plundered the hive was completely covered by them; they took a particular fancy to the hair of his head, and fastened on it by hundreds. I found forty-five species of these bees in different parts of the country; the largest was half an inch in length; the smallest were extremely minute, some kinds being not more than one-twelfth of an inch in size. These tiny fellows are often very troublesome in the woods, on account of their familiarity; for they settle on one's face and hands, and, in crawling about, get into the eyes and mouth, or up the nostrils.

The broad expansion of the hind shanks of bees is applied in some species to other uses besides the conveyance of clay and pollen. The female of the handsome golden and black Euglossa Surinamensis has this palette of very large size. This species builds its solitary nest also in crevices of walls or trees; but it closes up the chink with fragments of dried leaves and sticks, cemented together, instead of clay. It visits the cajú trees, and gathers with its hind legs a small quantity of the gum which exudes from their trunks. To this it adds the other materials required from the neighbouring bushes, and when laden flies off to its nest.

To the south my rambles never extended further than the banks of the Irurá, a stream which rises amongst the hills already spoken of, and running through a broad valley, wooded along the margins of the water-courses, falls into the Tapajos, at the head of the bay of Mapirí. All beyond, as before remarked, is terra incognita to the inhabitants of Santarem, The Brazilian settlers on the banks of the Amazons seem to have no taste for explorations by land, and I could find no person willing to accompany me on an excursion further towards the interior. Such a journey would be exceedingly difficult in this country, even if men could be obtained willing to undertake it. Besides, there were reports of a settlement of fierce runaway negroes on the Serra de Mururarú, and it was considered unsafe to go far in that direction, except with a large armed party. I visited the banks of the Irurá and the rich woods accompanying it, and two other streams in the same neighbourhood, one called the Panéma, and the other the Urumarí, once or twice a week during the whole time of my residence in Santarem, and made large collections of their natural productions. These forest brooks, with their clear cold waters brawling over their sandy or pebbly beds, through wild tropical glens, always had a great charm for me. The beauty of the moist, cool, and luxuriant glades was heightened by the contrast they afforded to the sterile country around them. The bare or scantily wooded hills which surround the valley are parched by the rays of the vertical sun. One of them, the Pico do Irurá, forms a nearly perfect cone, rising from a small grassy plain to a height of 500 or 600 feet, and its ascent is excessively fatiguing after the long walk from Santarem over the campos. I tried it one day, but did not reach the summit. A dense growth of coarse grasses clothed the steep sides of the hill, with here and there a stunted tree of kinds found in the plain beneath. In bared places, a red crumbly soil is exposed; and in one part a mass of rock, which appeared to me, from its compact texture and the absence of stratification, to be porphyritic; but I am not Geologist sufficient to pronounce on such questions. Mr. Wallace states that he found fragments of scoriæ, and believes the hill to be a volcanic cone. To the south and east of this isolated peak, the elongated ridges or table-topped hills attain a somewhat greater elevation.

The forest in the valley is limited to a tract a few hundred yards in width on each side of the different streams; in places where these run along the bases of the hills, the hill-sides facing the water are also richly wooded, although their opposite declivities are bare or nearly so. The trees are lofty and of great variety; amongst them are colossal examples of the Brazil nut tree (Bertholletia excelsa), and the Pikiá. This latter bears a large eatable fruit, curious in having a hollow chamber between the pulp and the kernel, beset with hard spines, which produce serious wounds if they enter the skin.

The eatable part appeared to me not much more palatable than a raw potato; but the inhabitants of Santarem are very fond of it, and undertake the most toilsome journeys on foot to gather a basketful. The tree which yields the tonka bean (Dipteryx odorata), used in Europe for scenting snuff, is also of frequent occurrence here. It grows to an immense height, and the fruit, which, although a legume, is of a rounded shape, and has but one seed, can be gathered only when it falls to the ground. A considerable quantity (from 1000 to 3000 pounds) is exported annually from Santarem, the produce of the whole region of the Tapajos. An endless diversity of trees and shrubs, some beautiful in flower and foliage, others bearing curious fruits, grow in this matted wilderness. It would be tedious to enumerate many of them. I was much struck with the variety of trees with large and diversely shaped fruits growing out of the trunk and branches, some within a few inches of the ground, like the cacao. Most of them are called by the natives Cupú, and the trees are of inconsiderable height. One of them, called Cupu-aï, bears a fruit of elliptical shape and of a dingy earthen colour, six or seven inches long, the shell of which is woody and thin, and contains a small number of seeds loosely enveloped in a juicy pulp of very pleasant flavour. The fruits hang like clayey ants' nests from the branches. Another kind more nearly resembles the cacao; this is shaped something like the cucumber, and has a green ribbed husk. It bears the name of Cacao de macaco, or monkey's chocolate, but the seeds are smaller than those of the common cacao. I tried once or twice to make chocolate from them. They contain plenty of oil of similar fragrance to that of the ordinary cacao-nut, and make up very well into paste; but the beverage has a repulsive clayey colour and an inferior flavour.

My excursions to the Irurá had always a pic-nic character. A few rude huts are scattered through the valley, but they are tenanted only for a few days in the year, when their owners come to gather and roast the mandioca of their small clearings. We used generally to take with us two boys—one negro, the other Indian—to carry our provisions for the day; a few pounds of beef or fried fish, farinha and bananas, with plates, and a kettle for cooking. José carried the guns, ammunition, and game-bags, and I the apparatus for entomologising—the insect net, a large leathern bag with compartments for corked boxes, phials, glass tubes, and so forth. It was our custom to start soon after sunrise, when the walk over the campos was cool and pleasant, the sky without a cloud, and the grass wet with dew. The paths are mere faint tracks; in our early excursions it was difficult to avoid missing our way. We were once completely lost, and wandered about for several hours over the scorching soil without recovering the road. A fine view is obtained of the country, from the rising ground about half-way across the waste. Thence to the bottom of the valley is a long, gentle, grassy slope, bare of trees. The strangely-shaped hills; the forest at their feet, richly varied with palms; the bay of Mapirí on the right, with the dark waters of the Tapajos and its white glistening shores, are all spread out before one, as if depicted on canvas. The extreme transparency of the atmosphere gives to all parts of the landscape such clearness of outline that the idea of distance is destroyed, and one fancies the whole to be almost within reach of the hand. Descending into the valley, a small brook has to be crossed, and then half a mile of sandy plain, whose vegetation wears a peculiar aspect, owing to the predominance of a stemless palm, the Curuá (Attalea spectabilis), whose large, beautifully pinnated, rigid leaves rise directly from the soil. The fruit of this species is similar to the coco-nut, containing milk in the interior of the kernel, but it is much inferior to it in size. Here, and indeed all along the road, we saw on most days in the wet season tracks of the jaguar. We never, however, met with the animal, although we sometimes heard his loud "hough" in the night, whilst lying in our hammocks at home, in Santarem, and knew he must be lurking somewhere near us.

My best hunting ground was a part of the valley sheltered on one side by a steep hill, whose declivity, like the swampy valley beneath, was clothed with magnificent forest. We used to make our halt in a small cleared place, tolerably free from ants and close to the water. Here we assembled after our toilsome morning's hunt in different directions through the woods, took our well-earned meal on the ground—two broad leaves of the wild banana serving us for a tablecloth—and rested for a couple of hours during the great heat of the afternoon. The diversity of animal productions was as wonderful as that of the vegetable forms in this rich locality. It was pleasant to lie down during the hottest part of the day, when my people lay asleep, and watch the movements of animals. Sometimes a troop of Anús (Crotophaga), a glossy black-plumaged bird, which lives in small societies in grassy places, would come in from the campos, one by one, calling to each other as they moved from tree to tree. Or a Toucan (Rhamphastos ariel) silently hopped or ran along and up the branches, peeping into chinks and crevices. Notes of solitary birds resounded from a distance through the wilderness. Occasionally a sulky Trogon would be seen, with its brilliant green back and rose-coloured breast, perched for an hour without moving, on a low branch. A number of large fat lizards, two feet long, of a kind called by the natives Jacuarú (Teius teguexim) were always observed in the still hours of mid-day scampering with great clatter over the dead leaves, apparently in chase of each other. The fat of this bulky lizard is much prized by the natives, who apply it as a poultice to draw palm spines or even grains of shot from the

The Jacuarú (Teius teguexim).

flesh. Other lizards of repulsive aspect, about three feet in length when full grown, splashed about and swam in the water: sometimes emerging to crawl into hollow trees on the banks of the stream, where I once found a female and a nest of eggs. The lazy flapping flight of large blue and black morpho butterflies high in the air, the hum of insects, and many inanimate sounds, contributed their share to the total impression this strange solitude produced. Heavy fruits from the crowns of trees which were mingled together at a giddy height overhead, fell now and then with a startling "plop" into the water. The breeze, not felt below, stirred in the top-most branches, setting the twisted and looped sipós in motion, which creaked and groaned in a great variety of notes. To these noises was added the monotonous ripple of the brook, which had its little cascade at every score or two yards of its course.

I seldom met with any of the larger animals in these excursions. We never saw a mammal of any kind on the campos; but tracks of three species were seen occasionally besides those of the jaguar; these belonged to a small tiger-cat, a deer, and an opossum; all of which animals must have been very rare, and probably nocturnal in their habits, with the exception of the deer. I saw in the woods, on one occasion, a small flock of monkeys, and once had an opportunity of watching the movements of a sloth. The latter was of the kind called by Cuvier Bradypus tridactylus, which is clothed with shaggy gray hair. The natives call it, in the Tupi language, Aï ybyreté (in Portuguese, Preguiça da terra firme), or sloth of the mainland, to distinguish it from the Bradypus infuscatus, which has a long, black and tawny stripe between the shoulders, and is called Aï ygapó (Preguiça das vargens), or sloth of the flooded lands. Some travellers in South America have described the sloth as very nimble in its native woods, and have disputed the justness of the name which has been bestowed upon it. The inhabitants of the Amazons region, however, both Indians and descendants of the Portuguese, hold to the common opinion, and consider the sloth as the type of laziness. It is very common for one native to call another, in reproaching him for idleness, "bicho do Embaüba" (beast of the Cecropia tree); the leaves of the Cecropia being the food of the sloth. It is a strange sight to watch the uncouth creature, fit production of these silent shades, lazily moving from branch to branch. Every movement betrays, not indolence exactly, but extreme caution. He never looses his hold from one branch without first securing himself to the next, and when he does not immediately find a bough to grasp with the rigid hooks into which his paws are so curiously transformed, he raises his body, supported on his hind legs, and claws around in search of a fresh foothold. After watching the animal for about half an hour, I gave him a charge of shot; he fell with a terrific crash, but caught a bough, in his descent, with his powerful claws, and remained suspended. Our Indian lad tried to climb the tree, but was driven back by swarms of stinging ants; the poor little fellow slid down in a sad predicament, and plunged into the brook to free himself. Two days afterwards I found the body of the sloth on the ground: the animal having dropped on the relaxation of the muscles a few hours after death. In one of our voyages, Mr. Wallace and I saw a sloth (B. infuscatus) swimming across a river, at a place where it was probably 300 yards broad. I believe it is not generally known that this animal takes to the water. Our men caught the beast, cooked, and ate him.

In returning from these trips we were sometimes benighted on the campos. We did not care for this on moonlit nights, when there was no danger of losing the path. The great heat felt in the middle hours of the day is much mitigated by four o'clock in the afternoon; a few birds then make their appearance; small flocks of ground doves run about the stony hillocks; parrots pass over and sometimes settle in the ilhas; pretty little finches of several species, especially one kind, streaked with olive-brown and yellow, and somewhat resembling our yellow-hammer, but, I believe, not belonging to the same genus, hop about the grass, enlivening the place with a few musical notes. The Carashué (Mimus) also then resumes its mellow blackbird-like song; and two or three species of humming-bird, none of which, however, are peculiar to the district, flit about from tree to tree. On the other hand, the little blue and yellow-striped lizards, which abound amongst the herbage during the scorching heats of midday, retreat towards this hour to their hiding-places; together with the day-flying insects and the numerous campo-butterflies. Some of these latter resemble greatly our English species found in heathy places, namely, a fritillary, Argynnis (Euptoieta) Hegesia, and two smaller kinds, which are deceptively like the little Nemeobius Lucina. After sunset the air becomes delightfully cool, and fragrant with the aroma of fruits and flowers. The nocturnal animals then come forth. A monstrous hairy spider, five inches in expanse, of a brown colour, with yellowish lines along its stout legs—which is very common here, inhabiting broad tubular galleries smoothly lined with silken web—may be then caught on the watch at the mouth of its burrow. It is only seen at night, and I think does not wander far from its den; the gallery is about two inches in diameter, and runs in a slanting direction, about two feet from the surface of the soil. As soon as it is night, swarms of goat-suckers suddenly make their appearance, wheeling about in a noiseless, ghostly manner, in chase of night-flying insects. They sometimes descend and settle on a low branch, or even on the pathway close to where one is walking, and then, squatting down on their heels, are difficult to distinguish from the surrounding soil. One kind has a long forked tail. In the day-time they are concealed in the wooded ilhas, where I very often saw them crouched and sleeping on the ground in the dense shade. They make no nest, but lay their eggs on the bare ground. Their breeding time is in the rainy season, and fresh eggs are found from December to June. Later in the evening, the singular notes of the goat-suckers are heard, one species crying Quao, Quao, another Chuck-co-co-cao; and these are repeated at intervals far into the night in the most monotonous manner. A great number of toads are seen on the bare sandy pathways soon after sunset. One of them was quite a colossus, about seven inches in length and three in height. This big fellow would never move out of the way until we were close to him. If we jerked him out of the path with a stick, he would slowly recover himself, and then turn round to have a good impudent stare. I have counted as many as thirty of these monsters within a distance of half a mile.

CHAPTER IX.

VOYAGE UP THE TAPAJOS.

Preparations for voyage—First day's sail—Loss of boat—Altar do Chaõ—Modes of obtaining fish—Difficulties with crew—Arrival at Aveyros—Excursions in the neighbourhood—White Cebus and habits and dispositions of Cebi monkeys—Tame parrot—Missionary settlement—Enter the River Cuparí—Adventure with Anaconda—Smoke-dried monkey—Boa constrictor—Village of Mundurucú Indians, and incursion of a wild tribe—Falls of the Cuparí—Hyacinthine macaw—Re-emerge into the broad Tapajos—Descent of river to Santarem.

June, 1852.—I will now proceed to relate the incidents of my principal excursion up the Tapajos, which I began to prepare for, after residing about six months at Santarem.

I was obliged, this time, to travel in a vessel of my own; partly because trading canoes large enough to accommodate a naturalist very seldom pass between Santarem and the thinly-peopled settlements on the river, and partly because I wished to explore districts at my ease, far out of the ordinary track of traders. I soon found a suitable canoe; a two-masted cuberta, of about six tons' burthen, strongly built of Itaüba or stone-wood, a timber of which all the best vessels in the Amazons country are constructed, and said to be more durable than teak. This I hired of a merchant at the cheap rate of 500 reis, or about one shilling and twopence per day. I fitted up the cabin, which, as usual in canoes of this class, was a square structure with its floor above the water-line, as my sleeping and working apartments. My chests, filled with store-boxes and trays for specimens, were arranged on each side, and above them were shelves and pegs to hold my little stock of useful books, guns, and game bags, boards and materials for skinning and preserving animals, botanical press and papers, drying cages for insects and birds, and so forth. A rush mat was spread on the floor, and my rolled-up hammock, to be used only when sleeping ashore, served for a pillow. The arched covering over the hold in the fore part of the vessel contained, besides a sleeping place for the crew, my heavy chests, stock of salt provisions and groceries, and an assortment of goods wherewith to pay my way amongst the half-civilised or savage inhabitants of the interior. The goods consisted of caschaça, powder and shot, a few pieces of coarse checked-cotton cloth and prints, fish-hooks, axes, large knives, harpoons, arrow-heads, looking-glasses, beads, and other small wares. José and myself were busy for many days arranging these matters. We had to salt the meat and grind a supply of coffee ourselves. Cooking utensils, crockery, water jars, a set of useful carpenter's tools, and many other things had to be provided. We put all the groceries and other perishable articles in tin canisters and boxes, having found that this was the only way of preserving them from damp and insects in this climate. When all was done, our canoe looked like a little floating workshop.

I could get little information about the river, except vague accounts of the difficulty of the navigation, and the famito or hunger which reigned on its banks. As I have before mentioned, it is about a thousand miles in length, and flows from south to north; in magnitude it stands the sixth amongst the tributaries of the Amazons. It is navigable, however, by sailing vessels only for about 160 miles above Santarem. The hiring of men to navigate the vessel was our greatest trouble. José was to be my helmsman, and we thought three other hands would be the fewest with which we could venture. But all our endeavours to procure these were fruitless. Santarem is worse provided with Indian canoemen than any other town on the river. I found on applying to the tradesmen to whom I had brought letters of introduction, and to the Brazilian authorities, that almost any favour would be sooner granted than the loan of hands. A stranger, however, is obliged to depend on them; for it is impossible to find an Indian or half-caste whom some one or other of the head-men do not claim as owing him money or labour. I was afraid at one time I should have been forced to abandon my project on this account. At length, after many rebuffs and disappointments, José contrived to engage one man, a mulatto, named Pinto, a native of the mining country of Interior Brazil, who knew the river well; and with these two I resolved to start, hoping to meet with others at the first village on the road.

We left Santarem on the 8th of June. The waters were then

at their highest point, and my canoe had been anchored close to the back door of our house. The morning was cool, and a brisk wind blew, with which we sped rapidly past the white-washed houses and thatched Indian huts of the suburbs. The charming little bay of Mapirí was soon left behind; we then doubled Point Maria Josepha, a headland formed of high cliffs of Tabatinga clay, capped with forest. This forms the limit of the river view from Santarem, and here we had our last glimpse, at a distance of seven or eight miles, of the city, a bright line of tiny white buildings resting on the dark water. A stretch of wild, rocky, uninhabited coast was before us, and we were fairly within the Tapajos.

Our course lay due west for about twenty miles. The wind increased as we neared Point Cururú, where the river bends from its northern course. A vast expanse of water here stretches to the west and south, and the waves, with a strong breeze, run very high. As we were doubling the point, the cable which held our montaria in tow astern, parted, and in endeavouring to recover the boat, without which we knew it would be difficult to get ashore on many parts of the coast, we were very near capsizing. We tried to tack down the river; a vain attempt with a strong breeze and no current. Our ropes snapped, the sails flew to rags, and the vessel, which we now found was deficient in ballast, heeled over frightfully. Contrary to José's advice, I ran the cuberta into a little bay, thinking to cast anchor there and wait for the boat coming up with the wind; but the anchor dragged on the smooth sandy bottom, and the vessel went broadside on to the rocky beach. With a little dexterous management, but not until after we had sustained some severe bumps, we managed to get out of this difficulty, clearing the rocky point at a close shave with our jib-sail. Soon after, we drifted into the smooth water of a sheltered bay, which leads to the charmingly situated village of Altar do Chaõ; and we were obliged to give up our attempt to recover the montaria.

The little settlement, Altar do Chaõ (Altar of the ground, or Earth altar), owes its singular name to the existence, at the entrance to the harbour, of one of those strange flat-topped hills which are so common in this part of the Amazons country, shaped like the high altar in Roman Catholic churches. It is an isolated one, and much lower in height than the similarly truncated hills and ridges near Almeyrim, being elevated probably not more than 300 feet above the level of the river. It is bare of trees, but covered in places with a species of fern. At the head of the bay is an inner harbour, which communicates by a channel with a series of lakes lying in the valleys between hills, and stretching far into the interior of the land. The village is peopled almost entirely by semi-civilised Indians, to the number of sixty or seventy families; and the scattered houses are arranged in broad streets on a strip of green sward, at the foot of a high, gloriously-wooded ridge.

I was so much pleased with the situation of this settlement, and the number of rare birds and insects which tenanted the forest, that I revisited it in the following year, and spent four months making collections. The houses in the village swarmed with vermin; bats in the thatch; fire-ants (formiga de fogo) under the floors; cockroaches and spiders on the walls. Very few of them had wooden doors and locks. Altar do Chaõ was originally a settlement of the aborigines, and was called Burarí. As in all the semi-civilised villages, where the original orderly and industrious habits of the Indian have been lost without anything being learnt from the whites to make amends, the inhabitants live in the greatest poverty. The scarcity of fish in the clear waters and rocky bays of the neighbourhood is no doubt partly the cause of the poverty and perennial hunger which reign here. When we arrived in the port, our canoe was crowded with the half-naked villagers—men, women, and children, who came to beg each a piece of salt pirarucu "for the love of God." They are not quite so badly off in the dry season. The shallow lakes and bays then contain plenty of fish, and the boys and women go out at night to spear them by torchlight; the torches being made of thin strips of green bark from the leaf-stalks of palms, tied in bundles. Many excellent kinds of fish are thus obtained, amongst them the Pescada, whose white and flaky flesh, when boiled, has the appearance and flavour of cod-fish; and the Tucunaré (Cichla temensis), a handsome species with a large prettily-coloured eye-like spot on its tail. Many small Salmonidæ are also met with, and a kind of sole, called Aramassá, which moves along the clear sandy bottom of the bay. At these times a species of sting-ray is common on the sloping beach, and bathers are frequently stung most severely by it. The weapon of this fish is a strong blade with jagged edges, about three inches long, growing from the side of the long fleshy tail. I once saw a woman wounded by it whilst bathing; she shrieked frightfully, and was obliged to be carried to her hammock, where she lay for a week in great pain; I have known strong men to be lamed for many months by the sting.

There was a mode of taking fish here which I had not before seen employed, but found afterwards to be very common on the Tapajos. This is by using a poisonous liana called Timbó (Paullinia pinnata). It will act only in the still waters of creeks and pools. A few rods, a yard in length, are mashed and soaked in the water, which quickly becomes discoloured with the milky deleterious juice of the plant. In about half an hour all the smaller fishes, over a rather wide space around the spot, rise to the surface, floating on their sides, and with the gills wide open. The poison acts evidently by suffocating the fishes; it spreads slowly in the water, and a very slight mixture seems sufficient to stupefy them. I was surprised, on beating the water in places where no fishes were visible in the clear depths, for many yards round, to find, sooner or later, sometimes twenty-four hours afterwards, a considerable number floating dead on the surface.

The climate is rather more humid than that of Santarem. I suppose this is to be attributed to the neighbouring country being densely wooded, instead of an open campo. In no part of the country did I enjoy more the moonlit nights than here in the dry season. After the day's work was done I used to go down to the shores of the bay, and lie at full length on the cool sand for two or three hours before bed-time. The soft pale light, resting on broad sandy beaches and palm-thatched huts, reproduced the effect of a mid-winter scene in the cold north when a coating of snow lies on the landscape. A heavy shower falls about once a week, and the shrubby vegetation never becomes parched up as at Santarem. Between the rains the heat and dryness increase from day to day: the weather on the first day after the rain is gleamy, with intervals of melting sunshine and passing clouds; the next day is rather drier, and the east wind begins to blow; then follow days of cloudless sky, with gradually increasing strength of breeze. When this has continued about a week, a light mistiness begins to gather about the horizon, clouds are formed, grumbling thunder is heard, and then, generally in the night-time, down falls the refreshing rain. The sudden chill caused by the rains produces colds, which are accompanied by the same symptoms as in our own climate; with this exception the place is very healthy.

June 17th.—The two young men returned without meeting

with my montaria, and I found it impossible here to buy a new one. The head-man could find me only one hand. This was a blunt-spoken but willing young Indian, named Manoel. He came on board this morning at eight o'clock, and we then got up our anchor and resumed our voyage.

The wind was light and variable all day, and we made only about fifteen miles by seven o'clock in the evening. The coast formed a succession of long shallow bays with sandy beaches, on which the waves broke in a long line of surf. Ten miles above Altar do Chaõ is a conspicuous headland, called Point Cajetúba. During a lull of the wind, towards midday, we ran the cuberta aground in shallow water and waded ashore, but the woods were scarcely penetrable, and not a bird was to be seen. The only thing observed worthy of note was the quantity of drowned winged ants along the beach; they were all of one species, the terrible formiga de fogo (Myrmica sævissima); the dead or half-dead bodies of which were heaped up in a line an inch or two in height and breadth, the line continuing without interruption for miles at the edge of the water. The countless thousands had been doubtless cast into the river while flying during a sudden squall the night before, and afterwards cast ashore by the waves. We found ourselves at seven o'clock near the mouth of a creek leading to a small lake, called Aramána-í; and the wind having died away we anchored, guided by the lights ashore, near the house of a settler, named Jeronymo, whom I knew, and who soon after showed us a a snug little harbour, where we could remain in safety for the night. The river here cannot be less than ten miles broad; it is quite clear of islands, and free from shoals at this season of the year. The opposite coast appeared in the day-time as a long thin line of forest, with dim grey hills in the background.

June 20th.—We had a light baffling wind off shore all day on the 20th, and made but fourteen or fifteen miles by six p.m.; when, the wind failing us, we anchored at the mouth of a narrow channel, called Tapaiúna, which runs between a large island and the mainland. About three o'clock we passed in front of Boim, a village on the opposite (western) coast. The breadth of the river is here six or seven miles: a confused patch of white on the high land opposite was all we saw of the village, the separate houses being undistinguishable on account of the distance. The coast along which we sailed to-day is a continuation of the low and flooded land of Paquiatúba.

June 21st.—The next morning we sailed along the Tapaiúna

channel, which is from 400 to 600 yards in breadth. We advanced but slowly, as the wind was generally dead against us, and stopped frequently to ramble ashore. Wherever the landing-place was sandy it was impossible to walk about on account of the swarms of the terrible fire-ant, whose sting is likened by the Brazilians to the puncture of a red-hot needle. There was scarcely a square inch of ground free from them. About three p.m. we glided into a quiet, shady creek, on whose banks an industrious white settler had located himself. I resolved to pass the rest of the day and night here, and endeavour to obtain a fresh supply of provisions, our stock of salt beef being now nearly exhausted. The situation of the house was beautiful; the little harbour being gay with water plants, Pontederiæ, now full of purple blossom, from which flocks of stilt-legged water-fowl started up screaming as we entered. The owner sent a boy with my men to show them the best place for fish up the creek, and in the course of the evening sold me a number of fowls, besides baskets of beans and farinha. The result of the fishing was a good supply of Jandiá, a handsome spotted Siluride fish, and Piránha, a kind of salmon. Piránhas are of several kinds, many of which abound in the waters of the Tapajos. They are caught with almost any kind of bait, for their taste is indiscriminate and their appetite most ravenous. They often attack the legs of bathers near the shore, inflicting severe wounds with their strong triangular teeth. At Paquiatúba and this place I added about twenty species of small fishes to my collection, caught by hook and line, or with the hand in shallow pools under the shade of the forest.

My men slept ashore, and on the coming aboard in the morning Pinto was drunk and insolent. According to José, who had kept himself sober, and was alarmed at the other's violent conduct, the owner of the house and Pinto had spent the greater part of the night together, drinking aguardente de beijú, a spirit distilled from the mandioca root. We knew nothing of the antecedents of this man, who was a tall, strong, self-willed fellow, and it began to dawn on us that this was not a very safe travelling companion in a wild country like this. I thought it better now to make the best of our way to the next settlement, Aveyros, and get rid of him. Our course to-day lay along a high rocky coast, which extended without a break for about eight miles. The height of the perpendicular rocks was from 100 to 150 feet; ferns and flowering shrubs grew in the crevices, and the summit supported a luxuriant growth of forest, like the rest of the river banks. The waves beat with loud roar at the foot of these inhospitable barriers. At two p.m. we passed the mouth of a small picturesque harbour, formed by a gap in the precipitous coast. Several families have here settled; the place is called Itá-puáma, or "standing rock," from a remarkable isolated cliff, which stands erect at the entrance to the little haven. A short distance beyond Itá-puáma we found ourselves opposite to the village of Pinhel, which is perched, like Boim, on high ground, on the western side of the river. The stream is here from six to seven miles wide. A line of low islets extends in front of Pinhel, and a little further to the south is a larger island, called Capitarí, which lies nearly in the middle of the river.

June 23rd.—The wind freshened at ten o'clock in the morning of the 23rd. A thick black cloud then began to spread itself over the sky a long way down the river, the storm which it portended, however, did not reach us, as the dark threatening mass crossed from east to west, and the only effect it had was to impel a column of cold air up the river, creating a breeze with which we bounded rapidly forward. The wind in the afternoon strengthened to a gale; we carried on with one foresail only, two of the men holding on to the boom to prevent the whole thing from flying to pieces. The rocky coast continued for about twelve miles above Itá-puáma, then succeeded a tract of low marshy land, which had evidently been once an island whose channel of separation from the mainland had become silted up. The island of Capitari and another group of islets succeeding it, called Jacaré, on the opposite side, helped also to contract at this point the breadth of the river, which was now not more than about three miles. The little cuberta almost flew along this coast, there being no perceptible current, past extensive swamps, margined with thick floating grasses. At length, on rounding a low point, higher land again appeared on the right bank of the river, and the village of Aveyros hove in sight, in the port of which we cast anchor late in the afternoon.

Aveyros is a small settlement, containing only fourteen or fifteen houses besides the church; but it is the place of residence of the authorities of a large district; the priest, Juiz de Paz, the subdelegado of police, and the Captain of the Trabalhadores. The district includes Pinhel, which we passed about twenty miles lower down on the left bank of the river. Five miles beyond Aveyros, and also on the left bank, is the missionary village of Santa Cruz, comprising thirty or forty families of baptized Mundurucú Indians, who are at present under the management of a Capuchin Friar, and are independent of the Captain of Trabalhadores of Aveyros. The river view from this point towards the south was very grand; the stream is from two to three miles broad, with green islets resting on its surface, and on each side a chain of hills stretches away in long perspective. I resolved to stay here for a few weeks to make collections. On landing, my first care was to obtain a house or room, that I might live ashore. This was soon arranged; the head man of the place, Captain Antonio, having received notice of my coming, so that before night all the chests and apparatus I required were housed and put in order for working.

I here dismissed Pinto, who again got drunk and quarrelsome a few hours after he came ashore. He left the next day, to my great relief, in a small trading canoe that touched at the place on its way to Santarem. The Indian Manoel took his leave at the same time, having engaged to accompany me only as far as Aveyros: I was then dependent on Captain Antonio for fresh hands. The captains of Trabalhadores are appointed by the Brazilian Government, to embody the scattered Indian labourers and canoe-men of their respective districts, to the end that they may supply passing travellers with men when required. A semi-military organisation is given to the bodies; some of the steadiest amongst the Indians themselves being nominated as sergeants, and all the members mustered at the principal village of their district twice a year. The captains, however, universally abuse their authority, monopolising the service of the men for their own purposes, so that it is only by favour that the loan of a canoe-hand can be wrung from them. I was treated by Captain Antonio with great consideration, and promised two good Indians when I should be ready to continue my voyage.

Little happened worth narrating during my forty days' stay at Aveyros. The time was spent in the quiet, regular pursuit of natural history: every morning I had my long ramble in the forest, which extended to the back-doors of the houses, and the afternoons were occupied in preserving and studying the objects collected. The priest was a lively old man, but rather a bore from being able to talk of scarcely anything except homœopathy, having been smitten with the mania during a recent visit to Santarem. He had a Portuguese

Homœopathic Dictionary, and a little leather case containing glass tubes filled with globules, with which he was doctoring the whole village. The weather, during the month of July, was uninterruptedly fine; not a drop of rain fell, and the river sank rapidly. The mornings, for two hours after sunrise, were very cold; we were glad to wrap ourselves in blankets on turning out of our hammocks, and walk about at a quick pace in the early sunshine. But in the afternoons the heat was sickening; for the glowing sun then shone full on the front of the row of whitewashed houses, and there was seldom any wind to moderate its effects. I began now to understand why the branch rivers of the Amazons were so unhealthy, whilst the main stream was pretty nearly free from diseases arising from malaria. The cause lies, without doubt, in the slack currents of the tributaries in the dry season, and the absence of the cooling Amazonian trade-wind, which purifies the air along the banks of the main river. The trade-wind does not deviate from its nearly straight westerly course, so that the branch streams, which run generally at right angles to the Amazons, and have a slack current for a long distance from their mouths, are left to the horrors of nearly stagnant air and water.

Aveyros may be called the head-quarters of the fire-ant, which might be fittingly termed the scourge of this fine river. The Tapajos is nearly free from the insect pests of other parts, mosquitoes, sand-flies, motúcas and piums; but the formiga de fogo is perhaps a greater plague than all the others put together. It is found only on sandy soils in open places, and seems to thrive most in the neighbourhood of houses and weedy villages such as Aveyros: it does not occur at all in the shades of the forest. I noticed it in most places on the banks of the Amazons, but the species is not very common on the main river, and its presence is there scarcely noticed, because it does not attack man, and the sting is not so virulent as it is in the same species on the banks of the Tapajos. Aveyros was deserted a few years before my visit on account of this little tormentor, and the inhabitants had only recently returned to their houses, thinking its numbers had decreased. It is a small species, of a shining reddish colour, not greatly differing from the common red stinging ant of our own country (Myrmica rubra), except that the pain and irritation caused by its sting are much greater. The soil of the whole village is undermined by it: the ground is perforated with the entrances to their subterranean galleries, and a little sandy dome occurs here and there, where the insects bring their young to receive warmth near the surface. The houses are overrun with them; they dispute every fragment of food with the inhabitants, and destroy clothing for the sake of the starch. All eatables are obliged to be suspended in baskets from the rafters, and the cords well soaked with copaüba balsam, which is the only means known of preventing them from climbing. They seem to attack persons out of sheer malice: if we stood for a few moments in the street, even at a distance from their nests, we were sure to be overrun and severely punished, for the moment an ant touched the flesh, he secured himself with his jaws, doubled in his tail, and stung with all his might. When we were seated on chairs in the evenings in front of the house to enjoy a chat with our neighbours, we had stools to support our feet, the legs of which, as well as those of the chairs, were well anointed with the balsam. The cords of hammocks are obliged to be smeared in the same way to prevent the ants from paying sleepers a visit.

The inhabitants declare that the fire-ant was unknown on the Tapajos before the disorders of 1835-6, and believe that the hosts sprang up from the blood of the slaughtered Cabanas or rebels. They have, doubtless, increased since that time, but the cause lies in the depopulation of the villages, and the rank growth of weeds in the previously cleared, well-kept spaces. I have already described the line of sediment formed, on the sandy shores lower down the river, by the dead bodies of the winged individuals of this species. The exodus from their nests of the males and females takes place at the end of the rainy season (June), when the swarms are blown into the river by squalls of wind, and subsequently cast ashore by the waves. I was told that this wholesale destruction of ant-life takes place annually, and that the same compact heap of dead bodies, which I saw only in part, extends along the banks of the river for twelve or fifteen miles.

The forest behind Aveyros yielded me little except insects, but in these it was very rich. It is not too dense, and broad sunny paths skirted by luxuriant beds of Lycopodiums, which form attractive sporting places for insects, extend from the village to a swampy hollow or ygapó, which lies about a mile inland. Of butterflies alone I enumerated fully 300 species, captured or seen in the course of forty days, within a half-hour's walk of the village. This is a greater number than is found in the whole of Europe. The only monkey I observed was the Callithrix moloch—one of the kinds called by the Indians

Whaiápu-saí. It is a moderate-sized species, clothed with long brown hair, and having hands of a whitish hue. Although nearly allied to the Cebi, it has none of their restless vivacity, but is a dull listless animal. It goes in small flocks of five or six individuals, running along the main boughs of the trees. One of the specimens which I obtained here was caught on a low fruit-tree at the back of our house, at sunrise one morning. This was the only instance of a monkey being captured in such a position that I ever heard of. As the tree was isolated, it must have descended to the ground from the neighbouring forest, and walked some distance to get at it. The species is sometimes kept in a tame state by the natives: it does not make a very amusing pet, and survives captivity only a short time.

I heard that the white Cebus, the Caiarára branca, a kind of monkey I had not yet seen, and wished very much to obtain, inhabited the forests on the opposite side of the river; so one day, on an opportunity being afforded by our host going over in a large boat, I crossed to go in search of it. We were about twenty persons in all, and the boat was an old rickety affair, with the gaping seams rudely stuffed with tow and pitch. In addition to the human freight we took three sheep with us, which Captain Antonio had just received from Santarem, and was going to add to his new cattle farm on the other side. Ten Indian paddlers carried us quickly across. The breadth of the river could not be less than three miles, and the current was scarcely perceptible. When a boat has to cross the main Amazons, it is obliged to ascend along the banks for half a mile or more to allow for drifting by the current; in this lower part of the Tapajos this is not necessary. When about half-way, the sheep, in moving about, kicked a hole in the bottom of the boat. The passengers took the matter very coolly, although the water spouted up alarmingly, and I thought we should inevitably be swamped. Captain Antonio took off his socks to stop the leak, inviting me and the Juiz de Paz, who was one of the party, to do the same, whilst two Indians baled out the water with large cuyas. We thus managed to keep afloat until we reached our destination, when the men patched up the leak for our return journey.

The landing-place lay a short distance within the mouth of a shady inlet, on whose banks, hidden amongst the dense woods, were the houses of a few Indian and mameluco settlers. The path to the cattle farm led first through a tract of swampy forest; it then ascended a slope and emerged on a fine sweep of prairie, varied with patches of timber. The wooded portion occupied the hollows, where the soil was of a rich chocolate-brown colour, and of a peaty nature. The higher grassy, undulating parts of the campo had a lighter and more sandy soil. Leaving our friends, I and José took our guns and dived into the woods in search of the monkeys. As we walked rapidly along I was very near treading on a rattlesnake, which lay stretched out nearly in a straight line on the bare sandy pathway. It made no movement to get out of the way, and I escaped the danger by a timely and sudden leap, being unable to check my steps in the hurried walk. We tried to excite the sluggish reptile by throwing handfuls of sand and sticks at it, but the only notice it took was to raise its ugly horny tail and shake its rattle. At length it began to move rather nimbly, when we dispatched it by a blow on the head with a pole, not wishing to fire on account of alarming our game.

We saw nothing of the white Caiarára; we met, however, with a flock of the common light-brown allied species (Cebus albifrons?), and killed one as a specimen. A resident on this side of the river told us that the white kind was found further to the south, beyond Santa Cruz. The light-brown Caiarára is pretty generally distributed over the forests of the level country. I saw it very frequently on the banks of the Upper Amazons, where it was always a treat to watch a flock leaping amongst the trees, for it is the most wonderful performer in this line of the whole tribe. The troops consist of thirty or more individuals, which travel in single file. When the foremost of the flock reaches the outermost branch of an unusually lofty tree, he springs forth into the air without a moment's hesitation, and alights on the dome of yielding foliage belonging to the neighbouring tree, maybe fifty feet beneath; all the rest following the example. They grasp, on falling, with hands and tail, right themselves in a moment, and then away they go along branch and bough to the next tree. The Caiarára owes its name in the Tupí language, macaw or large-headed (Acain, head, and Arára macaw), to the disproportionate size of the head compared with the rest of the body. It is very frequently kept as a pet in houses of natives. I kept one myself for about a year, which accompanied me in my voyages and became very familiar, coming to me always on wet nights to share my blanket. It is a most restless creature, but is not playful like most of the American monkeys; the restlessness of its disposition seeming to arise from great nervous irritability and discontent. The anxious, painful, and changeable expression of its countenance, and the want of purpose in its movements, betray this. Its actions are like those of a wayward child; it does not seem happy even when it has plenty of its favourite food, bananas; but will leave its own meal to snatch the morsels out of the hands of its companions. It differs in these mental traits from its nearest kindred, for another common Cebus, found in the same parts of the forest, the Prego monkey (Cebus cirrhifer?), is a much quieter and better-tempered animal; it is full of tricks, but these are generally of a playful character.

The Caiarára keeps the house in a perpetual uproar where it is kept: when alarmed, or hungry, or excited by envy, it screams piteously; it is always, however, making some noise or other, often screwing up its mouth and uttering a succession of loud notes resembling a whistle. My little pet, when loose, used to run after me, supporting itself for some distance on its hind legs, without, however, having been taught to do it. He offended me greatly, one day, by killing, in one of his jealous fits, another and much choicer pet—the nocturnal owl-faced monkey (Nyctipithecus trivirgatus). Some one had given this a fruit, which the other coveted, so the two got to quarrelling. The Nyctipithecus fought only with its paws, clawing out and hissing like a cat; the other soon obtained the mastery, and before I could interfere, finished his rival by cracking its skull with his teeth. Upon this I got rid of him.

After I had obtained the two men promised, stout young Indians, 17 or 18 years of age, one named Ricardo and the other Alberto, I paid a second visit to the western side of the river in my own canoe; being determined, if possible, to obtain specimens of the White Cebus. We crossed over first to the mission village, Santa Cruz, which consists of 30 or 40 wretched-looking mud huts, closely built together in three straight ugly rows on a high gravelly bank. The place was deserted, with the exception of two or three old men and women and a few children. A narrow belt of wood runs behind the village; beyond this is an elevated barren campo, with a clayey and gravelly soil. To the south the coast country is of a similar description; a succession of scantily-wooded hills, bare grassy spaces, and richly-timbered hollows. We traversed forest and campo in various directions, during three days, without meeting with monkeys, or indeed with any-

14

thing that repaid us the time and trouble. The soil of the district appeared too dry; at this season of the year I had noticed, in other parts of the country, that mammals and birds resorted to the more humid areas of forest; we therefore proceeded to explore carefully the low and partly swampy tract, along the coast to the north of Santa Cruz. We spent two days in this way, landing at many places, and penetrating a good distance in the interior. Although unsuccessful with regard to the White Cebus, the time was not wholly lost, as I added several small birds of species new to my collection. On the second evening we surprised a large flock, composed of about 50 individuals, of a curious eagle with a very long and slender hooked beak, the Rostrhamus hamatus. They were perched on the bushes which surrounded a shallow lagoon, separated from the river by a belt of floating grass: my men said they fed on toads and lizards found at the margins of pools. They formed a beautiful sight as they flew up and wheeled about at a great height in the air. We obtained only one specimen.

Before returning to Aveyros we paid another visit to the Jacaré inlet, leading to Captain Antonio's cattle farm, for the sake of securing further specimens of the many rare and handsome insects found there; landing at the port of one of the settlers. The owner of the house was not at home, and the wife, a buxom young woman, a dark mameluco, with clear though dark complexion and fine rosy cheeks, was preparing, in company with another stout-built Amazon, her rod and lines to go out fishing for the day's dinner. It was now the season for Tucunarés, and Senhora Joaquina showed us the fly baits used to take this kind of fish, which she had made with her own hands of parrots' feathers. The rods used are slender bamboos, and the lines made from the fibres of pine-apple leaves. It is not very common for the Indian and half-caste women to provide for themselves, in the way these spirited dames were doing, although they are all expert paddlers, and very frequently cross wide rivers in their frail boats without the aid of men. It is possible that parties of Indian women, seen travelling alone in this manner, may have given rise to the fable of a nation of Amazons, invented by the first Spanish explorers of the country. Senhora Joaquina invited me and José to a Tucunaré dinner for the afternoon, and then, shouldering their paddles and tucking up their skirts, the two dusky fisherwomen marched down to their canoe. We sent the two

Indians into the woods to cut palm leaves to mend the thatch of our cuberta, whilst I and José rambled through the woods which skirted the campo. On our return we found a most bountiful spread in the house of our hostess. A spotless white cloth was laid on the mat, with a plate for each guest, and a pile of fragrant newly made farinha by the side of it. The boiled Tucunarés were soon taken from the kettles and set before us. I thought the men must be happy husbands who owned such wives as these. The Indian and mameluco women certainly do make excellent managers; they are more industrious than the men, and most of them manufacture farinha for sale on their own account, their credit always standing higher with the traders on the river than that of their male connections. I was quite surprised at the quantity of fish they had taken, there being sufficient for the whole party, including several children, two old men from a neighbouring hut, and my Indians. I made our good-natured entertainers a small present of needles and sewing-cotton, articles very much prized, and soon after we re-embarked, and again crossed the river to Aveyros.

August 2nd.—Left Aveyros, having resolved to ascend a branch river, the Cuparí, which enters the Tapajos about eight miles above this village, instead of going forward along the main stream. I should have liked to visit the settlements of the Mundurucú tribe, which lie beyond the first cataract of the Tapajos, if it had been compatible with the other objects I had in view. But to perform this journey a lighter canoe than mine would have been necessary, and six or eight Indian paddlers, which in my case it was utterly impossible to obtain. There would be, however, an opportunity of seeing this fine race of people on the Cuparí, as a horde was located towards the head waters of this stream. The distance from Aveyros to the last civilised settlement on the Tapajos, Itaitúba, is about forty miles. The falls commence a short distance beyond this place. Ten formidable cataracts or rapids then succeed each other at intervals of a few miles; the chief of which are the Coaita, the Bubure, the Salto Grande (about thirty feet high), and the Montanha. The canoes of Cuyabá tradesmen which descend annually to Santarem are obliged to be unloaded at each of these, and the cargoes carried by land on the backs of Indians, whilst the empty vessels are dragged by ropes over the obstructions. The Cuparí was described to me as flowing through a rich moist clayey valley, covered with forests and abounding in game; whilst the banks of the Tapajos beyond Aveyros were barren sandy campos, with ranges of naked or scantily-wooded hills, forming a kind of country which I had always found very unproductive in Natural History objects in the dry season, which had now set in.

We entered the mouth of the Cuparí on the evening of the following day (August 3rd). It was not more than 100 yards wide, but very deep: we found no bottom in the middle with a line of eight fathoms. The banks were gloriously wooded; the familiar foliage of the cacao growing abundantly amongst the mass of other trees, reminding me of the forests of the main Amazons. We rowed for five or six miles, generally in a south-easterly direction, although the river had many abrupt bends, and stopped for the night at a settler's house, situated on a high bank, and accessible only by a flight of rude wooden steps fixed in the clayey slope. The owners were two brothers, half-breeds, who with their families shared the large roomy dwelling; one of them was a blacksmith, and we found him working with two Indian lads at his forge, in an open shed under the shade of mango trees. They were the sons of a Portuguese immigrant, who had settled here forty years previously, and married a Mundurucú woman. He must have been a far more industrious man than the majority of his countrymen who emigrate to Brazil now-a-days, for there were signs of former extensive cultivation at the back of the house, in groves of orange, lemon, and coffee trees, and a large plantation of cacao occupied the lower grounds.

The next morning one of the brothers brought me a beautiful opossum, which had been caught in the fowl-house a little before sunrise. It was not so large as a rat, and had soft brown fur, paler beneath and on the face, with a black stripe on each cheek. This made the third species of marsupial rat I had so far obtained: but the number of these animals is very considerable in Brazil, where they take the place of the shrews of Europe; shrew mice and, indeed, the whole of the insectivorous order of mammals, being entirely absent from Tropical America. One kind of these rat-like opossums is aquatic, and has webbed feet. The terrestrial species are nocturnal in their habits, sleeping during the day in hollow trees, and coming forth at night to prey on birds in their roosting places. It is very difficult to rear poultry in this country, on account of these small opossums; scarcely a night passing, in some parts, in which the fowls are not attacked by them.

August 5th.—The river reminds me of some parts of the Jaburú channel, being hemmed in by two walls of forest, rising to the height of at least 100 feet, and the outlines of the trees being concealed throughout by a dense curtain of leafy creepers. The impression of vegetable profusion and overwhelming luxuriance increases at every step; the deep and narrow valley of the Cuparí has a moister climate than the banks of the Tapajos. We have now frequent showers, whereas we left everything parched up by the sun at Aveyros.

After leaving the last sitio we advanced about eight miles, and then stopped at the house of Senhor Antonio Malagueita, a mameluco settler, whom we had been recommended to visit. His house and outbuildings were extensive, the grounds well weeded, and the whole wore an air of comfort and well-being which is very uncommon in this country. A bank of indurated white clay sloped gently up from the tree-shaded port to the house, and beds of kitchen-herbs extended on each side, with (rare sight!) rose and jasmine trees in full bloom. Senhor Antonio, a rather tall middle-aged man, with a countenance beaming with good nature, came down to the port as soon as we anchored. I was quite a stranger to him, but he had heard of my coming, and seemed to have made preparations. I never met with a heartier welcome. On entering the house, the wife, who had more of the Indian tint and features than her husband, was equally warm and frank in her greeting. I stayed here two days. We had together several long and successful rambles, along a narrow pathway which extended several miles into the forest. I here met with a new insect pest, one which the natives may be thankful is not spread more widely over the country; it was a large brown fly of the Tabanidæ family (genus Pangonia), with a proboscis half an inch long and sharper than the finest needle. It settled on our backs by twos and threes at a time, and pricked us through our thick cotton shirts, making us start and cry out with the sudden pain. I secured a dozen or two as specimens. As an instance of the extremely confined ranges of certain species, it may be mentioned that I did not find this insect in any other part of the country, except along half a mile or so of this gloomy forest road.

We were amused at the excessive and almost absurd tameness of a fine Mutum or Curassow turkey that ran about the house. It was a large glossy-black species (the Mitu tuberosa), having an orange-coloured beak, surmounted by a bean-shaped excrescence of the same hue. It seemed to consider itself as one of the family: attended at all the meals, passing from one person to another round the mat to be fed, and rubbing the sides of its head in a coaxing way against their cheeks or shoulders. At night it went to roost on a chest in a sleeping-room beside the hammock of one of the little girls, to whom it seemed particularly attached, following her wherever she went about the grounds. I found this kind of Curassow bird was very common in the forests of the Cuparí; but it is rare on the Upper Amazons, where an allied species, which has a round instead of a bean-shaped waxen excrescence on the beak (Crax globicera), is the prevailing kind. These birds in their natural state never descend from the tops of the loftiest trees, where they live in small flocks and build their nests. The Mitu tuberosa lays two rough-shelled white eggs; it is fully as large a bird as the common turkey, but the flesh when cooked is drier and not so well-flavoured. It is difficult to find the reason why these superb birds have not been reduced to domestication by the Indians, seeing that they so readily become tame. The obstacle offered by their not breeding in confinement, which is probably owing to their arboreal habits, might perhaps be overcome by repeated experiment; but for this the Indians probably had not sufficient patience or intelligence. The reason cannot lie in their insensibility to the value of such birds; for the common turkey, which has been introduced into the country, is much prized by them.

We had an unwelcome visitor whilst at anchor in the port of Antonio Malagueita. I was awoke a little after midnight, as I lay in my little cabin, by a heavy blow struck at the sides of the canoe close to my head, which was succeeded by the sound of a weighty body plunging in the water. I got up; but all was again quiet, except the cackle of fowls in our hen-coop, which hung over the side of the vessel, about three feet from the cabin door. I could find no explanation of the circumstance, and, my men being all ashore, I turned in again and slept till morning. I then found my poultry loose about the canoe, and a large rent in the bottom of the hen-coop, which was about two feet from the surface of the water: a couple of fowls were missing. Senhor Antonio said the depredator was a Sucurujú (the Indian name for the Anaconda, or great water serpent, Eunectes murinus), which had for months past been haunting this part of the river, and had carried off many ducks and fowls from the ports of various houses. I was inclined to doubt the fact of a serpent striking at its prey from the water, and thought an alligator more likely to be the culprit, although we had not yet met with alligators in the river. Some days afterwards the young men belonging to the different sitios agreed together to go in search of the serpent. They began in a systematic manner, forming two parties, each embarked in three or four canoes, and starting from points several miles apart, whence they gradually approximated, searching all the little inlets on both sides the river. The reptile was found at last, sunning itself on a log at the mouth of a muddy rivulet, and despatched with harpoons. I saw it the day after it was killed: it was not a very large specimen, measuring only eighteen feet nine inches in length, and sixteen inches in circumference at the widest part of the body. I measured skins of the Anaconda afterwards, twenty-one feet in length and two feet in girth. The reptile has a most hideous appearance, owing to its being very broad in the middle, and tapering abruptly at both ends. It is very abundant in some parts of the country, nowhere more so than in the Lago Grande, near Santarem, where it is often seen coiled up in the corners of farmyards, and is detested for its habit of carrying off poultry, young calves, or whatever animal it can get within reach of.

At Ega a large Anaconda was once near making a meal of a young lad about ten years of age, belonging to one of my neighbours. The father and his son went, as was their custom, a few miles up the Teffé to gather wild fruit; landing on a sloping sandy shore, where the boy was left to mind the canoe whilst the man entered the forest. The beaches of the Teffé form groves of wild guava and myrtle trees, and during most months of the year are partly overflown by the river. Whilst the boy was playing in the water under the shade of these trees, a huge reptile of this species stealthily wound its coils around him, unperceived until it was too late to escape. His cries brought the father quickly to the rescue, who rushed forward, and seizing the Anaconda boldly by the head, tore his jaws asunder. There appears to be no doubt that this formidable serpent grows to an enormous bulk, and lives to a great age, for I heard of specimens having been killed which measured forty-two feet in length, or double the size of the largest I had an opportunity of examining. The natives of the Amazons country universally believe in the existence of a monster water-serpent, said to be many score fathoms in length, which appears successively in different parts of the river. They call it the

Mai d'agoa—the mother, or spirit, of the water. This fable, which was doubtless suggested by the occasional appearance of Sucurujús of unusually large size, takes a great variety of forms, and the wild legends form the subject of conversation amongst old and young, over the wood fires in lonely settlements.

August 6th and 7th.—On leaving the sitio of Antonio Malagueita we continued our way along the windings of the river, generally in a south-east and south-south-east direction, but sometimes due north, for about fifteen miles, when we stopped at the house of one Paulo Christo, a mameluco whose acquaintance I had made at Aveyros. Here we spent the night and part of the next day, doing in the morning a good five hours' work in the forest, accompanied by the owner of the place. In the afternoon of the 7th we were again under weigh: the river makes a bend to the east-north-east for a short distance above Paulo Christo's establishment, it then turns abruptly to the south-west, running from that direction about four miles. The hilly country of the interior then commences, the first token of it being a magnificently wooded bluff, rising nearly straight from the water to a height of about 250 feet. The breadth of the stream hereabout was not more than sixty yards, and the forest assumed a new appearance, from the abundance of the Urucurí palm, a species which has a noble crown of broad fronds, with symmetrical rigid leaflets.

We reached, in the evening, the house of the last civilised settler on the river, Senhor Joaõ (John) Aracú, a wiry, active fellow and capital hunter, whom I wished to make a friend of and persuade to accompany me to the Mundurucú village and the falls of the Cuparí, some forty miles further up the river.

I stayed at the sitio of John Aracú until the 19th, and again, in descending, spent fourteen days at the same place. The situation was most favourable for collecting the natural products of the district. The forest was not crowded with underwood, and pathways led through it for many miles and in various directions. I could make no use here of our two men as hunters, so, to keep them employed, whilst José and I worked daily in the woods, I set them to make a montaria under John Aracú's directions. The first day a suitable tree was found for the shell of the boat, of the kind called Itaüba amarello, the yellow variety of the stone-wood. They felled it, and shaped out of the trunk a log nineteen feet in length; this they dragged from the forest, with the help of my host's men, over a road they had previously made with cylindrical pieces of wood to act as rollers. The distance was about half a mile, and the rope used for drawing the heavy load were tough lianas cut from the surrounding trees. This part of the work occupied about a week; the log had then to be hollowed out, which was done with strong chisels through a slit made down the whole length. The heavy portion of the task being then completed, nothing remained but to widen the opening, fit two planks for the sides, and the same number of semicircular boards for the ends, make the benches, and caulk the seams.

The expanding of the log thus hollowed out is a critical operation, and not always successful, many a good shell being spoilt by its splitting or expanding irregularly. It is first reared on tressels, with the slit downwards, over a large fire, which is kept up for seven or eight hours, the process requiring unremitting attention to avoid cracks and make the plank bend with the proper dip at the two ends. Wooden straddlers, made by cleaving pieces of tough elastic wood and fixing them with wedges, are inserted into the opening, their compass being altered gradually as the work goes on, but in different degree according to the part of the boat operated upon. Our casca turned out a good one: it took a long time to cool, and was kept in shape whilst it did so by means of wooden cross-pieces. When the boat was finished, it was launched with great merriment by the men, who hoisted coloured handkerchiefs for flags, and paddled it up and down the stream to try its capabilities. My people had suffered as much inconvenience from the want of a montaria as myself, so this was a day of rejoicing to all of us.

I was very successful at this place with regard to the objects of my journey. About twenty new species of fishes and a considerable number of small reptiles were added to my collection; but very few birds were met with worth preserving. A great number of the most conspicuous insects of the locality were new to me, and turned out to be species peculiar to this part of the Amazons valley. The most interesting acquisition was a large and handsome monkey, of a species I had not before met with—the white-whiskered Coaitá, or spider-monkey (Ateles marginatus). I saw a pair one day in the forest moving slowly along the branches of a lofty tree, and shot one of them; the next day John Aracú brought down another, possibly the companion. The species is of about the same size as the common black kind, of which I have given an account in a former chapter, and has a similar lean body, with limbs clothed with coarse black hair; but it differs in having the whiskers and a triangular patch on the crown of the head of a white colour. I thought the meat the best flavoured I had ever tasted. It resembled beef, but had a richer and sweeter taste. During the time of our stay in this part of the Cuparí we could get scarcely anything but fish to eat, and as this diet ill agreed with me, three successive days of it reducing me to a state of great weakness, I was obliged to make the most of our Coaitá meat. We smoke-dried the joints instead of salting them; placing them for several hours on a framework of sticks arranged over a fire, a plan adopted by the natives to preserve fish when they have no salt, and which they call "muquiar." Meat putrefies in this climate in less than twenty-four hours, and salting is of no use, unless the pieces are cut in thin slices and dried immediately in the sun. My monkeys lasted me about a fortnight, the last joint being an arm with the clenched fist, which I used with great economy, hanging it in the intervals between my frugal meals on a nail in the cabin. Nothing but the hardest necessity could have driven me so near to cannibalism as this, but we had the greatest difficulty in obtaining here a sufficient supply of animal food. About every three days the work on the montaria had to be suspended, and all hands turned out for the day to hunt and fish, in which they were often unsuccessful, for although there was plenty of game in the forest, it was too widely scattered to be available. Ricardo and Alberto occasionally brought in a tortoise or anteater, which served us for one day's consumption. We made acquaintance here with many strange dishes, amongst them Iguana eggs; these are of oblong form, about an inch in length, and covered with a flexible shell. The lizard lays about two score of them in the hollows of trees. They have an oily taste; the men ate them raw, beaten up with farinha, mixing a pinch of salt in the mess; I could only do with them when mixed with Tucupí sauce, of which we had a large jarful always ready for the tempering of unsavoury morsels.

One day as I was entomologizing alone and unarmed, in a dry Ygapó, where the trees were rather wide apart and the ground coated to the depth of eight or ten inches with dead leaves, I was near coming into collision with a boa-constrictor. I had just entered a little thicket to capture an insect, and whilst pinning it was rather startled by a rushing noise in the vicinity. I looked up to the sky, thinking a squall was coming on, but not a breath of wind stirred in the tree-tops. On stepping out of the bushes I met face to face a huge serpent coming down a slope, and making the dry twigs crack and fly with his weight as he moved over them. I had very frequently met with a smaller boa, the Cutim-boia, in a similar way, and knew from the habits of the family that there was no danger, so I stood my ground. On seeing me the reptile suddenly turned, and glided at an accelerated pace down the path. Wishing to take a note of his probable size and the colours and markings of his skin, I set off after him; but he increased his speed, and I was unable to get near enough for the purpose. There was very little of the serpentine movement in his course. The rapidly moving and shining body looked like a stream of brown liquid flowing over the thick bed of fallen leaves, rather than a serpent with skin of varied colours. He descended towards the lower and moister parts of the Ygapó. The huge trunk of an uprooted tree here lay across the road; this he glided over in his undeviating course, and soon after penetrated a dense swampy thicket, where of course I did not choose to follow him.

I suffered terribly from heat and mosquitoes as the river sank with the increasing dryness of the season, although I made an awning of the sails to work under, and slept at night in the open air, with my hammock slung between the masts. But there was no rest in any part; the canoe descended deeper and deeper into the gulley, through which the river flows between high clayey banks, as the water subsided, and with the glowing sun overhead we felt at midday as if in a furnace. I could bear scarcely any clothes in the daytime, between eleven in the morning and five in the afternoon, wearing nothing but loose and thin cotton trousers and a light straw hat, and could not be accommodated in John Aracú's house, as it was a small one and full of noisy children. One night we had a terrific storm. The heat in the afternoon had been greater than ever, and at sunset the sky had a brassy glare: the black patches of cloud which floated in it being lighted up now and then by flashes of sheet lightning. The mosquitoes at night were more than usually troublesome, and I had just sunk exhausted into a doze, towards the early hours of morning, when the storm began; a complete deluge of rain, with incessant lightning and rattling explosions of thunder. It lasted for eight hours; the grey dawn opening amidst the crash of the tempest. The rain trickled through the seams of the cabin roof on to my collections, the late hot weather having warped the boards, and it gave me immense trouble to secure them in the midst of the confusion. Altogether I had a bad night of it; but what with storms, heat, mosquitoes, hunger, and, towards the last, ill health, I seldom had a good night's rest on the Cuparí.

A small creek traversed the forest behind John Aracú's house, and entered the river a few yards from our anchoring place; I used to cross it twice a day, on going and returning from my hunting-ground. One day early in September, I noticed that the water was two or three inches higher in the afternoon than it had been in the morning. This phenomenon was repeated the next day, and in fact daily, until the creek became dry with the continued subsidence of the Cuparí, the time of rising shifting a little from day to day. I pointed out the circumstance to John Aracú, who had not noticed it before (it was only his second year of residence in the locality), but agreed with me that it must be the "maré." Yes, the tide! the throb of the great oceanic pulse felt in this remote corner, 530 miles distant from the place where it first strikes the body of fresh water at the mouth of the Amazons. I hesitated at first at this conclusion, but on reflecting that the tide was known to be perceptible at Obydos, more than 400 miles from the sea; that at high water in the dry season a large flood from the Amazons enters the mouth of the Tapajos, and that there is but a very small difference of level between that point and the Cuparí, a fact shown by the absence of current in the dry season; I could have no doubt that this conclusion was a correct one.

The fact of the tide being felt 530 miles up the Amazons, passing from the main stream to one of its affluents 380 miles from its mouth, and thence to a branch in the third degree, is a proof of the extreme flatness of the land which forms the lower part of the Amazonian valley. This uniformity of level is shown also in the broad lake-like expanses of water formed, near their mouths, by the principal affluents which cross the valley to join the main river.

August 21st.—John Aracú consented to accompany me to the falls, with one of his men, to hunt and fish for me. One of my objects was to obtain specimens of the hyacinthine macaw, whose range commences on all the branch rivers of the Amazons which flow from the south through the interior of Brazil, with the first cataracts. We started on the 19th, our direction on that day being generally south-west. On the 20th our course was southerly and south-easterly. This morning (August 21st) we arrived at the Indian settlement, the first house of which lies about thirty-one miles above the sitio of John Aracú. The river at this place is from sixty to seventy yards wide, and runs in a zigzag course between steep clayey banks, twenty to fifty feet in height. The houses of the Mundurucús, to the number of about thirty, are scattered along the banks for a distance of six or seven miles. The owners appear to have chosen all the most picturesque sites—tracts of level ground at the foot of wooded heights, or little havens with bits of white sandy beach—as if they had an appreciation of natural beauty. Most of the dwellings are conical huts, with walls of framework filled in with mud and thatched with palm leaves, the broad eaves reaching half-way to the ground. Some are quadrangular, and do not differ in structure from those of the semi-civilised settlers in other parts; others are open sheds or ranchos. They seem generally to contain not more than one or two families each.

At the first house we learnt that all the fighting men had this morning returned from a two days' pursuit of a wandering horde of savages of the Parárauáte tribe, who had strayed this way from the interior lands and robbed the plantations. A little further on we came to the house of the Tushaúa, or chief, situated on the top of a high bank, which we had to ascend by wooden steps. There were four other houses in the neighbourhood, all filled with people. A fine old fellow, with face, shoulders, and breast tattooed all over in a cross-bar pattern, was the first strange object that caught my eye. Most of the men lay lounging or sleeping in their hammocks. The women were employed in an adjoining shed making farinha, many of them being quite naked, and rushing off to the huts to slip on their petticoats when they caught sight of us. Our entrance aroused the Tushaúa from a nap; after rubbing his eyes he came forward and bade us welcome with the most formal politeness, and in very good Portuguese. He was a tall, broad-shouldered, well-made man, apparently about thirty years of age, with handsome regular features, not tattooed, and a quiet good-humoured expression of countenance. He had been several times to Santarem and once to Pará, learning the Portuguese language during these journeys. He was dressed in shirt and trousers made of blue-checked cotton cloth, and there was not the slightest trace of the savage in his appearance or demeanour. I was told that he had come into the chieftainship by inheritance, and that the Cuparí horde of Mundurucús, over which his fathers had ruled before him, was formerly much more numerous, furnishing 300 bows in time of war. They could now scarcely muster forty; but the horde has no longer a close political connection with the main body of the tribe, which inhabits the banks of the Tapajos, six days' journey from the Cuparí settlement.

I spent the remainder of the day here, sending Aracú and the men to fish, whilst I amused myself with the Tushaúa and his people. A few words served to explain my errand on the river; he comprehended at once why white men should admire, and travel to collect the beautiful birds and animals of his country, and neither he nor his people spoke a single word about trading, or gave us any trouble by coveting the things we had brought. He related to me the events of the preceding three days. The Parárauátes were a tribe of intractable savages, with whom the Mundurucús have been always at war. They had no fixed abode, and of course made no plantations, but passed their lives like the wild beasts, roaming through the forest, guided by the sun: wherever they found themselves at night-time, there they slept, slinging their bast hammocks, which are carried by the women, to the trees. They cross the streams which lie in their course in bark canoes, which they make on reaching the water, and cast away after landing on the opposite side. The tribe is very numerous, but the different hordes obey only their own chieftains. The Mundurucús of the upper Tapajos have an expedition on foot against them at the present time, and the Tushaúa supposed that the horde which had just been chased from his maloca were fugitives from that direction. There were about a hundred of them—including men, women, and children. Before they were discovered the hungry savages had uprooted all the macasheira, sweet potatoes, and sugar-cane, which the industrious Mundurucús had planted for the season, on the east side of the river. As soon as they were seen they made off, but the Tushaúa quickly got together all the young men of the settlement, about thirty in number, who armed themselves with guns, bows and arrows, and javelins, and started in pursuit. They tracked them, as before related, for two days through the forest, but lost their traces on the further bank of the Cuparitinga, a branch stream flowing from the north-east. The pursuers thought, at one time, they were close upon them, having found the inextinguished fire of their last encampment. The footmarks of the chief could be distinguished from the rest by their great size and the length of the stride. A small necklace made of scarlet beans was the only trophy of the expedition, and this the Tushaúa gave to me.

I saw very little of the other male Indians, as they were asleep in their huts all the afternoon. There were two other tattooed men lying under an open shed, besides the old man already mentioned. One of them presented a strange appearance, having a semicircular black patch in the middle of his face, covering the bottom of the nose and mouth ; crossed lines on his back and breast, and stripes down his arms and legs. It is singular that the graceful curved patterns used by the South Sea Islanders are quite unknown among the Brazilian red men ; they being all tattooed either in simple lines or patches. The nearest approach to elegance of design which I saw, was amongst the Tucúnas of the Upper Amazons, some of whom have a scroll-like mark on each cheek, proceeding from the corner of the mouth. The taste, as far as form is concerned, of the American Indian, would seem to be far less refined than that of the Tahitian and New Zealander.

To amuse the Tushaúa, I fetched from the canoe the two volumes of Knight's Pictorial Museum of Animated Nature. The engravings quite took his fancy, and he called his wives, of whom, as I afterwards learnt from Aracú, he had three or four, to look at them ; one of them was a handsome girl, decorated with necklace and bracelets of blue beads. In a short time others left their work, and I then had a crowd of women and children around me, who all displayed unusual curiosity for Indians. It was no light task to go through the whole of the illustrations, but they would not allow me to miss a page, making me turn back when I tried to skip. The pictures of the elephant, camels, orang-otangs, and tigers, seemed most to astonish them ; but they were interested in almost everything, down even to the shells and insects. They recognised the portraits of the most striking birds and mammals which are found in their own country ; the jaguar, howling monkeys, parrots, trogons, and toucans. The elephant was settled to be a large kind of tapir ; but they made but few remarks, and those in the Mundurucú language, of which I understood only two or three words. Their way of expressing surprise was a clicking sound made with the teeth, similar to the one we ourselves use, or a subdued exclamation, Hm ! hm ! Before I finished, from fifty to sixty had assembled ; there was no pushing or rudeness, the grown-up women letting the young girls and children stand before them, and all behaved in the most quiet and orderly manner possible.

The Mundurucús are perhaps the most numerous and formidable tribe of Indians now surviving in the Amazons region. They inhabit the shores of the Tapajos (chiefly the right bank), from 3° to 7° south latitude, and the interior of the country between that part of the river and the Madeira. On the Tapajos alone they can muster, I was told, 2,000 fighting men; the total population of the tribe may be about 20,000. They were not heard of until about ninety years ago, when they made war on the Portuguese settlements; their hosts crossing the interior of the country eastward of the Tapajos, and attacking the establishments of the whites in the province of Maranham. The Portuguese made peace with them in the beginning of the present century, the event being brought about by the common cause of quarrel entertained by the two peoples against the hated Múras. They have ever since been firm friends of the whites. It is remarkable how faithfully this friendly feeling has been handed down amongst the Mundurucús, and spread to the remotest of the scattered hordes. Wherever a white man meets a family, or even an individual of the tribe, he is almost sure to be reminded of this alliance. They are the most warlike of the Brazilian tribes, and are considered also the most settled and industrious; they are not, however, superior in this latter respect to the Jurís and Passés on the Upper Amazons, or the Uapés Indians near the head waters of the Rio Negro. They make very large plantations of mandioca, and sell the surplus produce, which amounts on the Tapajos to from 3,000 to 5,000 baskets (60 lb. each) annually, to traders who ascend the river from Santarem between the months of August and January. They also gather large quantities of salsaparilla, india-rubber, and Tonka beans, in the forests. The traders, on their arrival at the Campinas (the scantily wooded region inhabited by the main body of Mundurucús beyond the cataracts) have first to distribute their wares—cheap cotton cloths, iron hatchets, cutlery, small wares, and cashaça—amongst the minor chiefs, and then wait three or four months for repayment in produce.

A rapid change is taking place in the habits of these Indians through frequent intercourse with the whites, and those who dwell on the banks of the Tapajos now seldom tattoo their children. The principal Tushaúa of the whole tribe or nation, named Joaquim, was rewarded with a commission in the Brazilian army, in acknowledgment of the assistance he gave to the legal authorities during the rebellion of 1835-6. It would be a misnomer to call the Mundurucús of the Cuparí and many parts of the Tapajos, savages; their regular mode of life, agricultural habits, loyalty to their chiefs, fidelity to treaties, and gentleness of demeanour, give them a right to a better title. Yet they show no aptitude for the civilised life of towns, and, like the rest of the Brazilian tribes, seem incapable of any further advance in culture. In their former wars they exterminated two of the neighbouring peoples, the Júmas and the Jacarés; and make now an annual expedition against the Parárauátes, and one or two other similar wild tribes who inhabit the interior of the land, but are sometimes driven by hunger towards the banks of the great rivers to rob the plantations of the agricultural Indians. These campaigns begin in July, and last throughout the dry months; the women generally accompanying the warriors to carry their arrows and javelins. They had the diabolical custom, in former days, of cutting off the heads of their slain enemies, and preserving them as trophies around their houses. I believe this, together with other savage practices, has been relinquished in those parts where they have had long intercourse with the Brazilians, for I could neither see nor hear anything of these preserved heads. They used to sever the head with knives made of broad bamboo, and then, after taking out the brain and fleshy parts, soak it in bitter vegetable oil (andiroba), and expose it for several days over the smoke of a fire or in the sun. In the tract of country between the Tapajos and the Madeira, a deadly war has been for many years carried on between the Mundurucús and the Aráras. I was told by a Frenchman at Santarem, who had visited that part, that all the settlements there have a military organisation. A separate shed is built outside each village, where the fighting men sleep at night, sentinels being stationed to give the alarm with blasts of the Turé on the approach of the Aráras, who choose the night for their onslaughts.

Each horde of Mundurucús has its pajé or medicine man, who is the priest and doctor; fixes upon the time most propitious for attacking the ememy; exorcises evil spirits, and professes to cure the sick. All illness whose origin is not very apparent is supposed to be caused by a worm in the part affected. This the pajé pretends to extract; he blows on the seat of pain the smoke from a large cigar, made with an air of great mystery by rolling tobacco in folds of Tauarí, and then sucks the place, drawing from his mouth, when he has finished, what he pretends to be the worm. It is a piece of very clumsy conjuring. One of these pajés was sent for by a woman in John Aracú's family, to operate on a child who suffered much from pains in the head. Senhor John contrived to get possession of the supposed worm after the trick was performed in our presence, and it turned out to be a long white air-root of some plant. The pajé was with difficulty persuaded to operate whilst Senhor John and I were present. I cannot help thinking that he, as well as all others of the same profession, are conscious impostors, handing down the shallow secret of their divinations and tricks from generation to generation. The institution seems to be common to all tribes of Indians, and to be held to more tenaciously than any other.

I bought of the Tushaúa two beautiful feather sceptres, with their bamboo cases. These are of cylindrical shape, about three feet in length and three inches in diameter, and are made by gluing with wax the fine white and yellow feathers from the breast of the toucan on stout rods, the tops being ornamented with long plumes from the tails of parrots, trogons, and other birds. The Mundurucús are considered to be the most expert workers in feathers of all the South American tribes. It is very difficult, however, to get them to part with the articles, as they seem to have a sort of superstitious regard for them. They manufacture head-dresses, sashes, and tunics, besides sceptres; the feathers being assorted with a good eye to the proper contrast of colours, and the quills worked into strong cotton webs, woven with knitting sticks in the required shape. The dresses are worn only during their festivals, which are celebrated, not at stated times, but whenever the Tushaúa thinks fit. Dancing, singing, sports, and drinking, appear to be the sole objects of these occasional holidays. When a day is fixed upon, the women prepare a great quantity of tarobá, and the monotonous jingle is kept up, with little intermission, night and day, until the stimulating beverage is finished.

We left the Tushaúa's house early the next morning. The impression made upon me by the glimpse of Indian life in its natural state obtained here, and at another cluster of houses visited higher up, was a pleasant one, notwithstanding the disagreeable incident of the Parárauáte visit. The Indians are here seen to the best advantage; having relinquished many of their most barbarous practices, without being corrupted by too close contact with the inferior whites and half-breeds of the civilised settlements. The manners are simpler, the demeanour more gentle, cheerful, and frank, than amongst the Indians who live near the towns. I could not help contrasting their well-fed condition, and the signs of orderly, industrious habits, with the poverty and laziness of the semi-civilised people of Altar do Chaõ. I do not think that the introduction of liquors has been the cause of much harm to the Brazilian Indian. He has his drinking bout now and then, like the common working people of other countries. It was his habit in his original state, before Europeans visited his country; but he is always ashamed of it afterwards, and remains sober during the pretty long intervals. The harsh, slave-driving practices of the Portuguese and their descendants have been the greatest curses to the Indians; the Mundurucús of the Cuparí, however, have been now for many years protected against ill-treatment. This is one of the good services rendered by the missionaries, who take care that the Brazilian law in favour of the aborigines shall be respected by the brutal and unprincipled traders who go amongst them. I think no Indians could be in a happier position than these simple, peaceful, and friendly people on the banks of the Cuparí. The members of each family live together, and seem to be much attached to each other; and the authority of the chief is exercised in the mildest manner. Perpetual summer reigns around them; the land is of the highest fertility, and a moderate amount of light work produces them all the necessaries of their simple life. It is difficult to get at their notions on subjects that require a little abstract thought; but the mind of the Indian is in a very primitive condition. I believe he thinks of nothing except the matters that immediately concern his daily material wants. There is an almost total absence of curiosity in his mental disposition, consequently he troubles himself very little concerning the causes of the natural phenomena around him. He has no idea of a Supreme Being; but, at the same time, he is free from revolting superstitions—his religious notions going no farther than the belief in an evil spirit, regarded merely as a kind of hobgoblin, who is at the bottom of all his little failures, troubles in fishing, hunting, and so forth. With so little mental activity, and with feelings and passions slow of excitement, the life of these people is naturally monotonous and dull, and their virtues are, properly speaking, only negative;

but the picture of harmless homely contentment they exhibit is very pleasing, compared with the state of savage races in many other parts of the world.

The men awoke me at four o'clock with the sound of their oars on leaving the port of the Tushaúa. I was surprised to find a dense fog veiling all surrounding objects, and the air quite cold. The lofty wall of forest, with the beautiful crowns of Assai palms standing out from it on their slender, arching stems, looked dim and strange through the misty curtain. The sudden change a little after sunrise had quite a magical effect; for the mist rose up like the gauze veil before the transformation scene at a pantomime, and showed the glorious foliage in the bright glow of morning, glittering with dew drops. We arrived at the falls about ten o'clock. The river here is not more than forty yards broad, and falls over a low ledge of rock stretching in a nearly straight line across.

We had now arrived at the end of the navigation for large vessels—a distance from the mouth of the river, according to a rough calculation, of a little over seventy miles. I found it the better course now to send José and one of the men forward in the montaria with John Aracú, and remain myself with the cuberta and our other men, to collect in the neighbouring forest. We stayed here four days; one of the boats returning each evening from the upper river with the produce of the day's chase of my huntsmen. I obtained six good specimens of the hyacinthine macaw, besides a number of smaller birds, a species new to me of Guaríba, or howling monkey, and two large lizards. The Guaríba was an old male, with the hair much worn from his rump and breast, and his body disfigured with large tumours made by the grubs of a gad-fly (Æstrus). The back and tail were of a ruddy-brown colour; the limbs and underside of the body, black. The men ascended to the second falls, which form a cataract several feet in height, about fifteen miles beyond our anchorage. The macaws were found feeding in small flocks on the fruit of the Tucumá palm (Astryocaryum Tucumá), the excessively hard nut of which is crushed into pulp by the powerful beak of the bird. I found the craws of all the specimens filled with the sour paste to which the stone-like fruit had been reduced. Each bird took me three hours to skin, and I was occupied with these and my other specimens every evening until midnight, after my own laborious day's hunt; working on the roof of my cabin by the light of a lamp.

The place where the cuberta was anchored formed a little rocky haven, with a sandy beach sloping to the forest, within which were the ruins of an Indian Maloca, and a large weed-grown plantation. The port swarmed with fishes, whose movements it was amusing to watch in the deep, clear water. The most abundant were the Piránhas. One species which varied in length, according to age, from two to six inches, but was recognisable by a black spot at the root of the tail, was always the quickest to seize any fragment of meat thrown into the water. When nothing was being given to them, a few only were seen scattered about, their heads all turned one way in an attitude of expectation; but as soon as any offal fell from

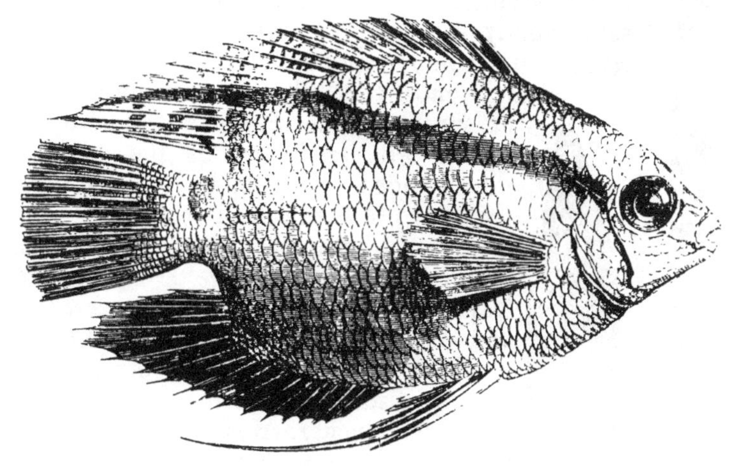

Acará (Mesonauta insignis).

the canoe, the water was blackened with the shoals that rushed instantaneously to the spot. Those who did not succeed in securing a fragment, fought with those who had been more successful, and many contrived to steal the coveted morsels from their mouths. When a bee or fly passed through the air near the water, they all simultaneously darted towards it as if roused by an electric shock. Sometimes a larger fish approached, and then the host of Piránhas took the alarm and flashed out of sight. The population of the water varied from day to day. Once a small shoal of a handsome black-banded fish, called by the natives Acará bandeira (Mesonauta insignis, of Günther), came gliding through at a slow pace, forming a very pretty sight. At another time, little troops of needle fish, eel-like animals with excessively long and slender toothed jaws, sailed through the field, scattering before them the hosts of smaller fry; and in the rear of the needle-fishes a strangely-shaped kind called Sarapó came wriggling along, one by one, with a slow movement. We caught with hook and line, baited with pieces of banana, several Curimatá (Anodus Amazonum), a most delicious fish, which, next to the Tucunaré and the Pescada, is most esteemed by the natives. The Curimatá seemed to prefer the middle of the stream, where the waters were agitated beneath the little cascade.

The weather was now settled and dry, and the river sank rapidly—six inches in twenty-four hours. In this remote and solitary spot I can say that I heard for the first and almost the only time the uproar of life at sunset, which Humboldt describes as having witnessed towards the sources of the Orinoco, but which is unknown on the banks of the larger rivers. The noises of animals began just as the sun sank behind the trees after a sweltering afternoon, leaving the sky above of the intensest shade of blue. Two flocks of howling monkeys, one close to our canoe, the other about a furlong distant, filled the echoing forest with their dismal roaring. Troops of parrots, including the hyacinthine macaw we were in search of, began then to pass over; the different styles of cawing and screaming of the various species making a terrible discord. Added to these noises were the songs of strange Cicadas, one large kind perched high on the trees around our little haven setting up a most piercing chirp; it began with the usual harsh jarring tone of its tribe, but this gradually and rapidly became shriller, until it ended in a long and loud note resembling the steam-whistle of a locomotive engine. Half-a-dozen of these wonderful performers made a considerable item in the evening concert. I had heard the same species before at Pará, but it was

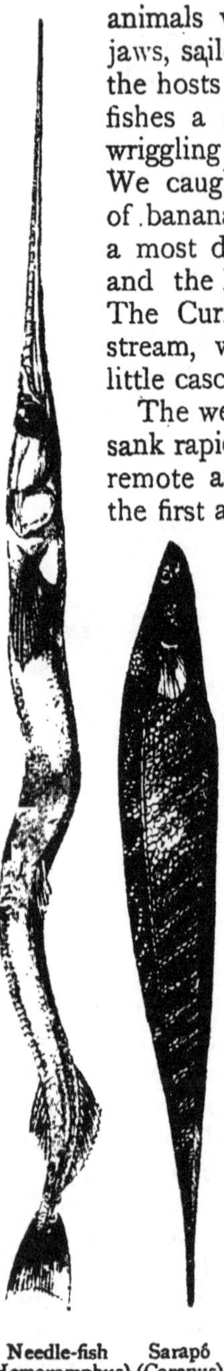

Needle-fish (Hemaramphus). Sarapó (Carapus).

there very uncommon: we obtained here one of them for my collection by a lucky blow with a stone. The uproar of beasts, birds, and insects lasted but a short time: the sky quickly lost its intense hue, and the night set in. Then began the tree-frogs—quack-quack, drum-drum, hoo-hoo; these accompanied by a melancholy night-jar, kept up their monotonous cries until very late.

My men encountered on the banks of the stream a Jaguar and a black Tiger, and were very much afraid of falling in with the Parárauátes, so that I could not, after their return on the fourth day, induce them to undertake another journey. We began our descent of the river in the evening of the 26th of August. At night forest and river were again enveloped in mist, and the air before sunrise was quite cold. There is a considerable current from the falls to the house of John Aracú, and we accomplished the distance, with its aid and by rowing, in seventeen hours.

September 21st.—At five o'clock in the afternoon we emerged from the confined and stifling gully through which the Cuparí flows, into the broad Tapajos, and breathed freely again. How I enjoyed the extensive view after being so long pent up: the mountainous coasts, the gray distance, the dark waters tossed by a refreshing breeze! Heat, mosquitoes, insufficient and bad food, hard work and anxiety, had brought me to a very low state of health; and I was now anxious to make all speed back to Santarem.

We touched at Aveyros, to embark some chests I had left there, and to settle accounts with Captain Antonio: finding nearly all the people sick with fever and vomit, against which the Padre's homœopathic globules were of no avail. The Tapajos had been pretty free from epidemics for some years past, although it was formerly a very unhealthy river. A sickly time appeared to be now returning; in fact, the year following my visit (1853) was the most fatal one ever experienced in this part of the country. A kind of putrid fever broke out, which attacked people of all races alike. The accounts we received at Santarem were most distressing: my Cuparí friends especially suffered very severely. John Aracú and his family all fell victims, with the exception of his wife: my kind friend Antonio Malagueita also died, and a great number of people in the Mundurucú village.

The descent of the Tapajos in the height of the dry season, which was now close at hand, is very hazardous on account of the strong winds, absence of current, and shoaly water far away from the coasts. The river towards the end of September is about thirty feet shallower than in June; and in many places ledges of rock are laid bare, or covered with only a small depth of water. I had been warned of these circumstances by my Cuparí friends, but did not form an adequate idea of what we should have to undergo. Canoes, in descending, only travel at night, when the terral, or light land-breeze, blows off the eastern shore. In the day-time a strong wind rages from down river, against which it is impossible to contend, as there is no current, and the swell raised by its sweeping over scores of miles of shallow water is dangerous to small vessels. The coast for the greater part of the distance affords no shelter: there are, however, a number of little harbours, called *esperas*, which the canoe-men calculate upon, carefully arranging each night voyage so as to reach one of them before the wind begins the next morning.

We left Aveyros in the evening of the 21st, and sailed gently down with the soft land-breeze, keeping about a mile from the eastern shore. It was a brilliant moonlit night, and the men worked cheerfully at the oars, when the wind was slack; the terral wafting from the forest a pleasant perfume like that of mignonette. At midnight we made a fire and got a cup of coffee, and at three o'clock in the morning reached the sitio of Ricardo's father, an Indian named André, where we anchored and slept.

September 22nd.—Old André with his squaw came aboard this morning. They brought three Tracajás, a turtle, and a basketful of Tracajá eggs, to exchange with me for cotton cloth and cashaça. Ricardo, who had been for some time very discontented, having now satisfied his longing to see his parents, cheerfully agreed to accompany me to Santarem. The loss of a man at this juncture would have been very annoying, with Captain Antonio ill at Aveyros, and not a hand to be had anywhere in the neighbourhood; but if we had not called at André's sitio, we should not have been able to have kept Ricardo from running away at the first landing-place. He was a lively, restless lad, and although impudent and troublesome at first, had made a very good servant; his companion, Alberto, was of quite a different disposition, being extremely taciturn, and going through all his duties with the quietest regularity.

We left at 11 a.m., and progressed a little before the wind

began to blow from down river, when we were obliged again to cast anchor. The terral began at six o'clock in the evening, and we sailed with it past the long line of rock-bound coast near Itapuáma. At ten o'clock a furious blast of wind came from a cleft between the hills, catching us with the sails close-hauled, and throwing the canoe nearly on its beam-ends, when we were about a mile from the shore. José had the presence of mind to slacken the sheet of the mainsail, whilst I leapt forward and lowered the sprit of the foresail; the two Indians standing stupefied in the prow. It was what the canoe-men call a *trovoada secca*, or white squall. The river in a few minutes became a sheet of foam; the wind ceased in about half an hour, but the terral was over for the night, so we pulled towards the shore to find an anchoring place.

We reached Tapaiuna by midnight on the 23rd, and on the morning of the 24th arrived at the Retiro, where we met a shrewd Santarem trader, whom I knew, Senhor Chico Honorio, who had a larger and much better provided canoe than our own. The wind was strong from below all day, so we remained at this place in his company. He had his wife with him, and a number of Indians, male and female. We slung our hammocks under the trees, and breakfasted and dined together, our cloth being spread on the sandy beach in the shade; after killing a large quantity of fish with *timbó*, of which we had obtained a supply at Itapuáma. At night we were again under weigh with the land breeze. The water was shoaly to a great distance off the coast, and our canoe having the lighter draught went ahead, our leadsman crying out the soundings to our companion: the depth was only one fathom, half a mile from the coast. We spent the next day (25th) at the mouth of a creek called Piní, which is exactly opposite the village of Boim, and on the following night advanced about twelve miles. Every point of land had a long spit of sand stretching one or two miles towards the middle of the river, which it was necessary to double by a wide circuit. The terral failed us at midnight when we were near an *espera*, called Maraï, the mouth of a shallow creek.

September 26th.—I did not like the prospect of spending the whole dreary day at Maraï, where it was impossible to ramble ashore, the forest being utterly impervious, and the land still partly under water. Besides, we had used up our last stick of firewood to boil our coffee at sunrise, and could not get a fresh supply at this place. So there being a dead calm on the river in the morning, I gave orders at ten o'clock to move out of the harbour, and try with the oars to reach Paquiatúba, which was only five miles distant. We had doubled the shoaly point which stretches from the mouth of the creek, and were making way merrily across the bay, at the head of which was the port of the little settlement, when we beheld to our dismay, a few miles down the river, the signs of the violent day breeze coming down upon us—a long, rapidly advancing line of foam with the darkened water behind it. Our men strove in vain to gain the harbour; the wind overtook us, and we cast anchor in three fathoms, with two miles of shoaly water between us and the land on our lee. It came with the force of a squall: the heavy billows washing over the vessel and drenching us with the spray. I did not expect that our anchor would hold; I gave out, however, plenty of cable, and watched the result at the prow; José placing himself at the helm, and the men standing by the jib and foresail, so as to be ready if we dragged, to attempt the passage of the Maraï spit, which was now almost dead to leeward. Our little bit of iron, however, held its place; the bottom being fortunately not so sandy as in most other parts of the coast; but our weak cable then began to cause us anxiety. We remained in this position all day without food, for everything was tossing about in the hold; provision-chests, baskets, kettles, and crockery. The breeze increased in strength towards the evening, when the sun set fiery red behind the misty hills on the western shore, and the gloom of the scene was heightened by the strange contrasts of colour; the inky water and the lurid gleam of the sky. Heavy seas beat now and then against the prow of our vessel with a force that made her shiver. If we had gone ashore in this place, all my precious collections would have been inevitably lost; but we ourselves could have scrambled easily to land, and re-embarked with Senhor Honorio, who had remained behind in the Piní, and would pass in the course of two or three days. When night came, I lay down exhausted with watching and fatigue, and fell asleep, as my men had done some time before. About nine o'clock I was awoke by the montaria bumping against the sides of the vessel, which had veered suddenly round, and the full moon, previously astern, then shone full in the cabin. The wind had abruptly ceased, giving place to light puffs from the eastern shore, and leaving a long swell rolling into the shoaly bay.

After this I resolved not to move a step beyond Paquiatúba,

without an additional man, and one who understood the navigation of the river at this season. We reached the landing-place at ten o'clock, and anchored within the mouth of the creek. In the morning I walked through the beautiful shady alleys of the forest, which were water-paths in June when we touched here in ascending the river, to the house of Inspector Cypriano. After an infinite deal of trouble I succeeded in persuading him to furnish me with another Indian. There are about thirty families established in this place, but the able-bodied men had been nearly all drafted off within the last few weeks by the Government, to accompany a military expedition against runaway negroes, settled in villages in the interior. Senhor Cypriano was a pleasant-looking and extremely civil young Mameluco. He accompanied us, on the night of the 28th, five miles down the river to Point Jaguararí, where the man lived whom he intended to send with me. I was glad to find my new hand a steady, middle-aged and married Indian; his name was of very good promise, Angelo Custodio (Guardian Angel).

Point Jaguararí forms at this season of the year a high sand-bank, which is prolonged as a narrow spit, stretching about three miles towards the middle of the river. We rounded this with great difficulty in the night of the 29th; reaching before daylight a good shelter behind a similar sandbank at Point Acarátingarí, a headland situated not more than five miles in a straight line from our last anchoring place. We remained here all day: the men beating *timbó* in a quiet pool between the sandbank and the mainland, and obtaining a great quantity of fish, from which I selected six species new to my collection. We made rather better progress the two following nights, but the terral now always blew strongly from the north-north-east after midnight, and thus limited the hours during which we could navigate, forcing us to seek the nearest shelter to avoid being driven back faster than we came.

On the 2nd of October we reached Point Cajetúba, and had a pleasant day ashore. The river scenery in this neighbourhood is of the greatest beauty. A few houses of settlers are seen at the bottom of the broad bay of Aramána-í at the foot of a range of richly-timbered hills, the high beach of snow-white sand stretching in a bold curve from point to point. The opposite shores of the river are ten or eleven miles distant, but towards the north is a clear horizon of water and sky. The country near Point Cajetúba is similar to the neighbourhood of Santarem: namely, campos with scattered trees. We gathered a large quantity of wild fruit: Cajú, Umirí, and Aápiránga. The Umirí berry (Humirium floribundum) is a black drupe similar in appearance to the Damascene plum, and not greatly unlike it in taste. The Aápiránga is a bright vermilion-coloured berry, with a hard skin and a sweet viscid pulp enclosing the seeds. Between the point and Altar do Chao was a long stretch of sandy beach with moderately deep water: our men therefore took a rope ashore, and towed the cuberta at merry speed until we reached the village. A long, deeply-laden canoe with miners from the interior provinces here passed us. It was manned by ten Indians, who propelled the boat by poles; the men, five on each side, trotting one after the other along a plank arranged for the purpose from stem to stern.

It took us two nights to double Point Cururú, where, as already mentioned, the river bends from its northerly course beyond Altar do Chao. A confused pile of rocks, on which many a vessel heavily laden with farinha has been wrecked, extends at the season of low water from the foot of a high bluff far into the stream. We were driven back on the first night (October 3rd) by a squall. The light terral was carrying us pleasantly round the spit, when a small black cloud which lay near the rising moon suddenly spread over the sky to the northward: the land breeze then ceased, and furious blasts began to blow across the river. We regained, with great difficulty, the shelter of the point. It blew almost a hurricane for two hours, during the whole of which time the sky over our heads was beautifully clear and starlit. Our shelter at first was not very secure, for the wind blew away the lashings of our sails, and caused our anchor to drag. Angelo Custodio, however, seized a rope which was attached to the foremast, and leapt ashore; had he not done so, we should probably have been driven many miles backwards up the storm-tossed river. After the cloud had passed, the regular east wind began to blow, and our further progress was effectually stopped for the night. The next day we all went ashore, after securing well the canoe, and slept from eleven o'clock till five, under the shade of trees.

The distance between Point Cururú and Santarem was accomplished in three days, against the same difficulties of contrary and furious winds, shoaly water, and rocky coasts. I was thankful at length to be safely housed, with the whole of my collections, made under so many privations and perils, landed without the loss or damage of a specimen. The men, after unloading the canoe and delivering it to its owner, came to receive their payment. They took part in goods and part in money, and after a good supper, on the night of the 7th October, shouldered their bundles and set off to walk by land some eighty miles to their homes. I was rather surprised at the good feeling exhibited by these poor Indians at parting. Angelo Custodio said that whenever I should wish to make another voyage up the Tapajos, he would be always ready to serve me as pilot. Alberto was undemonstrative as usual; but Ricardo, with whom I had had many sharp quarrels, actually shed tears when he shook hands and bid me the final "adeos."

CHAPTER X.

THE UPPER AMAZONS—VOYAGE TO EGA.

Departure from Barra—First day and night on the Upper Amazons—Desolate appearance of river in the flood season—Cucáma Indians—Mental condition of Indians—Squalls—Manatee—Forest—Floating pumice-stones from the Andes—Falling banks—Ega and its inhabitants—Daily life of a Naturalist at Ega—The four seasons of the Upper Amazons.

I MUST now take the reader from the picturesque, hilly country of the Tapajos, and its dark, streamless waters, to the boundless wooded plains, and yellow turbid current of the Upper Amazons or Solimoens. I will resume the narrative of my first voyage up the river, which was interrupted at the Barra of the Rio Negro, in the seventh chapter, to make way for the description of Santarem and its neighbourhood.

I embarked at Barra on the 26th of March, 1850, three years before steamers were introduced on the upper river, in a cuberta which was returning to Ega, the first and only town of any importance in the vast solitudes of the Solimoens, from Santarem, whither it had been sent with a cargo of turtle oil in earthenware jars. The owner, an old white-haired Portuguese trader of Ega, named Daniel Cardozo, was then at Barra, attending the assizes as juryman, a public duty performed without remuneration, which took him six weeks away from his business. He was about to leave Barra himself, in a small boat, and recommended me to send forward my heavy baggage in the cuberta and make the journey with him. He would reach Ega, 370 miles distant from Barra, in twelve or fourteen days! whilst the large vessel would be thirty or forty days on the road. I preferred, however, to go in company with my luggage, looking forward to the many opportunities I should have of landing and making collections on the banks of the river.

I shipped the collections made between Pará and the Rio Negro in a large cutter which was about descending to the capital, and after a heavy day's work got all my chests aboard the Ega canoe by eight o'clock at night. The Indians were then all embarked, one of them being brought dead drunk by his companions, and laid to sober himself all night on the wet boards of the tombadilha. The cabo, a spirited young white, named Estulano Alves Carneiro, who has since risen to be a distinguished citizen of the new province of the Upper Amazons, soon after gave orders to get up the anchor. The men took to the oars, and in a few hours we crossed the broad mouth of the Rio Negro; the night being clear, calm, and starlit, and the surface of the inky waters smooth as a lake.

When I awoke the next morning, we were progressing by espia along the left bank of the Solimoens. The rainy reason had now set in over the region through which the great river flows; the sand-banks and all the lower lands were already under water, and the tearing current, two or three miles in breadth, bore along a continuous line of uprooted trees and islets of floating plants. The prospect was most melancholy; no sound was heard but the dull murmur of the waters; the coast along which we travelled all day was encumbered every step of the way with fallen trees, some of which quivered in the currents which set around projecting points of land. Our old pest, the Motúca, began to torment us as soon as the sun gained power in the morning. White egrets were plentiful at the edge of the water, and humming-birds, in some places, were whirring about the flowers overhead. The desolate appearance of the landscape increased after sunset, when the moon rose in mist.

This upper river, the Alto-Amazonas, or Solimoens, is always spoken of by the Brazilians as a distinct stream. This is partly owing, as before remarked, to the direction it seems to take at the fork of the Rio Negro; the inhabitants of the country, from their partial knowledge, not being able to comprehend the whole river system in one view. It has, however, many peculiarities to distinguish it from the lower course of the river. The trade-wind, or sea-breeze, which reaches, in the height of the dry season, as far as the mouth of the Rio Negro, 900 or 1000 miles from the Atlantic, never blows on the upper river. The atmosphere is therefore more stagnant and sultry, and the winds that do prevail are of irregular direction and short duration. A great part of the land on the borders of the Lower Amazons is hilly; there are extensive campos, or open plains, and long stretches of sandy soil clothed with thinner forests. The climate, in consequence, is comparatively dry, many months in succession during the fine season passing without rain. All this is changed on the Solimoens. A fortnight of clear sunny weather is a rarity: the whole region through which the river and its affluents flow, after leaving the easternmost ridges of the Andes, which Pöppig describes as rising like a wall from the level country, 240 miles from the Pacific, is a vast plain, about 1000 miles in length, and 500 or 600 in breadth, covered with one uniform, lofty, impervious, and humid forest. The soil is nowhere sandy, but always either a stiff clay, alluvium, or vegetable mould, which latter, in many places, is seen in water-worn sections of the river banks to be twenty or thirty feet in depth. With such a soil and climate, the luxuriance of vegetation, and the abundance and beauty of animal forms which are already so great in the region nearer the Atlantic, increase on the upper river. The fruits, both wild and cultivated, common to the two sections of the country, reach a progressively larger size in advancing westward, and some trees which blossom only once a year at Pará and Santarem, yield flower and fruit all the year round at Ega. The climate is healthy, although one lives here as in a permanent vapour bath. I must not, however, give here a lengthy description of the region, whilst we are yet on its threshold. I resided and travelled on the Solimoens altogether for four years and a half. The country on its borders is a magnificent wilderness where civilized man, as yet, has scarcely obtained a footing; the cultivated ground from the Rio Negro to the Andes amounting only to a few score acres. Man, indeed, in any condition, from his small numbers, makes but an insignificant figure in these vast solitudes. It may be mentioned that the Solimoens is 2130 miles in length, if we reckon from the source of what is usually considered the main stream (Lake Lauricocha, near Lima); but 2500 miles by the route of the Ucayali, the most considerable and practicable fork of the upper part of the river. It is navigable at all seasons by large steamers, for upwards of 1400 miles from the mouth of the Rio Negro.

On the 28th we passed the mouth of Ariaüi, a narrow inlet

which communicates with the Rio Negro, emerging in front of Barra. Our vessel was nearly drawn into this, by the violent current which set from the Solimoens. The towing-cable was lashed to a strong tree about thirty yards ahead, and it took the whole strength of crew and passengers to pull across. We passed the Guariba, a second channel connecting the two rivers, on the 30th, and on the 31st sailed past a straggling settlement called Manacápurú, situated on a high rocky bank. Many citizens of Barra have *sitios*, or country-houses, in this place, although it is eighty miles distant from the town by the nearest road. Beyond Manacápurú all traces of high land cease; both shores of the river, henceforward for many hundred miles, are flat, except in places where the Tabatinga formation appears, in clayey elevations of from twenty to forty feet above the line of highest water. The country is so completely destitute of rocky or gravelly beds, that not a pebble is seen during many weeks' journey. Our voyage was now very monotonous. After leaving the last house at Manacápurú we travelled nineteen days without seeing a human habitation, the few settlers being located on the banks of inlets or lakes some distance from the shores of the main river. We met only one vessel during the whole of the time, and this did not come within hail, as it was drifting down in the middle of the current in a broad part of the river, two miles from the bank along which we were laboriously warping our course upwards.

After the first two or three days we fell into a regular way of life aboard. Our crew was composed of ten Indians of the Cucáma nation, whose native country is a portion of the borders of the upper river, in the neighbourhood of Nauta, in Peru. The Cucámas speak the Tupi language, using, however, a harsher accent than is common amongst the semi-civilized Indians from Ega downwards. They are a shrewd, hard-working people, and are the only Indians who willingly and in a body engage themselves to navigate the canoes of traders. The pilot, a steady and faithful fellow named Vicente, told me that he and his companions had now been fifteen months absent from their wives and families, and that on arriving at Ega they intended to take the first chance of a passage to Nauta. There was nothing in the appearance of these men to distinguish them from canoe-men in general. Some were tall and well built, others had squat figures with broad shoulders and excessively thick arms and legs. No two of them were at all similar in the shape of the head: Vicente had an oval visage, with fine regular features, whilst a little dumpy fellow, the wag of the party, was quite a Mongolian in breadth and prominence of cheek, spread of nostrils, and obliquity of eyes; but these two formed the extremes as to face and figure. None of them were tattooed or disfigured in any way; and they were all quite destitute of beard. The Cucámas are notorious on the river for their provident habits. The desire of acquiring property is so rare a trait in Indians, that the habits of these people are remarked on with surprise by the Brazilians. The first possession which they strive to acquire, on descending the river into Brazil, which all the Peruvian Indians look upon as a richer country than their own, is a wooden trunk with lock and key; in this they stow away carefully all their earnings converted into clothing, hatchets, knives, harpoon heads, needles and thread, and so forth. Their wages are only fourpence or sixpence a day, which is often paid in goods charged a hundred per cent. above Pará prices, so that it takes them a long time to fill their chest.

It would be difficult to find a better-behaved set of men on a voyage than these poor Indians. During our thirty-five days' journey they lived and worked together in the most perfect good fellowship. I never heard an angry word pass amongst them. Senhor Estulano let them navigate the vessel in their own way, exerting his authority only now and then when they were inclined to be lazy. Vicente regulated the working hours. These depended on the darkness of the nights. In the first and second quarters of the moon they kept it up with *espia*, or oars, until towards midnight; in the third and fourth quarters they were allowed to go to sleep soon after sunset, and aroused at three or four o'clock in the morning to resume their work. On cool, rainy days we all bore a hand at the *espia*, trotting with bare feet on the sloppy deck in Indian file, to the tune of some wild boatman's chorus. We had a favourable wind for two days only out of the thirty-five, by which we made about forty miles; the rest of our long journey was accomplished literally by pulling our way from tree to tree. When we encountered a *remanso* near the shore, we got along very pleasantly for a few miles by rowing: but this was a rare occurrence. During leisure hours the Indians employed themselves in sewing. Vicente was a good hand at cutting out shirts and trousers, and acted as master tailor to the whole party, each of whom had a thick steel thimble and a stock of needles and thread of his own. Vicente made for me a set of blue-check cotton shirts during the passage.

The goodness of these Indians, like that of most others amongst whom I lived, consisted perhaps more in the absence of active bad qualities, than in the possession of good ones; in other words, it was negative rather than positive. Their phlegmatic, apathetic temperament, coldness of desire and deadness of feeling, want of curiosity and slowness of intellect, make the Amazonian Indians very uninteresting companions anywhere. Their imagination is of a dull, gloomy quality, and they seemed never to be stirred by the emotions—love, pity, admiration, fear, wonder, joy, enthusiasm. These are characteristics of the whole race. The good fellowsh p of our Cucámas seemed to arise, not from warm sympathy, but simply from the absence of eager selfishness in small matters. On the morning when the favourable wind sprang up, one of the crew, a lad of about seventeen years of age, was absent ashore at the time of starting, having gone alone in one of the montarias to gather wild fruit. The sails were spread and we travelled for several hours at great speed, leaving the poor fellow to paddle after us against the strong current. Vicente, who might have waited a few minutes at starting, and the others, only laughed when the hardship of their companion was alluded to. He overtook us at night, having worked his way with frightful labour the whole day without a morsel of food. He grinned when he came on board, and not a dozen words were said on either side.

Their want of curiosity is extreme. One day we had an unusually sharp thunder-shower. The crew were lying about the deck, and after each explosion all set up a loud laugh; the wag of the party exclaiming, "There's my old uncle hunting again!" an expression showing the utter emptiness of mind of the spokesman. I asked Vicente what he thought was the cause of lightning and thunder. He said, "Timaá ichoquá,"—I don't know. He had never given the subject a moment's thought! It was the same with other things. I asked him who made the sun, the stars, the trees? He didn't know, and had never heard the subject mentioned amongst his tribe. The Tupí language, at least as taught by the old Jesuits, has a word—Tupána—signifying God. Vicente sometimes used this word, but he showed by his expressions that he did not attach the idea of a Creator to it. He seemed to think it meant some deity, or visible image, which the whites worshipped in the churches he had seen in the villages. None of the Indian tribes on the Upper Amazons have an idea of a Supreme Being, and consequently have no word to express it in their own languages. Vicente thought the river on which we were travelling encircled the whole earth, and that the land was an island like those seen in the stream, but larger. Here a gleam of curiosity and imagination in the Indian mind is revealed: the necessity of a theory of the earth and water has been felt, and a theory has been suggested. In all other matters not concerning the common wants of life the mind of Vicente was a blank, and such I always found to be the case of the Indian in his natural state. Would a community of any race of men be otherwise, were they isolated for centuries in a wilderness like the Amazonian Indians, associated in small numbers, wholly occupied in procuring a mere subsistence, and without a written language, or a leisured class to hand down acquired knowledge from generation to generation?

One day a smart squall gave us a good lift onward; it came with a cold, fine, driving rain, which enveloped the desolate landscape as with a mist: the forest swayed and roared with the force of the gale, and flocks of birds were driven about in alarm over the tree tops. On another occasion a similar squall came from an unfavourable quarter: it fell upon us quite unawares, when we had all our sails out to dry, and blew us broadside foremost on the shore. The vessel was fairly lifted on to the tall bushes which lined the banks, but we sustained no injury beyond the entanglement of our rigging in the branches. The days and nights usually passed in a dead calm, or with light intermittent winds from up river, and consequently full against us. We landed twice a day, to give ourselves and the Indians a little rest and change, and to cook our two meals—breakfast and dinner. There was another passenger beside myself—a cautious, middle-aged Portuguese, who was going to settle at Ega, where he had a brother long since established. He was accommodated in the fore-cabin, or arched covering over the hold. I shared the cabin-proper with Senhores Estulano and Manoel, the latter a young half-caste, son-in-law to the owner of the vessel, under whose tuition I made good progress in learning the Tupí language during the voyage.

Our men took it in turns, two at a time, to go out fishing, for which purpose we carried a spare montaria. The master had brought from Barra, as provisions, nothing but stale salt pirarecú—half-rotten fish, in large, thin, rusty slabs—farinha, coffee, and treacle. In these voyages passengers are expected to provide for themselves, as no charge is made except for freight of the heavy luggage or cargo they take with them. The Portuguese and myself had brought a few luxuries, such as beans, sugar, biscuits, tea, and so forth; but we found ourselves almost obliged to share them with our two companions and the pilot, so that before the voyage was one-third finished the small stock of most of these articles was exhausted. In return we shared in whatever the men brought. Sometimes they were quite unsuccessful, for fish is extremely difficult to procure in the season of high water, on account of the lower lands, lying between the inlets and infinite chain of pools and lakes, being flooded from the main river, thus increasing tenfold the area over which the finny population has to range. On most days, however, they brought two or three fine fish, and once they harpooned a manatee, or Vacca marina. On this last-mentioned occasion we made quite a holiday; the canoe was stopped for six or seven hours, and all turned out into the forest to help to skin and cook the animal. The meat was cut into cubical slabs, and each person skewered a dozen or so of these on a long stick. Fires were made, and the spits stuck in the ground and slanted over the flames to roast. A drizzling rain fell all the time, and the ground around the fires swarmed with stinging ants, attracted by the entrails and slime which were scattered about. The meat has somewhat the taste of very coarse pork; but the fat, which lies in thick layers between the lean parts, is of a greenish colour, and of a disagreeable, fishy flavour. The animal was a large one, measuring nearly ten feet in length, and nine in girth at the broadest part. The manatee is one of the few objects which excite the dull wonder and curiosity of the Indians, notwithstanding its commonness. The fact of its suckling its young at the breast, although an aquatic animal resembling a fish, seems to strike them as something very strange. The animal, as it lay on its back, with its broad rounded head and muzzle, tapering body, and smooth, thick, lead-coloured skin, reminded me of those Egyptian tombs which are made of dark, smooth stone, and shaped to the human figure.

Notwithstanding the hard fare, the confinement of the canoe, the trying weather,—frequent and drenching rains, with gleams of fiery sunshine,—and the woful desolation of the river scenery, I enjoyed the voyage on the whole. We were not much troubled by mosquitoes, and therefore passed the nights very pleasantly, sleeping on deck, wrapped in blankets or old sails. When the rains drove us below, we were less comfortable, as there was only just room in the small cabin for three of us to lie close together, and the confined air was stifling. I became inured to the Piums in the course of the first week; all the exposed parts of my body, by that time, being so closely covered with black punctures that the little bloodsuckers could not very easily find an unoccupied place to operate upon. Poor Miguel, the Portuguese, suffered horribly from these pests, his ankles and wrists being so much inflamed that he was confined to his hammock, slung in the hold, for weeks. At every landing-place I had a ramble in the forest, whilst the red skins made the fire and cooked the meal. The result was a large daily addition to my collection of insects, reptiles, and shells. Sometimes the neighbourhood of our gipsy-like encampment was a tract of dry and spacious forest, pleasant to ramble in; but more frequently it was a rank wilderness, into which it was impossible to penetrate many yards, on account of uprooted trees, entangled webs of monstrous woody climbers, thickets of spiny bamboos, swamps, or obstacles of one kind or other. The drier lands were sometimes beautified to the highest degrees by groves of the Urucurí palm (Attalea excelsa), which grew by thousands under the crowns of the lofty ordinary forest trees; their smooth columnar stems being all of nearly equal height (forty or fifty feet), and their broad, finely-pinnated leaves interlocking above to form arches and woven canopies of elegant and diversified shapes. The fruit of this palm ripens on the upper river in April, and during our voyage I saw immense quantities of it strewn about under the trees in places where we encamped. It is similar in size and shape to the date, and has a pleasantly-flavoured juicy pulp. The Indians would not eat it; I was surprised at this, as they greedily devoured many other kinds of palm fruit, whose sour and fibrous pulp was much less palatable. Vicente shook his head when he saw me one day eating a quantity of the Urucurí plums. I am not sure they were not the cause of a severe indigestion under which I suffered for many days afterwards.

In passing slowly along the interminable wooded banks week after week, I observed that there were three tolerably distinct kinds of coast and corresponding forest constantly recurring on this upper river. First, there were the low and most recent alluvial deposits,—a mixture of sand and mud, covered with tall, broad-leaved grasses, or with the arrow-grass before described, whose feathery-topped flower-stem rises to a height of fourteen or fifteen feet. The only large trees which grow in these places are the Cecropiæ. Many of the smaller and newer islands were of this description. Secondly, there were the moderately high banks, which are only partially overflowed when the flood season is at its height; these are wooded with a magnificent varied forest, in which a great variety of palms and broad-leaved Marantaceæ form a very large proportion of the vegetation. The general foliage is of a vivid light-green hue; the water frontage is sometimes covered with a diversified mass of greenery; but where the current sets strongly against the friable earthy banks, which at low water are twenty-five to thirty feet high, these are cut away, and expose a section of forest, where the trunks of trees loaded with epiphytes appear in massy colonnades. One might safely say that three-fourths of the land bordering the Upper Amazons, for a thousand miles, belong to this second class. The third description of coast is the higher, undulating, clayey land, which appears only at long intervals, but extends sometimes for many miles along the borders of the river. The coast at these places is sloping, and composed of red or variegated clay. The forest is of a different character from that of the lower tracts: it is rounder in outline, more uniform in its general aspect; palms are much less numerous and of peculiar species—the strange bulging-stemmed species, Iriartea ventricosa, and the slender glossy-leaved Bacába-í (Ænocarpus minor), being especially characteristic; and, in short, animal life, which imparts some cheerfulness to the other parts of the river, is seldom apparent. This "terra firme," as it is called, and a large portion of the fertile lower land, seemed well adapted for settlement; some parts were originally peopled by the aborigines, but these have long since become extinct or amalgamated with the white immigrants. I afterwards learnt that there were not more than eighteen or twenty families settled throughout the whole country from Manacápurú to Quarý, a distance of 240 miles; and these, as before observed, do not live on the banks of the main stream, but on the shores of inlets and lakes.

The fishermen twice brought me small rounded pieces of very porous pumice-stone, which they had picked up floating on the surface of the main current of the river. They were to me objects of great curiosity, as being messengers from the distant volcanoes of the Andes: Cotopaxi, Llanganete, or Sangay, which rear their peaks amongst the rivulets that feed some of the early tributaries of the Amazons, such as the Macas, the Pastaza, and the Napo. The stones must have already travelled a distance of 1,200 miles. I afterwards found them rather common; the Brazilians use them for cleaning rust from their guns, and firmly believe them to be solidified river foam. A friend once brought me, when I lived at Santarem, a large piece which had been found in the middle of the stream below Monte Alegre, about 900 miles further down the river; having reached this distance, pumice-stones would be pretty sure of being carried out to sea, and floated thence with the north-westerly Atlantic current, to shores many thousand miles distant from the volcanoes which ejected them. They are sometimes stranded on the banks in different parts of the river. Reflecting on this circumstance since I arrived in England, the probability of these porous fragments serving as vehicles for the transportation of seeds of plants, eggs of insects, spawn of fresh-water fish, and so forth, has suggested itself to me. Their rounded, water-worn appearance showed that they must have been rolled about for a long time in the shallow streams near the sources of the rivers at the feet of the volcanoes, before they leapt the waterfalls and embarked on the currents which lead direct for the Amazons. They may have been originally cast on the land and afterwards carried to the rivers by freshets; in which case the eggs and seeds of land insects and plants might be accidentally introduced, and safely enclosed with particles of earth in their cavities. As the speed of the current in the rainy season has been observed to be from three to five miles an hour, they might travel an immense distance before the eggs or seeds were destroyed. I am ashamed to say that I neglected the opportunity,

Bulging-stemmed Palm: Pashiúba barrigudo (Iriartea ventricosa).

whilst on the spot, of ascertaining whether this was actually the case. The attention of Naturalists has only lately been turned to the important subject of occasional means of wide dissemination of species of animals and plants. Unless such be shown to exist, it is impossible to solve some of the most difficult problems connected with the distribution of plants and animals. Some species, with most limited powers of locomotion, are found in opposite parts of the earth, without existing in the intermediate regions; unless it can be shown that these may have migrated or been accidentally transported from one point to the other, we shall have to come to the strange conclusion that the same species had been created in two separate districts.

Canoemen on the Upper Amazons live in constant dread of the "terras cahidas," or landslips, which occasionally take place along the steep earthy banks, especially when the waters are rising. Large vessels are sometimes overwhelmed by these avalanches of earth and trees. I should have thought the accounts of them exaggerated, if I had not had an opportunity during this voyage of seeing one on a large scale. One morning I was awoke before sunrise by an unusual sound resembling the roar of artillery. I was lying alone on the top of the cabin; it was very dark, and all my companions were asleep, so I lay listening. The sounds came from a considerable distance, and the crash which had aroused me was succeeded by others much less formidable. The first explanation which occurred to me was that it was an earthquake; for, although the night was breathlessly calm, the broad river was much agitated and the vessel rolled heavily. Soon after, another loud explosion took place, apparently much nearer than the former one; then followed others. The thundering peal rolled backwards and forwards, now seeming close at hand, now far off; the sudden crashes being often succeeded by a pause, or a long-continued dull rumbling. At the second explosion, Vicente, who lay snoring by the helm, awoke and told me it was a "terra cahida;" but I could scarcely believe him. The day dawned after the uproar had lasted about an hour, and we then saw the work of destruction going forward on the other side of the river, about three miles off. Large masses of forest, including trees of colossal size, probably 200 feet in height, were rocking to and fro, and falling headlong one after the other into the water. After each avalanche the wave which it caused returned on the crumbly bank with tremendous force, and caused the fall of other masses by undermining them. The line of coast over which the landslip extended was a mile or two in length; the end of it, however, was hid from our view by an intervening island. It was a grand sight; each downfall created a cloud of spray; the concussion in one place causing other masses to give way a long distance from it, and thus the crashes continued, swaying to any fro, with little prospect of a termination. When we glided out of sight, two hours after sunrise, the destruction was still going on.

On the 22nd we threaded the Paraná-mirím of Arauana-í, one of the numerous narrow by-waters which lie conveniently for canoes away from the main river, and often save a considerable circuit round a promontory or island. We rowed for half a mile through a magnificent bed of Victoria water-lilies, the flower-buds of which were just beginning to expand. Beyond the mouth of the Catuá, a channel leading to one of the great lakes so numerous in the plains of the Amazons, which we passed on the 25th, the river appeared greatly increased in breadth. We travelled for three days along a broad reach, which both up and down river presented a blank horizon of water and sky: this clear view was owing to the absence of islands, but it renewed one's impressions of the magnitude of the stream, which here, 1,200 miles from its mouth, showed so little diminution of width. Further westward a series of large islands commences, which divides the river into two and sometimes three channels, each about a mile in breadth. We kept to the southernmost of these, travelling all day on the 30th of April along a high and rather sloping bank.

In the evening we arrived at a narrow opening, which would be taken, by a stranger navigating the main channel, for the outlet of some insignificant stream: it was the mouth of the Teffé, on whose banks Ega is situated, the termination of our voyage. After having struggled for thirty-five days with the muddy currents and insect pests of the Solimoens, it was unspeakably refreshing to find one's-self again in a dark-water river, smooth as a lake, and free from Pium and Motuca. The rounded outline, small foliage, and sombre-green of the woods, which seemed to rest on the glassy waters, made a pleasant contrast to the tumultuous piles of rank, glaring, light-green vegetation, and torn, timber-strewn banks, to which we had been so long accustomed on the main river. The men rowed lazily until nightfall, when, having done a laborious day's work, they discontinued and went to sleep, intending to make for Ega in the morning. It was not thought worth while to secure the vessel to the trees or cast anchor, as there was no current. I sat up for two or three hours after my companions had gone to rest, enjoying the solemn calm of the night. Not a breath of air stirred; the sky was of a deep blue, and the stars seemed to stand forth in sharp relief; there was no sound of life in the woods, except the occasional melancholy note of some nocturnal bird. I reflected on my own wandering life: I had now reached the end of the third stage of my journey, and was now more than half-way across the continent. It was necessary for me, on many accounts, to find a rich locality for Natural History explorations, and settle myself in it for some months or years. Would the neighbourhood of Ega turn out to be suitable? and should I, a solitary stranger on a strange errand, find a welcome amongst its people?

Our Indians resumed their oars at sunrise the next morning (May 1st), and after an hour's rowing along the narrow channel, which varies in breadth from 100 to 500 yards, we doubled a low wooded point, and emerged suddenly on the so-called Lake of Ega: a magnificent sheet of water, five miles broad—the expanded portion of the Teffé. It is quite clear of islands, and curves away to the west and south, so that its full extent is not visible from this side. To the left, on a gentle grassy slope at the point of junction of a broad tributary with the Teffé, lay the little settlement: a cluster of a hundred or so of palm-thatched cottages and white-washed red-tiled houses, each with its neatly-enclosed orchard of orange, lemon, banana, and guava trees. Groups of palms, with their tall slender shafts and feathery crowns, overtopped the buildings and lower trees. A broad grass-carpeted street led from the narrow strip of white sandy beach to the rudely-built barn-like church, with its wooden crucifix on the green before it, in the centre of the town. Cattle were grazing before the houses, and a number of dark-skinned natives were taking their morning bath amongst the canoes of various sizes which were anchored or moored to stakes in the port. We let off rockets and fired salutes, according to custom, in token of our safe arrival, and shortly afterwards went ashore.

I made Ega my head-quarters during the whole of the time I remained on the Upper Amazons (four years and a half). My excursions into the neighbouring region extended sometimes as far as 300 and 400 miles from the place. An account of these excursions will be given in subsequent chapters; in the intervals between them I led a quiet, uneventful life in the settlement; following my pursuit in the same peaceful, regular way as a Naturalist might do in a European village. For many weeks in succession my journal records little more than the notes made on my daily captures. I had a dry and spacious cottage, the principal room of which was made a workshop and study; here a large table was placed, and my little library of reference arranged on shelves in rough wooden boxes. Cages for drying specimens were suspended from the rafters by cords well anointed, to prevent ants from descending, with a bitter vegetable oil: rats and mice were kept from them by inverted *cuyas*, placed half-way down the cords. I always kept on hand a large portion of my private collection, which contained a pair of each species and variety, for the sake of comparing the old with the new acquisitions. My cottage was whitewashed inside and out about once a year by the proprietor, a native trader; the floor was of earth; the ventilation was perfect, for the outside air, and sometimes the rain as well, entered freely through gaps at the top of the walls under the eaves, and through wide crevices in the doorways. Rude as the dwelling was, I look back with pleasure on the many happy months I spent in it. I rose generally with the sun, when the grassy streets were wet with dew, and walked down to the river to bathe: five or six hours of every morning were spent in collecting in the forest, whose borders lay only five minutes' walk from my house: the hot hours of the afternoon, between three and six o'clock, and the rainy days, were occupied in preparing and ticketing the specimens, making notes, dissecting, and drawing. I frequently had short rambles by water in a small montaria, with an Indian lad to paddle. The neighbourhood yielded me, up to the last day of my residence, an uninterrupted succession of new and different forms in the different classes of the animal kingdom, but especially insects.

There were, of course, many drawbacks to the amenities of the place as a residence for a European; but these were not of a nature that my readers would perhaps imagine. There was scarcely any danger from wild animals: it seems almost ridiculous to refute the idea of danger from the natives, in a country where even incivility to an unoffending stranger is a rarity. A Jaguar, however, paid us a visit one night. It was considered an extraordinary event, and so much uproar was made by the men who turned out with guns and bows and arrows, that the animal scampered off and was heard of no more. Alligators were rather troublesome in the dry season. During these months there was almost always one or two lying in wait near of the bathing-place for anything that might turn up at the edge the water; dog, sheep, pig, child, or drunken Indian. When this visitor was about, every one took extra care whilst bathing. I used to imitate the natives in not advancing far from the bank, and in keeping my eye fixed on that of the monster, which stares with a disgusting leer along the surface of the water; the body being submerged to the level of the eyes, and the top of the head, with part of the dorsal crest, the only portions visible. When a little motion was perceived in the water behind the reptile's tail, bathers were obliged to beat a quick retreat. I was never threatened myself, but I often saw the crowds of women and children scared whilst bathing, by the beast making a movement towards them; a general scamper to the shore and peals of laughter were always the result in these cases. The men can always destroy these alligators when they like to take the trouble to set out with montarias and harpoons for the purpose; but they never do it unless one of the monsters, bolder than usual, puts some one's life in danger. This arouses them, and they then track the enemy with the greatest pertinacity; when half killed, they drag it ashore and dispatch it amid loud execrations. Another, however, is sure to appear some days or weeks afterwards, and take the vacant place on the station. Besides alligators, the only animals to be feared are the poisonous serpents. These are certainly common enough in the forest, but no fatal accident happened during the whole time of my residence.

I suffered most inconvenience from the difficulty of getting news from the civilised world down river, from the irregularity of receipt of letters, parcels of books and periodicals, and towards the latter part of my residence from ill health arising from bad and insufficient food. The want of intellectual society, and of the varied excitement of European life, was also felt most acutely, and this, instead of becoming deadened by time, increased until it became almost insupportable. I was obliged, at last, to come to the conclusion that the contemplation of Nature alone is not sufficient to fill the human heart and mind. I got on pretty well when I received a parcel from England by the steamer once in two or four months. I used to be very economical with my stock of reading, lest it should be finished before the next arrival, and leave me utterly destitute. I went over the periodicals, the "Athenæum," for instance, with great deliberation, going through every number three times; the first time devouring the more interesting articles; the second, the whole of the remainder; and the third, reading all the advertisements from beginning to end. If four months (two steamers) passed without a fresh parcel, I felt discouraged in the extreme. I was worst off in the first year, 1850, when twelve months elapsed without letters or remittances. Towards the end of this time my clothes had worn to rags: I was barefoot, a great inconvenience in tropical forests, notwithstanding statements to the contrary that have been published by travellers; my servant ran away, and I was robbed of nearly all my copper money. I was obliged then to descend to Pará, but returned, after finishing the examination of the middle part of the Lower Amazons and the Tapajos, in 1855, with my Santarem assistant, and better provided for making collections on the upper river. This second visit was in pursuit of the plan before mentioned, of exploring in detail the whole valley of the Amazons, which I formed in Pará in the year 1851.

During so long a residence I witnessed, of course, many changes in the place. Some of the good friends who made me welcome on my first arrival died, and I followed their remains to their last resting-place in the little rustic cemetery on the borders of the surrounding forest. I lived there long enough, from first to last, to see the young people grow up, attended their weddings, and the christenings of their children, and, before I left, saw them old married folks with numerous families. In 1850, Ega was only a village, dependent on Pará, 1,400 miles distant, as the capital of the then undivided province. In 1852, with the creation of the new province of the Amazons, it became a city; returned its members to the provincial parliament at Barra; had its assizes, its resident judges, and rose to be the chief town of the *comarca* or country. A year after this, namely, in 1853, steamers were introduced on the Solimoens; and from 1855 one ran regularly every two months between the Rio Negro and Nauta in Peru, touching at all the villages, and accomplishing the distance in ascending, about 1,200 miles, in eighteen days. The trade and population, however, did not increase with these changes. The people became more "civilised," that is, they began to dress according to the latest Parisian fashions, instead of going about in stockingless feet, wooden clogs, and shirt sleeves; acquired a taste for money-getting and office-holding; became divided into parties, and lost part of their former simplicity of manners. But the place remained, when I left in 1859, pretty nearly what it was when I first arrived in 1850—a semi-Indian village, with much in the ways and notions of its people more like those a small country town in Northern Europe than a South of American settlement. The place is healthy, and almost free from insect pests; perpetual verdure surrounds it; the soil is of marvellous fertility, even for Brazil; the endless rivers and labyrinths of channels teem with fish and turtle; a fleet of steamers might anchor at any season of the year in the lake, which has uninterrupted water communication straight to the Atlantic. What a future is in store for the sleepy little tropical village!

After speaking of Ega as a city, it will have a ludicrous effect to mention that the total number of its inhabitants is only about 1,200. It contains just 107 houses, about half of which are miserably built mud-walled cottages, thatched with palm-leaves. A fourth of the population are almost always absent, trading or collecting produce on the rivers. The neighbourhood within a radius of thirty miles, and including two other small villages, contains probably 2,000 more people. The settlement is one of the oldest in the country, having been founded in 1688 by Father Samuel Fritz, a Bohemian Jesuit, who induced several of the docile tribes of Indians, then scattered over the neighbouring region, to settle on the site. From 100 to 200 acres of sloping ground around the place were afterwards cleared of timber; but such is the encroaching vigour of vegetation in this country, that the site would quickly relapse into jungle if the inhabitants neglected to pull up the young shoots as they arose. There is a stringent municipal law which compels each resident to weed a given space around his dwelling. Every month, whilst I resided here, an inspector came round with his wand of authority, and fined every one who had not complied with the regulation. The Indians of the surrounding country have never been hostile to the European settlers. The rebels of Pará and the Lower Amazons, in 1835-6, did not succeed in rousing the natives of the Solimoens against the whites. A party of forty of them ascended the river for that purpose, but on arriving at Ega, instead of meeting with sympathisers as in other places, they were surrounded by a small body of armed residents, and shot down without mercy. The military commandant at the time, who was the prime mover in this orderly resistance to anarchy, was a courageous and loyal negro, named José Patricio, an officer known throughout the Upper Amazons for his unflinching honesty and love of order, whose acquaintance I had the pleasure of making at St. Paulo in 1858. Ega was the head-quarters of the great scientific commission, which met in the years from 1781 to 1791, to settle the boundaries between the Spanish and Portuguese territories in South America. The chief commissioner for Spain, Don Francisco Requena, lived some time in the village with his family. I found only one person at Ega, my old friend Romaõ de Oliveira, who recollected, or had any knowledge of this important time, when a numerous staff of astronomers, surveyors, and draughtsmen, explored much of the surrounding country, with large bodies of soldiers and natives.

Many of the Ega Indians, including all the domestic servants, are savages who have been brought from the neighbouring rivers; the Japurá, the Issá, and the Solimoens. I saw here individuals of at least sixteen different tribes; most of whom had been bought, when children, of the native chiefs. This species of slave dealing, although forbidden by the laws of Brazil, is winked at by the authorities, because without it there would be no means of obtaining servants. They all become their own masters when they grow up, and never show the slightest inclination to return to utter savage life. But the boys generally run away and embark on the canoes of traders; and the girls are often badly treated by their mistresses, the jealous, passionate, and ill-educated Brazilian women. Nearly all the enmities which arise amongst residents at Ega and other places are caused by disputes about Indian servants. No one who has lived only in old settled countries, where service can be readily bought, can imagine the difficulties and annoyances of a land where the servant class are ignorant of the value of money, and hands cannot be obtained except by coaxing them from the employ of other masters.

Great mortality takes place amongst the poor captive children on their arrival at Ega. It is a singular circumstance, that the Indians residing on the Japurá and other tributaries always fall ill on descending to the Solimoens, whilst the reverse takes place with the inhabitants of the banks of the main river, who never fail of taking intermittent fever when they first ascend these branch rivers, and of getting well when they return. The finest tribes of savages who inhabit the country near Ega are the Jurís and Passés: these are now, however, nearly extinct, a few families only remaining on the banks of the retired creeks connected with the Teffé, and on other branch rivers between the Teffé and the Jutahí. They are a peaceable, gentle, and industrious people, devoted to agriculture and fishing, and have always been friendly to the whites. I shall have occasion to speak again of the Passés, who are a slenderly-built and superior race of Indians, distinguished by a large square tattooed patch in the middle of their faces. The principal cause of their decay in numbers seems to be a disease which always appears amongst them when a village is visited by people from the civilised settlements—a slow fever, accompanied by the symptoms of a common cold, "defluxo," as the Brazilians term it, ending probably in consumption. The disorder has been known to break out when the visitors were entirely free from it; the simple contact of civilised men, in some mysterious way, being sufficient to create it. It is generally fatal to the Jurís and Passés: the first question the poor patient Indians now put to an advancing canoe is, "Do you bring defluxo?"

My assistant, José, in the last year of our residence at Ega, "resgatou" (ransomed, the euphemism in use for purchased) two Indian children, a boy and a girl, through a Japurá trader. The boy was about twelve years of age, and of an unusually dark colour of skin: he had, in fact, the tint of a Cáfuzo, the offspring of Indian and negro. It was thought he had belonged to some perfectly wild and houseless tribe, similar to the Pará-rauátes of the Tapajos, of which there are several in different parts of the interior of South America. His face was of regular, oval shape, but his glistening black eyes had a wary, distrustful expression, like that of a wild animal; and his hands and feet were small and delicately formed. Soon after his arrival, finding that none of the Indian boys and girls in the houses of our neighbours understood his language, he became sulky and reserved; not a word could be got from him until many weeks afterwards, when he suddenly broke out with complete phrases of Portuguese. He was ill of swollen liver and spleen, the result of intermittent fever, for a long time after coming into our hands. We found it difficult to cure him, owing to his almost invincible habit of eating earth, baked clay, pitch, wax, and other similar substances. Very many children on the upper parts of the Amazons have this strange habit; not only Indians, but negroes and whites. It is not, therefore, peculiar to the famous Otomacs of the Orinoco, described by Humboldt, or to Indians at all, and seems to originate in a morbid craving, the result of a meagre diet of fish, wild-fruits, and mandioca meal. We gave our little savage the name of Sebastian. The use of these Indian children is to fill water-jars from the river, gather fire-wood in the forest, cook, assist in paddling the montaria in excursions, and so forth. Sebastian was often my companion in the woods, where he was very useful in finding the small birds I shot, which sometimes fell in the thickets amongst confused masses of fallen branches and dead leaves. He was wonderfully expert at catching lizards with his hands, and at climbing. The smoothest stems of palm-trees offered little difficulty to him: he would gather a few lengths of tough flexible lianas; tie them in a short endless band to support his feet with, in embracing the slippery shaft, and then mount upwards by a succession of slight jerks. It was very amusing, during the first few weeks, to witness the glee and pride with which he would bring to me the bunches of fruit he had gathered from almost inaccessible trees. He avoided the company of boys of his own race, and was evidently proud of being the servant of a real white man. We brought him down with us to Pará: but he showed no emotion at any of the strange sights of the capital; the steam-vessels, large ships and houses, horses and carriages, the pomp of church ceremonies, and so forth. In this he exhibited the usual dulness of feeling and poverty of thought of the Indian; he had, nevertheless, very keen perceptions, and was quick at learning any mechanical art. José, who had resumed, some time before I left the country, his old trade of goldsmith, made him his apprentice, and he made very rapid progress; for after about three months' teaching he came to me one day with radiant countenance, and showed me a gold ring of his own making.

The fate of the little girl, who came with a second batch of children all ill of intermittent fever, a month or two after Sebastian, was very different. She was brought to our house, after landing, one night in the wet season, when the rain was pouring in torrents, thin and haggard, drenched with wet, and shivering with ague. An old Indian who brought her to the door, said briefly, "ecui encommenda" (here's your little parcel, or order), and went away. There was very little of the savage in her appearance, and she was of a much lighter colour than the boy. We found she was of the Miránha tribe, all of whom are distinguished by a slit, cut in the middle of each wing of the nose, in which they wear on holiday occasions a large button made of pearly river-shell. We took the greatest care of our little patient; had the best nurses in the town, fomented her daily, gave her quinine and the most nourishing food; but it was all of no avail; she sank rapidly; her liver was enormously swollen, and almost as hard to the touch as stone. There was something uncommonly pleasing in her ways, and quite unlike anything I had yet seen in Indians. Instead of being dull and taciturn, she was always smiling and full of talk. We had an old woman of the same tribe to attend her, who explained what she said to us. She often begged to be taken to the river to bathe; asked for fruit, or coveted articles she saw in the room for playthings. Her native name was Oria. The last week or two she could not rise from the bed we had made for her in a dry corner of the room: when she wanted lifting, which was very often, she would allow no one to help her but me, calling me by the name of "Caríwa" (white man), the only word of Tupí she seemed to know. It was inexpressibly touching to hear her, as she lay, repeating by the hour the verses which she had been taught to recite with her companions in her native village: a few sentences repeated over and over again with a rhythmic accent, and relating to objects and incidents connected with the wild life of her tribe. We had her baptized before she died, and when this latter event happened, in opposition to the wishes of the big people of Ega, I insisted on burying her with the same honours as a child of the whites; that is, as an "anjinho" (little angel), according to the pretty Roman Catholic custom of the country. We had the corpse clothed in a robe of fine calico, crossed her hands on her breast over a "palma" of flowers, and made also a crown of flowers for her head. Scores of helpless children like our poor Oria die at Ega, or on the road; but generally not the slightest care is taken of them during their illness. They are the captives made during the merciless raids of one section of the Miránha tribe on the territories of another, and sold to the Ega traders. The villages of the attacked hordes are surprised, and the men and women killed or driven into the thickets without having time to save their children. There appears to be no doubt that the Miránhas are cannibals, and therefore the purchase of these captives probably saves them from a worse fate. The demand for them at Ega operates, however, as a direct cause of the supply, stimulating the unscrupulous chiefs, who receive all the profits, to undertake these murderous expeditions.

It is remarkable how quickly the savages of the various nations, which each have their own, to all appearance, widely different language, learn Tupí on their arrival at Ega, where it is the common idiom. This perhaps may be attributed chiefly to the grammatical forms of all the Indian tongues being the same, although the words are different. As far as I could learn, the feature is common to all, of placing the preposition *after* the noun, making it, in fact, a *post*-position, thus: "he is come the village *from*;" "go him *with*, the plantation *to*," and so forth. The ideas to be expressed in their limited sphere of life and thought are few; consequently the stock of words is extremely small; besides, all Indians have the same way of thinking, and the same objects to talk about; these circumstances also contribute to the ease with which they learn each other's language. Hordes of the same tribe living on the same branch rivers, speak mutually unintelligible languages; this happens with the Miránhas on the Japurá, and with the Collínas on the Jurúa; whilst Tupí is spoken with little corruption along the banks of the main Amazons for a distance of 2,500 miles. The purity of Tupí is kept up by frequent communication amongst the natives, from one end to the other of the main river; how complete and long-continued must be the isolation in which the small groups of savages have lived in other parts, to have caused so complete a segregation of dialects! It is probable that the strange inflexibility of the Indian organisation, both bodily and mental, is owing to the isolation in which each small tribe has lived, and to the narrow round of life and thought, and close intermarriages for countless generations, which are the necessary results. Their fecundity is of a low degree, for it is very rare to find an Indian family having so many as four children, and we have seen how great is their liability to sickness and death on removal from place to place.

I have already remarked on the different way in which the climate of this equatorial region affects Indians and negroes. No one could live long amongst the Indians of the Upper Amazons, without being struck with their constitutional dislike to the heat. Europeans certainly withstand the high temperature better than the original inhabitants of the country; I always found I could myself bear exposure to the sun or unusually hot weather quite as well as the Indians, although not well fitted by nature for a hot climate. Their skin is always hot to the touch, and they perspire little. No Indian resident of Ega can be induced to stay in the village (where the heat is felt more than in the forest or on the river) for many days together. They bathe many times a day, but do not plunge in the water, taking merely a *sitz-bath*, as dogs may be seen doing in hot climates, to cool the lower parts of the body. The women and children, who often remain at home, whilst the men are out for many days together fishing, generally find some excuse for trooping off to the shade of the forest in the hot hours of the afternoon. They are restless and discontented in fine dry weather, but cheerful in cool days, when the rain is pouring down on their naked backs. When suffering under fever, nothing but strict watching can prevent them from going down to bathe in the river, or eating immoderate quantities of juicy fruits, although these indulgences are frequently the cause of death. They are very subject to disorders of the liver, dysentery, and other diseases of hot climates; and when any epidemic is about, they fall ill quicker, and suffer more than negroes or even whites. How different all this is with the negro, the true child of tropical climes! The impression gradually forced itself on my mind that the red Indian lives as a stranger or immigrant in these hot regions, and that his constitution was not originally adapted, and has not since become perfectly adapted, to the climate.

The Indian element is very prominent in the amusements of the Ega people. All the Roman Catholic holidays are kept up with great spirit; rude Indian sports being mingled with the ceremonies introduced by the Portuguese. Besides these, the aborigines celebrate their own ruder festivals: the people of different tribes combining: for, in most of their features, the merry-makings were originally alike in all the tribes. The Indian idea of a holiday is bonfires, processions, masquerading, especially the mimicry of different kinds of animals, plenty of confused drumming and fifing, monotonous dancing, kept up hour after hour without intermission, and, the most important point of all, getting gradually and completely drunk. But he attaches a kind of superstitious significance to these acts, and thinks that the amusements appended to the Roman Catholic holidays, as celebrated by the descendants of the Portuguese, are also an essential part of the religious ceremonies. But in this respect the uneducated whites and half-breeds are not a bit more enlightened than the poor dull-souled Indian. All look upon a religious holiday as an amusement, in which the priest takes the part of director or chief actor.

Almost every unusual event, independent of saints' days, is made the occasion of a holiday by the sociable, easy-going people of the white and mameluco classes; funerals, christenings, weddings, the arrival of strangers, and so forth. The custom of "waking" the dead is also kept up. A few days after I arrived I was awoke in the middle of a dark moist night by Cardozo, to sit up with a neighbour whose wife had just died. I found the body laid out on a table, with crucifix and lighted wax candles at the head, and the room full of women and girls squatted on stools or on their haunches. The men were seated round the open door, smoking, drinking coffee, and telling stories; the bereaved husband exerting himself much to keep the people merry during the remainder of the night. The Ega people seem to like an excuse for turning night into day; it is so cool and pleasant, and they can sit about during these hours in the open air, clad as usual in simple shirt and trowsers, without streaming with perspiration.

The patron saint is Santa Theresa; the festival at whose anniversary lasts, like most of the others, ten days. It begins very quietly with evening litanies sung in the church, which are attended by the greater part of the population, all clean and gaily dressed in calicoes and muslins; the girls wearing jasmines and other natural flowers in the hair, no other head-dress being worn by females of any class. The evenings pass pleasantly; the church is lighted up with wax candles, and illuminated on the outside by a great number of little oil lamps—rude clay cups, or halves of the thick rind of the bitter orange, which are fixed all over the front. The congregation seem very attentive, and the responses to the litany of Our Lady, sung by a couple of hundred fresh female voices, ring agreeably through the still village. Towards the end of the festival the fun commences. The managers of the feast keep open houses, and dancing, drumming, tinkling of wire guitars, and unbridled drinking by both sexes, old and young, are kept up for a couple of days and a night with little intermission. The ways of the people at these merry-makings, of which there are many in the course of the year, always struck me as being not greatly different from those seen at an old-fashioned village wake in retired parts of England. The old folks look on and get very talkative over their cups; the children are allowed a little extra indulgence in sitting up; the dull, reserved fellows become loquacious, shake one another by the hand or slap each other on the back, discovering, all at once, what capital friends they are. The cantankerous individual gets quarrelsome, and the amorous unusually loving. The Indian, ordinarily so taciturn, finds the use of his tongue, and gives the minutest details of some little dispute which he had with his master years ago, and which every one else had forgotten; just as I have known lumpish labouring men in England do, when half-fuddled. One cannot help reflecting, when witnessing these traits of manners, on the similarity of human nature everywhere, when classes are compared whose state of culture and conditions of life are pretty nearly the same.

The Indians play a conspicuous part in the amusements at St. John's eve, and at one or two other holidays which happen about that time of the year—the end of June. In some of the sports the Portuguese element is visible, in others the Indian; but it must be recollected that masquerading, recitative singing, and so forth, are common originally to both peoples. A large number of men and boys disguise themselves to represent different grotesque figures, animals, or persons. Two or three dress themselves up as giants, with the help of a tall framework. One enacts the part of the Caypór, a kind of sylvan deity similar to the Curupíra which I have before mentioned. The belief in this being seems to be common to all the tribes of the Tupí stock. According to the figure they dressed up at Ega, he is a bulky, misshapen monster, with red skin and long shaggy red hair hanging half-way down his back. They believe that he has subterranean campos and hunting grounds in the forest, well stocked with pacas and deer. He is not at all an object of worship, nor of fear, except to children, being considered merely as a kind of hobgoblin. Most of the masquers make themselves up as animals—bulls, deer, magoary storks, jaguars, and so forth, with the aid of light frameworks, covered with old cloth dyed or painted, and shaped according to the object represented. Some of the imitations which I saw were capital. One ingenious fellow arranged an old piece of canvas in the form of a tapir, placed himself under it, and crawled about on all fours. He constructed an elastic nose to resemble that of the tapir, and made, before the doors of the principal residents, such a good imitation of the beast grazing, that peals of laughter greeted him wherever he went. Another man walked about solitarily, masked as a jabirú crane (a large animal standing about four feet high), and mimicked the gait and habits of the bird uncommonly well. One year an Indian lad imitated me, to the infinite amusement of the townsfolk. He came the previous day to borrow of me an old blouse and straw hat. I felt rather taken in when I saw him, on the night of the performance, rigged out as an entomologist, with an insect net, hunting bag, and pincushion. To make the imitation complete, he had borrowed the frame of an old pair of spectacles, and went about with it straddled over his nose. The jaguar now and then made a raid amongst the crowd of boys who were dressed as deer, goats, and so forth. The masquers kept generally together, moving from house to house, and the performances were directed by an old musician, who sang the orders and explained to the spectators what was going forward in a kind of recitative, accompanying himself on a wire guitar. The mixture of Portuguese and Indian customs is partly owing to the European immigrants in these parts having been uneducated men, who, instead of introducing European civilisation, have descended almost to the level of the Indians, and adopted some of their practices. The performances take place in the evening, and occupy five or six hours; bonfires are lighted along the grassy streets, and the families of the better class are seated at their doors, enjoying the wild but good-humoured fun.

We lived at Ega, during most part of the year, on turtle. The great fresh-water turtle of the Amazons grows on the upper river to an immense size, a full-grown one measuring nearly three feet in length by two in breadth, and is a load for the strongest Indian. Every house has a little pond, called a curral (pen), in the back-yard to hold a stock of the animals through the season of dearth—the wet months; those who have a number of Indians in their employ send them out for a month when the waters are low to collect a stock, and those who have not, purchase their supply; with some difficulty, however, as they are rarely offered for sale. The price of turtles, like that of all other articles of food, has risen greatly with the introduction of steam-vessels. When I arrived in 1850, a middle-sized one could be bought pretty readily for ninepence; but when I left in 1859, they were with difficulty obtained at eight and nine shillings each. The abundance of turtles, or rather the facility with which they can be found and caught, varies with the amount of annual subsidence of the waters. When the river sinks less than the average, they are scarce; but when more, they can be caught in plenty, the bays and shallow lagoons in the forest having then only a small depth of water. The flesh is very tender, palatable, and wholesome; but it is very cloying: every one ends, sooner or later, by becoming thoroughly surfeited. I became so sick of turtle in the course of two years that I could not bear the smell of it, although at the same time nothing else was to be had, and I was suffering actual hunger. The native women cook it in various ways. The entrails are chopped up and made into a delicious soup called *sarapatel*, which is generally boiled in the concave upper shell of the animal used as a kettle. The tender flesh of the breast is partially minced with farinha, and the breast shell then roasted over the fire, making a very pleasant dish. Steaks cut from the breast and cooked with the fat form another palatable dish. Large sausages are made of the thick-coated stomach, which is filled with minced meat and boiled. The quarters cooked in a kettle of Tucupí sauce form another variety of food. When surfeited with turtle in all other shapes, pieces of the lean part roasted on a spit and moistened only with vinegar make an agreeable change. The smaller kind of turtle, the tracajá, which makes its appearance in the main river, and lays its eggs a month earlier than the large species, is of less utility to the inhabitants, although its flesh is superior, on account of the difficulty of keeping it alive; it survives captivity but a very few days, although placed in the same ponds in which the large turtle keeps well for two or three years.

Those who cannot hunt and fish for themselves, and whose stomachs refuse turtle, are in a poor way at Ega. Fish, including many kinds of large and delicious salmonidæ, is abundant in the fine season; but each family fishes only for itself, and has no surplus for sale. An Indian fisherman remains out just long enough to draw what he thinks sufficient for a couple of days' consumption. Vacca marina is a great resource in the wet season; it is caught by harpooning, which requires much skill, or by strong nets made of very thick hammock twine, and placed across narrow inlets. Very few Europeans are able to eat the meat of this animal. Although there is a large quantity of cattle in the neighbourhood of the town, and pasture is abundant all the year round, beef can be had only when a beast is killed by accident. The most frequent cause of death is poisoning by drinking raw Tucupí, the juice of the mandioca root. Bowls of this are placed on the ground in the sheds where the women prepare farinha; it is generally done carelessly, but sometimes intentionally, through spite, when stray oxen devastate the plantations of the poorer people. The juice is almost certain to be drunk if cattle stray near the place, and death is the certain result. The owners kill a beast which shows symptoms of having been poisoned, and retail the beef in the town. Although every one knows it cannot be wholesome, such is the scarcity of meat and the uncontrollable desire to eat beef, that it is eagerly bought, at least by those residents who come from other provinces where beef is the staple article of food. Game of all kinds is scarce in the forest near the town, except in the months of June and July, when immense numbers of a large and handsome bird, Cuvier's toucan (Ramphastos Cuvieri) make their appearance. They come in well-fed condition, and are shot in such quantities that every family has the strange treat of stewed and roasted toucans daily for many weeks. Curassow birds are plentiful on the banks of the Solimoens, but to get a brace or two requires the sacrifice of several days for the trip. A tapir, of which the meat is most delicious and nourishing, is sometimes killed by a fortunate hunter. I have still a lively recollection of the pleasant effects which I once experienced from a diet of fresh tapir meat for a few days, after having been brought to a painful state of bodily and mental depression by a month's scanty rations of fish and farinha.

We sometimes had fresh bread at Ega, made from American flour brought from Pará, but it was sold at ninepence a pound. I was once two years without tasting wheaten bread, and attribute partly to this the gradual deterioration of health which I suffered on the Upper Amazons. Mandioca meal is a poor, weak substitute for bread; it is deficient in gluten, and consequently cannot be formed into a leavened mass or loaf, but is obliged to be roasted in hard grains in order to keep any length of time. Cakes are made of the half-roasted meal, but they become sour in a very few hours. A superior kind of meal is manufactured at Ega of the sweet mandioca (Manihot Aypi); it is generally made with a mixture of the starch of the root, and is therefore a much more wholesome article of food than the ordinary sort which, on the Amazons, is made of the pulp after the starch has been extracted by soaking in water. When we could get neither bread nor biscuit, I found tapioca soaked in coffee the best native substitute. We were seldom without butter, as every canoe brought one or two casks on each return voyage from Pará, where it is imported in considerable quantity from Liverpool. We obtained tea in the same way; it being served as a fashionable luxury at wedding and christening parties; the people were at first strangers to this article, for they used to stew it in a saucepan, mixing it up with coarse raw sugar, and stirring it with a spoon. Sometimes we had milk, but this was only when a cow calved; the yield from each cow was very small, and lasted only for a few weeks in each case, although the pasture is good, and the animals are sleek and fat.

Fruit of the ordinary tropical sorts could generally be had. I was quite surprised at the variety of the wild kinds, and of the delicious flavour of some of them. Many of these are utterly unknown in the regions nearer the Atlantic; being the peculiar productions of this highly favoured, and little known, interior country. Some have beeen planted by the natives in their clearings. The best was the *Jabutí-púhe*, or tortoise foot; a scaled fruit probably of the Anonaceous order. It is about the size of an ordinary apple; when ripe, the rind is moderately thin, and encloses, with the seeds, a quantity of custardy pulp of a very rich flavour. Next to this stands the Cumá (Collophora sp.) of which there are two species, not unlike, in appearance, small round pears; but the rind is rather hard, and contains a gummy milk, and the pulpy part is almost as delicious as that of the Jabutí-púhe. The Cumá tree is of moderate height, and grows rather plentifully in the more elevated and drier situations. A third kind is the Pamá, which is a stone fruit, similar in colour and appearance to the cherry, but of oblong shape. The tree is one of the loftiest in the forest, and has never, I believe, been selected for cultivation. To get at the fruit the natives are obliged to climb to the height of about a hundred feet, and cut off the heavily laden branches. I have already mentioned the Umarí and the Wishí: both these are now cultivated. The fatty, bitter pulp which surrounds the large stony seeds of these fruits is eaten mixed with farinha, and is very nourishing. Another cultivated fruit is the Purumá (Puruma cecropiæfolia, Martius), a round juicy berry, growing in large bunches and resembling grapes in taste. Another smaller kind, called Purumá-i, grows wild in the forest close to Ega, and has not yet been planted. The most singular of all these fruits is the Uikí, which is of oblong shape, and grows ap-

Uikí Fruit.

parently crosswise on the end of its stalk. When ripe, the thick green rind opens by a natural cleft across the middle, and discloses an oval seed the size of a Damascene plum, but of a vivid crimson colour. This bright hue belongs to a thin coating of pulp, which, when the seeds are mixed in a plate of stewed bananas, gives to the mess a pleasant rosy tint; and a rich creamy taste and consistence. *Mingau* (porridge) of bananas flavoured and coloured with Uikí is a favourite dish at Ega. The fruit, like most of the others here mentioned, ripens in January. Many smaller fruits, such as Wajurú (probably a species of Achras), the size of a gooseberry, which grows singly and contains a sweet gelatinous pulp, enclosing two large shining black seeds; Cashipári-arapaá, an oblong scarlet berry; two kinds of Bacurí, the Bacurí-siúma and the B. curúa, sour fruits of a bright lemon colour when ripe, and a great number of others, are of less importance as articles of food.

The celebrated "Peach palm," *Pupunha* of the Tupí nations (Guilielma speciosa), is a common tree at Ega. The name, I suppose, is in allusion to the colour of the fruit, and not to its flavour, for it is dry and mealy, and in taste may be compared to a mixture of chestnuts and cheese. Vultures devour it eagerly, and come in quarrelsome flocks to the trees when it is ripe. Dogs will also eat it: I do not recollect seeing cats do the same, although they go voluntarily to the woods to eat Tucumá, another kind of palm fruit. The tree, as it grows in clusters beside the palm-thatched huts, is a noble ornament, being, when full-grown, from fifty to sixty feet in height, and often as straight as a scaffold-pole. A bunch of fruit when ripe is a load for a strong man, and each tree bears several of them. The Pupunha grows wild nowhere on the Amazons. It is one of those few vegetable productions (including three kinds of mandioca and the American species of banana) which the Indians have cultivated from time immemorial, and brought with them in their original migration to Brazil. It is only, however, the more advanced tribes who have kept up the cultivation. The superiority of the fruit on the Solimoens to that grown on the Lower Amazons, and in the neighbourhood of Pará, is very striking. At Ega it is generally as large as a full-sized peach, and when boiled almost as mealy as a potato; whilst at Pará it is no bigger than a walnut, and the pulp is fibrous. Bunches of sterile or seedless fruits sometimes occur in both districts. It is one of the principal articles of food at

Ega when in season, and is boiled and eaten with treacle or salt. A dozen of the seedless fruits make a good nourishing meal for a grown-up person. It is the general belief that there is more nutriment in Pupunha than in fish or Vacca marina.

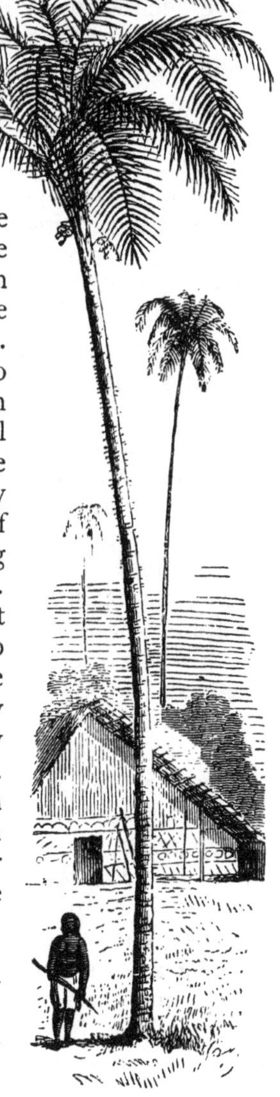

Pupunha Palm.

The seasons in the Upper Amazons region offer some points of difference from those of the lower river and the district of Pará, which two sections of the country we have already seen also differ considerably. The year at Ega is divided according to the rises and falls of the river, with which coincide the wet and dry periods. All the principal transactions of life of the inhabitants are regulated by these yearly recurring phenomena. The peculiarity of this upper region consists in there being two rises and two falls within the year. The great annual rise commences about the end of February, and continues to the middle of June, during which the rivers and lakes, confined during the dry periods to their ordinary beds, gradually swell, and overflow all the lower lands. The inundation progresses gently, inch by inch, and is felt everywhere, even in the interior of the forests, of the higher lands, miles away from the river; as these are traversed by numerous gullies, forming, in the fine season, dry spacious dells, which become gradually transformed by the pressure of the flood into broad creeks, navigable by small boats, under the shade of trees. All the countless swarms of turtle of various species then leave the main river for the inland pools: sand-banks go under water, and the flocks of wading birds then migrate northerly to the upper waters of the tributaries which flow from that direction, or to the Orinoco; which streams during the wet period of the Amazons are enjoying the cloudless skies of their dry season. The families of fishermen who have been employed, during the previous four or five months, in harpooning and salting pirarucú and shooting turtle in the great lakes, now return to the towns and villages; their temporarily constructed fishing establishments becoming gradually submerged, with the sand islets or beaches on which they were situated. This is the season, however, in which the Brazil nut and wild cacao ripen, and many persons go out to gather these harvests, remaining absent generally throughout the months of March and April. The rains during this time are not continuous; they fall very heavily at times, but rarely last so long at a stretch as twenty-four hours, and many days intervene of pleasant, sunny weather. The sky, however, is generally overcast and gloomy, and sometimes a drizzling rain falls.

About the first week in June the flood is at its highest; the water being then about forty-five feet above its lowest point; but it varies in different years to the extent of about fifteen feet. The "enchente," or flow, as it is called by the natives, who believe this great annual movement of the waters to be of the same nature as the tide towards the mouth of the Amazons, is then completed, and all begin to look forward to the "vasante," or ebb. The provision made for the dearth of the wet season is by this time pretty nearly exhausted; fish is difficult to procure, and many of the less provident inhabitants have become reduced to a diet of fruits and farinha porridge.

The fine season begins with a few days of brilliant weather—furious hot sun, with passing clouds. Idle men and women, tired of the dulness and confinement of the flood season, begin to report, on returning from their morning bath, the cessation of the flow: *as agoas estaõ paradas*, "the waters have stopped." The muddy streets, in a few days, dry up; groups of young fellows are now seen seated on the shady sides of the cottages, making arrows and knitting fishing-nets with tucúm twine; others are busy patching up and caulking their canoes, large and small: in fact, preparations are made on all sides for the much longed-for "veraõ" or summer, and the "migration," as it is called, of fish and turtle; that is, their descent from the inaccessible pools in the forest to the main river. Towards the middle of July the sandbanks begin to reappear above the surface of the waters, and with this change come flocks of sandpipers and gulls, which latter make known the advent of the fine season, as the cuckoo does of the European spring; uttering almost incessantly their plaintive cries as they fly about over the shallow waters of sandy shores. Most of the gaily-plumaged birds have now finished moulting, and begin to be more active in the forest.

The fall continues to the middle of October, with the interruption of a partial rise called "repiquet," of a few inches in the midst of very dry weather in September, caused by the swollen contribution of some large affluent higher up the river. The amount of subsidence also varies considerably, but it is never so great as to interrupt navigation by large vessels. The greater it is the more abundant is the season. Every one is prosperous when the waters are low; the shallow bays and pools being then crowded with the concentrated population of fish and turtle. All the people, men, women, and children, leave the villages, and spend the few weeks of glorious weather rambling over the vast undulating expanses of sand in the middle of the Solimoens, fishing, hunting, collecting eggs of turtle and plovers, and thoroughly enjoying themselves. The inhabitants pray always for a "vasante grande," or great ebb.

From the middle of October to the beginning of January, the second wet season prevails. The rise is sometimes not more than about fifteen feet, but it is, in some years, much more extensive, laying the large sand islands under water before the turtle eggs are hatched. In one year, whilst I resided at Ega, this second annual inundation reached to within ten feet of the highest water point, as marked by the stains on the trunks of trees by the river side.

The second dry season comes on in January, and lasts throughout February. The river sinks sometimes to the extent of a few feet only, but one year (1856) I saw it ebb to within about five feet of its lowest point in September. This is called the summer of the Umarí, "Veraõ do Umarí," after the fruit of this name already described, which ripens at this seaon. When the fall is great, this is the best time to catch turtles. In the year above mentioned, nearly all the residents who had a canoe, and could work a paddle, went out after them in the month of February, and about 2,000 were caught in the course of a few days. It appears that they had been arrested, in their migration towards the interior pools of the forest, by the sudden drying up of the water-courses, and so had become easy prey.

Thus the Ega year is divided into four seasons; two of dry weather and falling waters, and two of the reverse. Besides this variety, there is, in the month of May, a short season of very cold weather, a most surprising circumstance in this otherwise uniformly sweltering climate. This is caused by the continuance of a cold wind, which blows from the south over the humid forests that extend, without interruption, from north of the equator to the eighteenth parallel of latitude in Bolivia. I had, unfortunately, no thermometer with me at Ega; the only one I brought with me from England having been lost at Pará. The temperature is so much lowered that fishes die in the river Teffé, and are cast in considerable quantities on its shores. The wind is not strong; but it brings cloudy weather, and lasts from three to five or six days in each year. The inhabitants all suffer much from the cold, many of them wrapping themselves up with the warmest clothing they can get (blankets are here unknown), and shutting themselves indoors with a charcoal fire lighted. I found, myself, the change of temperature most delightful, and did not require extra clothing. It was a bad time, however, for my pursuit, as birds and insects all betook themselves to places of concealment, and remained inactive. The period during which this wind prevails is called the "tempo da friagem," or the season of coldness. The phenomenon, I presume, is to be accounted for by the fact that in May it is winter in the southern temperate zone, and that the cool currents of air travelling thence northwards towards the equator, become only moderately heated in their course, owing to the intermediate country being a vast partially-flooded plain covered with humid forests.

CHAPTER XI.

EXCURSIONS IN THE NEIGHBOURHOOD OF EGA.

The River Teffé—Rambles through groves on the beach—Excursion to the house of a Passé chieftain—Character and customs of the Passé tribe—First excursion to the sand islands of the Solimoens—Habits of great river-turtle—Second excursion—Turtle-fishing in the inland pools—Third excursion—Hunting rambles with natives in the forest—Return to Ega.

I WILL now proceed to give some account of the more interesting of my shorter excursions in the neighbourhood of Ega. The incidents of the longer voyages, which occupied each several months, will be narrated in a separate chapter.

The settlement, as before described, is built on a small tract of cleared land at the lower or eastern end of the lake, six or seven miles from the main Amazons, with which the lake communicates by a narrow channel. On the opposite shore of the broad expanse stands a small village, called Nogueira, the houses of which are not visible from Ega, except on very clear days; the coast on the Nogueira side is high, and stretches away into the grey distance towards the south-west. The upper part of the river Teffé is not visited by the Ega people, on account of its extreme unhealthiness, and its barrenness in salsaparilla and other wares. To Europeans it would seem a most surprising thing that the people of a civilised settlement, 170 years old, should still be ignorant of the course of the river on whose banks their native place, for which they proudly claim the title of city, is situated. It would be very difficult for a private individual to explore it, as the necessary number of Indian paddlers could not be obtained. I knew only one person who had ascended the Teffé to any considerable distance, and he was not able to give me a distinct account of the river. The only tribe known to live on its banks are the Catauishís, a people who perforate their lips all round, and wear rows of slender sticks in the holes: their territory lies between the Purús and the Juruá, embracing both shores of the Teffé. A large navigable stream, the Bararuá, enters the lake from the west, about thirty miles above Ega; the breadth of the lake is much contracted a little below the mouth of this tributary, but it again expands further south, and terminates abruptly where the Teffé proper, a narrow river with a strong current, forms its head water.

The whole of the country for hundreds of miles is covered with picturesque but pathless forests, and there are only two roads along which excursions can be made by land from Ega. One is a narrow hunter's track, about two miles in length, which traverses the forest in the rear of the settlement. The other is an extremely pleasant path along the beach to the west of the town. This is practicable only in the dry season, when a flat strip of white sandy beach is exposed at the foot of the high wooded banks of the lake, covered with trees, which, as there is no underwood, form a spacious shady grove. I rambled daily, during many weeks of each successive dry season, along this delightful road. The trees, many of which are myrtles and wild guavas, with smooth yellow stems, were in flower at this time; and the rippling waters of the lake, under the cool shade, everywhere bordered the path. The place was the resort of kingfishers, green and blue tree-creepers, purple-headed tanagers, and humming-birds. Birds generally, however, were not numerous. Every tree was tenanted by Cicadas, the reedy notes of which produced that loud, jarring, insect music which is the general accompaniment of a woodland ramble in a hot climate. One species was very handsome, having wings adorned with patches of bright green and scarlet. It was very common; sometimes three or four tenanting a single tree, clinging as usual to the branches. On approaching a tree thus peopled, a number of little jets of a clear liquid would be seen squirted from aloft. I have often received the well-directed discharge full on my face; but the liquid is harmless, having a sweetish taste, and is ejected by the insect from the anus, probably in self-defence, or from fear. The number and variety of gaily-tinted butterflies, sporting about in this grove on sunny days, were so great that the bright moving flakes of colour gave quite a character to the physiognomy of the place. It was impossible to walk far without disturbing flocks of them from the damp sand at the edge of the water, where they congregated to imbibe the moisture. They were of almost all colours, sizes, and shapes: I noticed here altogether eighty species, belonging to twenty-two different genera. It is a singular fact that, with very few exceptions, all the individuals of these various species thus sporting in sunny places were of the male sex; their partners, which are much more soberly dressed and immensely less numerous than the males, being confined to the shades of the woods. Every afternoon, as the sun was getting low, I used to notice these gaudy sunshine-loving swains trooping off to the forest, where I suppose they would find their sweethearts and wives. The most abundant, next to the very common sulphur-yellow and orange-coloured kinds, were about a dozen species of Eunica, which are of large size, and are conspicuous from their liveries of glossy dark-blue and purple. A superbly-adorned creature, the Callithea Markii, having wings of a thick texture, coloured sapphire-blue and orange, was only an occasional visitor. On certain days, when the weather was very calm, two small gilded green species (Symmachia Trochilus and Colubris) literally swarmed on the sands, their glittering wings lying wide open on the flat surface. The beach terminates, eight miles beyond Ega, at the mouth of a rivulet; the character of the coast then changes, the river banks being masked by a line of low islets amid a labyrinth of channels.

In all other directions my very numerous excursions were by water; the most interesting of those made in the immediate neighbourhood were to the houses of Indians on the banks of retired creeks; an account of one of these trips will suffice.

On the 23rd of May, 1850, I visited, in company with Antonio Cardozo, the Delegado, a family of the Passé tribe, who live near the head waters of the Igarapé, which flows from the south into the Teffé, entering it at Ega. The creek is more than a quarter of a mile broad near the town, but a few miles inland it gradually contracts, until it becomes a mere rivulet flowing through a broad dell in the forest. When the river rises, it fills this dell; the trunks of the lofty trees then stand many feet deep in the water, and small canoes are able to travel the distance of a day's journey under the shade, regular paths or alleys being cut through the branches and lower trees. This is the general character of the country of the Upper Amazons; a land of small elevation and abruptly undulated, the hollows forming narrow valleys in the dry months, and deep navigable creeks in the wet months. In retired nooks on the margins of these shady rivulets, a few families or small hordes of aborigines still linger in nearly their primitive state, the relics of their once numerous tribes. The family wė intended to visit on this trip, was that of Pedro-uassú (Peter the Great, or tall Peter), an old chieftain or Tushaúa of the Passés.

We set out at sunrise, in a small igarité, manned by six young Indian paddlers. After travelling about three miles along the broad portion of the creek—which, being surrounded by woods, had the appearance of a large pool—we came to a part where our course seemed to be stopped by an impenetrable hedge of trees and bushes. We were some time before finding the entrance, but when fairly within the shades, a remarkable scene presented itself. It was my first introduction to these singular water-paths. A narrow and tolerably straight alley stretched away for a long distance before us: on each side were the tops of bushes and young trees, forming a kind of border to the path, and the trunks of the tall forest trees rose at irregular intervals from the water, their crowns interlocking far over our heads, and forming a thick shade. Slender air roots hung down in clusters, and looping sipós dangled from the lower branches; bunches of grass, tillandsiæ, and ferns, sat in the forks of the larger boughs, and the trunks of trees near the water had adhering to them round dried masses of fresh-water sponges. There was no current perceptible, and the water was stained of a dark olive-brown hue, but the submerged stems could be seen through it to a great depth. We travelled at good speed for three hours along this shady road; the distance of Pedro's house from Ega being about twenty miles. When the paddlers rested for a time, the stillness and gloom of the place became almost painful: our voices waked dull echoes as we conversed, and the noise made by fishes occasionally whipping the surface of the water was quite startling. A cool, moist, clammy air pervaded the sunless shade.

The breadth of the wooded valley, at the commencement, is probably more than half a mile, and there is a tolerably clear view for a considerable distance on each side of the water-path through the irregular colonnade of trees: other paths also, in this part, branch off right and left from the principal road, leading to the scattered houses of Indians on the mainland. The dell contracts gradually towards the head of the rivulet. and the forest then becomes denser; the water-path also diminishes in width, and becomes more winding on account of the closer growth of the trees. The boughs of some are stretched forth at no great height over one's head, and are seen to be loaded with epiphytes; one orchid I noticed particularly, on account of its bright yellow flowers growing at the end of flower-stems several feet long. Some of the trunks, especially those of palms, close beneath their crowns, were clothed with a thick mass of glossy shield-shaped Pothos plants, mingled with ferns. Arrived at this part we were, in fact, in the heart of the virgin forest. We heard no noises of animals in the trees, and saw only one bird, the sky-blue chatterer, sitting alone on a high branch. For some distance the lower vegetation was so dense that the road runs under an arcade of foliage, the branches having been cut way only sufficiently to admit of the passage of a small canoe. These thickets are formed chiefly of bamboos, whose slender foliage and curving stems arrange themselves in elegant feathery bowers: but other social plants, slender green climbers with tendrils so eager in aspiring to grasp the higher boughs that they seem to be endowed almost with animal energy, and certain low trees having large elegantly-veined leaves, contribute also to the jungly masses. Occasionally we came upon an uprooted tree lying across the path, its voluminous crown still held by thick cables of sipó, connecting it with standing trees: a wide circuit had to be made in these cases, and it was sometimes difficult to find the right path again.

At length we arrived at our journey's end. We were then in a very dense and gloomy part of the forest: we could see, however, the dry land on both sides of the creek, and to our right a small sunny opening appeared, the landing-place to the native dwellings. The water was deep close to the bank, and a clean pathway ascended from the shady port to the buildings, which were about a furlong distant. My friend Cardozo was godfather to a grandchild of Pedro-uassú, whose daughter had married an Indian settled in Ega. He had sent word to the old man that he intended to visit him: we were therefore expected.

As we landed, Pedro-uassú himself came down to the port to receive us; our arrival having been announced by the barking of dogs. He was a tall and thin old man, with a serious but benignant expression of countenance, and a manner much freer from shyness and distrust than is usual with Indians. He was clad in a shirt of coarse cotton cloth, dyed with murishí, and trowsers of the same material turned up to the knee. His features were sharply delineated—more so than in any Indian face I had yet seen; the lips thin and the nose rather high and compressed. A large, square, blue-black tattooed patch occupied the middle of his face, which, as well as the other exposed parts of his body, was of a light reddish-tan colour, instead of the usual coppery-brown hue. He walked with an upright, slow gait, and on reaching us saluted Cardozo with the air of a man who wished it to be understood that he was dealing with an equal. My friend introduced me, and I was welcomed in the same grave, ceremonious manner. He seemed to have many questions to ask: but they were chiefly about Senhora Felippa, Cardozo's Indian housekeeper at Ega, and were purely complimentary. This studied politeness is quite natural to Indians of the advanced agricultural tribes. The language used was Tupí: I heard no other spoken all the day. It must be borne in mind that Pedro-uassú had never had much intercourse with whites: he was, although baptized, a primitive Indian, who had always lived in retirement; the ceremony of baptism having been gone through, as it generally is by the aborigines, simply from a wish to stand well with the whites.

Arrived at the house, we were welcomed by Pedro's wife: a thin, wrinkled, active old squaw, tattooed in precisely the same way as her husband. She had also sharp features, but her manner was more cordial and quicker than that of her husband; she talked much and with great inflection of voice; whilst the tones of the old man were rather drawling and querulous. Her clothing was a long petticoat of thick cotton cloth, and a very short chemise, not reaching to her waist. I was rather surprised to find the grounds around the establishment in neater order than in any sitio, even of civilised people, I had yet seen on the Upper Amazons: the stock of utensils and household goods of all sorts was larger, and the evidences of regular industry and plenty more numerous than one usually perceives in the farms of civilised Indians and whites. The buildings were of the same construction as those of the humbler settlers in all other parts of the country. The family lived in a large, oblong, open shed built under the shade of trees. Two smaller buildings, detached from the shed and having mud walls with low doorways, contained apparently the sleeping apartments of different members of the large household. A small mill for grinding sugar-cane, having two cylinders of hard notched wood; wooden troughs, and kettles for boiling the *guarápa* (cane juice), to make treacle, stood under a separate shed, and near it was a large enclosed mud-house for poultry. There was another hut and shed a short distance off, inhabited by a family dependent on Pedro, and a narrow pathway through the luxuriant woods led to more dwellings of the same kind. There was an abundance of fruit trees around the place, including the never-failing banana, with its long, broad, soft green leaf-blades, and groups of full-grown Pupúnhas, or peach palms. There was also a large number of cotton and coffee trees. Amongst the utensils I noticed baskets of different shapes, made of flattened maranta stalks, and dyed various colours. The making of these is an original art of the Passés, but I believe it is also practised by other tribes, for I saw several in the houses of semi-civilised Indians on the Tapajos.

There were only three persons in the house besides the old couple, the rest of the people being absent; several came in, however, in the course of the day. One was a daughter of Pedro's, who had an oval tattoed spot over her mouth; the second was a young grandson; and the third the son-in-law from Ega, Cardozo's *compadre.* The old woman was occupied, when we entered, in distilling spirits from cará, an eatable root similar to the potato, by means of a clay still, which had been manufactured by herself. The liquor had a reddish tint, but not a very agreeable flavour. A cup of it, warm from the still, however, was welcome after our long journey. Cardozo liked it, emptied his cup, and replenished it in a very short time. The old lady was very talkative, and almost fussy in her desire to please her visitors. We sat in tucúm hammocks, suspended between the upright posts of the shed. The young woman with the blue mouth—who, although married, was as shy as any young maiden of her race—soon became employed in scalding and plucking fowls for the dinner, near the fire on the ground at the other end of the dwelling. The son-in-law, Pedro-uassú, and Cardozo now began a long conversation on the subject of their deceased wife, daughter, and *comadre.** It appeared she had died of consumption—"tisica," as they called it, a word adopted by the Indians from the Portuguese. The widower repeated over and over again, in nearly the same words, his account of her illness, Pedro chiming in like a chorus, and Cardozo moralising and condoling. I thought the *cauim* (grog) had a good deal to do with the flow of talk and warmth of feeling of all three: the widower drank and wailed until he became maundering, and finally fell asleep.

I left them talking, and went a long ramble into the forest,

* Co-mother; the term expressing the relationship of a mother to the godfather of her child.

Pedro sending his grandson, a smiling well-behaved lad of about fourteen years of age, to show me the paths, my companion taking with him his *Zarabatana*, or blow-gun. This instrument is used by all the Indian tribes on the Upper Amazons. It is generally nine or ten feet long, and is made of two separate lengths of wood, each scooped out so as to form one-half of the tube. To do this with the necessary accuracy requires an enormous amount of patient labour, and considerable mechanical ability, the tools used being simply the incisor teeth of the Páca and Cutía. The two half-tubes when finished are secured together by a very close and tight spirally-wound strapping, consisting of long flat strips of Jacitára, or the wood of the climbing palm-tree; and the whole is smeared afterwards with black wax, the production of a Melipona bee. The pipe tapers towards the muzzle, and a cup-shaped mouth-piece, made of wood, is fitted in the broad end. A full-sized *Zarabatana* is heavy, and can only be used by an adult Indian who has had

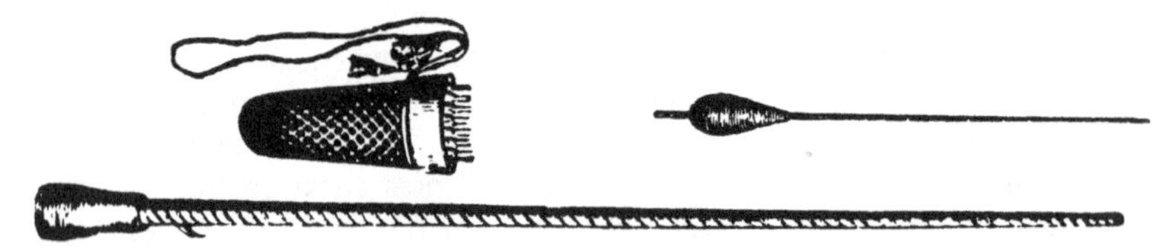

Blow-gun, quiver, and arrow.

great practice. The young lads learns to shoot with smaller and lighter tubes. When Mr. Wallace and I had lessons at Barra in the use of the blow-gun, of Julio, a Jurí Indian, then in the employ of Mr. Hauxwell, an English bird-collector, we found it very difficult to hold steadily the long tubes. The arrows are made from the hard rind of the leaf-stalks of certain palms, thin strips being cut, and rendered as sharp as needles by scraping the ends with a knife or the tooth of an animal. They are winged with a little oval mass of samaüma silk (from the seed-vessels of the silk-cotton tree, Eriodendron samaüma), cotton being too heavy. The ball of samaüma should fit to a nicety the bore of the blowgun; when it does so, the arrow can be propelled with such force by the breath, that it makes a noise, almost as loud as a pop-gun, on flying from the muzzle. My little companion was armed with a quiver full of these little missiles, a small number of which, sufficient for the day's sport, were tipped with the fatal Urarí poison. The quiver was an ornamental affair, the broad rim being made of highly-polished wood of a rich cherry-red colour (the Moira-piránga, or red-wood of the Japurá). The body was formed of neatly-plaited strips of Maranta stalks, and the belt by which it was suspended from the shoulder was decorated with cotton fringes and tassels.

We walked about two miles along a well-trodden pathway, through high caäpoeira (second-growth forest). A large proportion of the trees were Melastomas, which bore a hairy yellow fruit, nearly as large and as well flavoured as our gooseberry. The season, however, was nearly over for them. The road was bordered every inch of the way by a thick bed of elegant Lycopodiums. An artificial arrangement of trees and bushes could scarcely have been made to wear so finished an appearance as this naturally decorated avenue. The path at length terminated at a plantation of mandioca, the largest I had yet seen since I left the neighbourhood of Pará. There were probably ten acres of cleared land, and part of the ground was planted with Indian corn, water-melons, and sugar-cane. Beyond this field there was only a faint hunter's track, leading towards the untrodden interior. My companion told me he had never heard of there being any inhabitants in that direction (the south). We crossed the forest from this place to another smaller clearing, and then walked, on our road home, through about two miles of caäpoeira of various ages, the sites of old plantations. The only fruits of our ramble were a few rare insects and a Japú (Cassicus cristatus), a handsome bird with chestnut and saffron-coloured plumage, which wanders through the tree-tops in large flocks. My little companion brought this down from a height which I calculated at thirty yards. The blow-gun, however, in the hands of an expert adult Indian, can be made to propel arrows so as to kill at a distance of fifty and sixty yards. The aim is most certain when the tube is held vertically, or nearly so. It is a far more useful weapon in the forest than a gun, for the report of a firearm alarms the whole flock of birds or monkeys feeding on a tree, whilst the silent poisoned dart brings the animals down one by one, until the sportsman has a heap of slain by his side. None but the stealthy Indian can use it effectively. The poison, which must be fresh to kill speedily, is obtained only of the Indians who live beyond the cataracts of the rivers flowing from the north, especially the Rio Negro and the Japurá. Its principal ingredient is the wood of the Strychnos toxifera, a tree which does not grow in the humid forests of the river plains. A most graphic account of the Urarí, and of an expedition undertaken in search of the tree in Guiana, has been given by Sir Robert Schomburgk.*

When we returned to the house after mid-day, Cardoza was still sipping cauím, and now looked exceedingly merry. It was fearfully hot: the good fellow sat in his hammock with a cuya full of grog in his hands; his broad honest face all of a glow, and the perspiration streaming down his uncovered breast, the unbuttoned shirt having slipped half-way over his broad shoulders. Pedro-uassú had not drunk much; he was noted, as I afterwards learnt, for his temperance. But he was standing up as I had left him two hours previously, talking to Cardozo in the same monotonous tones, the conversation apparently not having flagged all the time. I had never heard so much talking amongst Indians. The widower was asleep: the stirring, managing old lady with her daughter were preparing dinner. This, which was ready soon after I entered, consisted of boiled fowls and rice, seasoned with large green peppers and lemon juice, and piles of new, fragrant farinha and raw bananas. It was served on plates of English manufacture on a tupé, or large plaited rush mat, such as is made by the natives pretty generally on the Amazons. Three or four other Indians, men and women of middle age, now made their appearance, and joined in the meal. We all sat round on the floor: the women, according to custom, not eating until after the men had done. Before sitting down our host apologised, in his usual quiet, courteous manner, for not having knives and forks; Cardozo and I ate by the aid of wooden spoons, the Indians using their fingers. The old man waited until we were all served before he himself commenced. At the end of the meal one of the women brought us water in a painted clay basin of Indian manufacture, and a clean but coarse cotton napkin, that we might wash our hands.

The horde of Passés of which Pedro-uassú was Tushaúa or chieftain, was at this time reduced to a very small number of individuals. The disease mentioned in the last chapter had for several generations made great havoc amongst them; many also had entered the service of whites at Ega, and, of late years, intermarriages with whites, half-castes, and civilised Indians had been frequent. The old man bewailed the fate of his race to Cardozo with tears in his eyes. "The people of my nation," he said, "have always been good friends to the Caríwas (whites), but before my grandchildren are old like me the name

* Annals and Magazine of Natural History, Vol. vii. p. 411.

of Passé will be forgotten." In so far as the Passés have amalgamated with European immigrants or their descendants, and become civilised Brazilian citizens, there can scarcely be ground for lamenting their extinction as a nation; but it fills one with regret to learn how many die prematurely of a disease which seems to arise on their simply breathing the same air as the whites. The original territory of the tribe must have been of large extent, for Passés are said to have been found by the early Portuguese colonists on the Rio Negro; an ancient settlement on that river, Barcellos, having been peopled by them when it was first established; and they formed also part of the original population of Fonte-boa on the Solimoens. Their hordes were therefore spread over a region 400 miles in length from east to west. It is probable, however, that they have been confounded by the colonists with other neighbouring tribes who tattoo their faces in a similar manner. The extinct tribe of Yurimaúas, or Sorimóas, from which the river Solimoens derives its name, according to traditions extant at Ega, resembled the Passés in their slender figures and friendly disposition. These tribes (with others lying between them) peopled the banks of the main river and its by-streams from the mouth of the Rio Negro to Peru. True Passés existed in their primitive state on the banks of the Issá, 240 miles to the west of Ega, within the memory of living persons. The only large body of them now extant are located on the Japurá, at a place distant about 150 miles from Ega: the population of this horde, however, does not exceed, from what I could learn, 300 or 400 persons. I think it probable that the lower part of the Japurá and its extensive delta lands formed the original home of this gentle tribe of Indians.

The Passés are always spoken of in this country as the most advanced of all the Indian nations in the Amazons region. Under what influences this tribe has become so strongly modified in mental, social, and bodily features it is hard to divine. The industrious habits, fidelity, and mildness of disposition of the Passés, their docility and, it may be added, their personal beauty, especially of the children and women, made them from the first very attractive to the Portuguese colonists. They were, consequently, enticed in great number from their villages and brought to Barra and other settlements of the whites. The wives of governors and military officers from Europe were always eager to obtain children for domestic servants; the girls being taught to sew, cook, weave ham mocks, manufacture pillow-laçe, and so forth. They have been generally treated with kindness, especially by the educated families in the settlements. It is pleasant to have to record that I never heard of a deed of violence perpetrated, on the one side or the other, in the dealings between European settlers and this noble tribe of savages.

We started on our return to Ega at half-past four o'clock in the afternoon. Our generous entertainers loaded us with presents. There was scarcely room for us to sit in the canoe, as they had sent down ten large bundles of sugar-cane, four baskets of farinha, three cedar planks, a small hamper of coffee, and two heavy bunches of bananas. After we were embarked the old lady came with a parting gift for me—a huge bowl of smoking hot banana porridge. I was to eat it on the road "to keep my stomach warm." Both stood on the bank as we pushed off, and gave us their adeos, "Ikuána Tupána eirúm" (Go with God): a form of salutation taught by the old Jesuit missionaries. We had a most uncomfortable passage, for Cardozo was quite tipsy, and had not attended to the loading of the boat. The cargo had been placed too far forward, and to make matters worse my heavy friend obstinately insisted on sitting astride on the top of the pile, instead of taking his place near the stern; singing from his perch a most indecent love-song, and disregarding the inconvenience of having to bend down almost every minute to pass under the boughs and hanging sipós as we sped rapidly along. The canoe leaked, but not at first alarmingly. Long before sunset darkness began to close in under these gloomy shades, and our steersman could not avoid now and then running the boat into the thicket. The first time this happened a piece was broken off the square prow (rodella); the second time we got squeezed between two trees. A short time after this latter accident, being seated near the stern, with my feet on the bottom of the boat, I felt rather suddenly the cold water above my ankles. A few minutes more and we should have sunk, for a seam had been opened forward under the pile of sugar-cane. Two of us began to bale, and by the most strenuous efforts managed to keep afloat without throwing overboard our cargo. The Indians were obliged to paddle with extreme slowness to avoid shipping water, as the edge of our prow was nearly level with the surface; but Cardozo was now persuaded to change his seat. The sun set, the quick twilight passed, and the moon soon after began to glimmer through the thick canopy of foliage. The prospect of being swamped in this hideous solitude was by no means pleasant, although I calculated on the chance of swimming to a tree and finding a nice snug place in the fork of some large bough wherein to pass the night. At length, after four hours' tedious progress, we suddenly emerged on the open stream, where the moonlight glittered in broad sheets on the gently rippling waters. A little extra care was now required in paddling. The Indians plied their strokes with the greatest nicety; the lights of Ega (the oil lamps in the houses) soon appeared beyond the black wall of forest, and in a short time we leapt safely ashore.

A few months after the excursion just narrated, I accompanied Cardozo in many wanderings on the Solimoens, during which we visited the praias (sand islands), the turtle pools in the forests, and the by-streams and lakes of the great desert river. His object was mainly to superintend the business of digging up turtle eggs on the sand-banks, having been elected commandante for the year, by the municipal council of Ega, of the "praia real" (royal sand island) of Shimuní, the one lying nearest to Ega. There are four of these royal praias within the Ega district (a distance of 150 miles from the town), all of which are visited annually by the Ega people for the purpose of collecting eggs and extracting oil from their yolks. Each has its commander, whose business is to make arrangements for securing to every inhabitant an equal chance in the egg harvest, by placing sentinels to protect the turtles whilst laying, and so forth. The pregnant turtles descend from the interior pools to the main river in July and August, before the outlets dry up, and then seek in countless swarms their favourite sand islands; for it is only a few praias that are selected by them out of the great number existing. The young animals remain in the pools throughout the dry season. These breeding-places of turtles then lie twenty to thirty or more feet above the level of the river, and are accessible only by cutting roads through the dense forest.

We left Ega on our first trip, to visit the sentinels whilst the turtles were yet laying, on the 26th of September. Our canoe was a stoutly-built igarité, arranged for ten paddlers, and having a large arched toldo at the stern, under which three persons could sleep pretty comfortably. Emerging from the Teffé, we descended rapidly on the swift current of the Solimoens to the south-eastern or lower end of the large wooded island of Bariá, which here divides the river into two great channels. We then paddled across to Shimuní, which lies in the middle of the north-easterly channel, reaching the commencement of the praia an hour before sunset. The island proper is about three miles long and half a mile broad: the forest with which it is covered rises to an immense and uniform height, and presents all round a compact, impervious front. Here and there a singular tree, called Pao mulatto (mulatto wood), with polished dark-green trunk, rose conspicuously amongst the mass of vegetation. The sand-bank, which lies at the upper end of the island, extends several miles, and presents an irregular, and in some parts, strongly-waved surface, with deep hollows and ridges. When upon it, one feels as though treading an almost boundless field of sand: for towards the south-east, where no forest-line terminates the view, the white rolling plain stretches away to the horizon. The north-easterly channel of the river, lying between the sands and the further shore of the river, is at least two miles in breadth; the middle one, between the two islands, Shimuní and Bariá, is not much less than a mile.

We found the two sentinels lodged in a corner of the praia, where it commences at the foot of the towering forest-wall of the island; having built for themselves a little rancho with poles and palm-leaves. Great precautions are obliged to be taken to avoid disturbing the sensitive turtles, who, previous to crawling ashore to lay, assemble in great shoals off the sand-bank. The men, during this time, take care not to show themselves, and warn off any fisherman who wishes to pass near the place. Their fires are made in a deep hollow near the borders of the forest, so that the smoke may not be visible. The passage of a boat through the shallow waters where the animals are congregated, or the sight of a man or a fire on the sand-bank, would prevent the turtles from leaving the water that night to lay their eggs, and if the causes of alarm were repeated once or twice they would forsake the praia for some other quieter place. Soon after we arrived our men were sent with the net to catch a supply of fish for supper. In half an hour four or five large basketfuls of Acarí were brought in. The sun set soon after our meal was cooked; we were then obliged to extinguish the fire and remove our supper materials to the sleeping ground, a spit of sand about a mile off; this course being necessary on account of the mosquitoes which swarm at night on the borders of the forest.

One of the sentinels was a taciturn, morose-looking, but sober and honest Indian, named Daniel; the other was a noted character of Ega, a little wiry mameluco, named Carepíra (Fish-hawk); known for his waggery, propensity for strong drink, and indebtedness to Ega traders. Both were intrepid canoemen and huntsmen, and both perfectly at home anywhere in these fearful wastes of forest and water. Carepíra had his son with him, a quiet little lad of about nine years of age. These men in a few minutes constructed a small shed with four upright poles and leaves of the arrow-grass, under which I and Cardozo slung our hammocks. We did not go to sleep, however, until after midnight: for when supper was over we lay about on the sand with a flask of rum in our midst, and whiled away the still hours in listening to Carepíra's stories.

I rose from my hammock by daylight, shivering with cold; a praia, on account of the great radiation of heat in the night from the sand, being towards the dawn the coldest place that can be found in this climate. Cardozo and the men were already up watching the turtles. The sentinels had erected for this purpose a stage about fifty feet high, on a tall tree near their station, the ascent to which was by a roughly-made ladder of woody lianas. They are enabled, by observing the turtles from this watch-tower, to ascertain the date of successive deposits of eggs, and thus guide the commandante in fixing the time for the general invitation to the Ega people. The turtles lay their eggs by night, leaving the water, when nothing disturbs them, in vast crowds, and crawling to the central and highest part of the praia. These places are, of course, the last to go under water when, in unusually wet seasons, the river rises before the eggs are hatched by the heat of the sand. One could almost believe, from this, that the animals used forethought in choosing a place; but it is simply one of those many instances in animals where unconscious habit has the same result as conscious prevision. The hours between midnight and dawn are the busiest. The turtles excavate with their broad webbed paws deep holes in the fine sand: the first comer, in each case, making a pit about three feet deep, laying its eggs (about 120 in number) and covering them with sand; the next making its deposit at the top of that of its predecessor, and so on until every pit is full. The whole body of turtles frequenting a praia does not finish laying in less than fourteen or fifteen days, even when there is no interruption. When all have done, the area (called by the

Brazilians *taboleiro*), over which they have excavated, is distinguishable from the rest of the praia only by signs of the sand having been a little disturbed.

On rising I went to join my friends. Few recollections of my Amazonian rambles are more vivid and agreeable than that of my walk over the white sea of sand on this cool morning. The sky was cloudless; the just-risen sun was hidden behind the dark mass of woods on Shimuní, but the long line of forest to the west, on Bariá, with its plumy decorations of palms, was lighted up with his yellow, horizontal rays. A faint chorus of singing birds reached the ears from across the water, and flocks of gulls and plovers were crying plaintively over the swelling banks of the praia, where their eggs lay in nests made in little hollows of the sand. Tracks of stray turtles were visible on the smooth white surface of the praia. The animals which thus wander from the main body are lawful prizes of the sentinels; they had caught in this way two before sunrise, one of which we had for dinner. In my walk I disturbed several pairs of the chocolate and drab-coloured wild-goose (Anser jubatus) which set off to run along the edge of the water. The enjoyment one feels in rambling over these free, open spaces, is no doubt enhanced by the novelty of the scene, the change being very great from the monotonous landscape of forest which everywhere else presents itself.

On arriving at the edge of the forest I mounted the sentinel's stage, just in time to see the turtles retreating to the water on the opposite side of the sand-bank, after having laid their eggs. The sight was well worth the trouble of ascending the shaky ladder. They were about a mile off, but the surface of the sands was blackened with the multitudes which were waddling towards the river; the margin of the praia was rather steep, and they all seemed to tumble head first down the declivity into the water.

I spent the morning of the 27th collecting insects in the woods of Shimuní; assisting my friend in the afternoon to beat a large pool for Tracajás, Cardozo wishing to obtain a supply for his table at home. The pool was nearly a mile long, and lay on one side of the island between the forests and the sand-bank. The sands are heaped up very curiously around the margins of these isolated sheets of water; in the present case they formed a steeply-inclined bank, from five to eight feet in height. What may be the cause of this formation I cannot imagine. The pools always contain a quantity of imprisoned fish, turtles, Tracajás, and Aiyussás.* The turtles and Aiyussás crawl out voluntarily in the course of a few days, and escape to the main river, but the Tracajas remain and become an easy prey to the natives. The ordinary mode of obtaining them is to whip the water in every part with rods for several hours during the day; this treatment having the effect of driving the animals out. They wait, however, until the night following the beating before making their exit. Our Indians were occupied for many hours in this work, and when night came they and the sentinels were placed at intervals along the edge of the water, to be ready to capture the runaways. Cardozo and I, after supper, went and took our station at one end of the pool.

We did not succeed, after all our trouble, in getting many Tracajás. This was partly owing to the intense darkness of the night, and partly, doubtless, to the sentinels having already nearly exhausted the pool, notwithstanding their declarations to the contrary. In waiting for the animals it was necessary to keep silence: not a pleasant way of passing the night; speaking only in whispers, and being without fire in a place liable to be visited by a prowling jaguar. Cardozo and I sat on a sandy slope with our loaded guns by our side, but it was so dark we could scarcely see each other. Towards midnight a storm began to gather around us. The faint wind which had breathed from over the water, since the sun went down, ceased; thick clouds piled themselves up until every star was obscured, and gleams of watery lightning began to play in the midst of the black masses. I hinted to Cardozo that I thought we had now had enough of watching, and suggested a cigarette. Just then a quick pattering movement was heard on the sands, and grasping our guns, we both started to our feet. Whatever it might have been it seemed to pass by, and a few moments afterwards a dark body appeared to be moving in another direction on the opposite slope of the sandy ravine where we lay. We prepared to fire, but luckily took the precaution of first shouting "Quem vai lá?" (Who goes there?) It turned out to be the taciturn sentinel, Daniel, who asked us mildly whether we had heard a "raposa" pass our way. The raposa is a kind of wild dog, with very long tapering muzzle, and black and white speckled hair. Daniel could distinguish all kinds of

* Specimens of this species of turtle are named in the British Museum collection, Podocnemis expansa.

animals in the dark by their footsteps. It now began to thunder, and our position was getting very uncomfortable. Daniel had not seen anything of the other Indians, and thought it was useless waiting any longer for Tracajás; we therefore sent him to call in the whole party, and made off ourselves, as quickly as we could, for the canoe. The rest of the night was passed most miserably; as indeed were very many of my nights on the Solimoens. A furious squall burst upon us; the wind blew away the cloths and mats we had fixed up at the ends of the arched awning of the canoe to shelter ourselves, and the rain beat right through our sleeping-place. There we lay, Cardozo and I, huddled together, and wet through, waiting for the morning.

A cup of strong and hot coffee put us to rights at sunrise; but the rain was still coming down, having changed to a steady

Surubim (Pimelodus tigrinus).

drizzle. Our men were all returned from the pool, having taken only four Tracajás. The business which had brought Cardozo hither being now finished, we set out to return to Ega, leaving the sentinels once more to their solitude on the sands. Our return route was by the rarely frequented north-easterly channel of the Solimoens, through which flows part of the waters of its great tributary stream, the Jupurá. We travelled for five hours along the desolate, broken, timber-strewn shore of Bariá. The channel is of immense breadth, the opposite coast being visible only as a long low line of forest. At three o'clock in the afternoon we doubled the upper end of the island, and then crossed towards the mouth of the Teffé by a broad transverse channel running between Bariá and another island called Quanarú. There is a small sand-bank at the north-westerly point of Bariá called Jacaré; we stayed here to dine, and afterwards fished with the net. A fine rain was still falling, and we had capital sport, in three hauls taking more fish than our canoe would conveniently hold. They were of two kinds only, the Surubim and the Piraepiéüa (species of Pimelodus), very handsome fishes, four feet in length, with flat spoon-shaped heads, and prettily-spotted and striped skins.

On our way from Jacaré to the mouth of the Teffé we had a little adventure with a black tiger or jaguar. We were paddling rapidly past a long beach of dried mud, when the Indians became suddenly excited, shouting "Ecuí Jauareté; Jauarí-pixúna!" (Behold the jaguar, the black jaguar!) Looking ahead we saw the animal quietly drinking at the water's edge. Cardozo ordered the steersman at once to put us ashore. By the time we were landed the tiger had seen us, and was retracing his steps towards the forest. On the spur of the moment, and without thinking of what we were doing, we took our guns (mine was a double-barrel, with one charge of B B and one of dust-shot) and gave chase. The animal increased his speed, and reaching the forest border dived into the dense mass of broad-leaved grass which formed its frontage. We peeped through the gap he had made, but, our courage being by this time cooled, we did not think it wise to go into the thicket after him. The black tiger appears to be more abundant than the spotted form of jaguar in the neighbourhood of Ega. The most certain method of finding it is to hunt, assisted by a string of Indians shouting and driving the game before them, in the narrow *restingas* or strips of dry land in the forest, which are isolated by the flooding of their neighbourhood in the wet season. We reached Ega by eight o'clock at night.

On the 6th of October we left Ega on a second excursion; the principal object of Cardozo being, this time, to search certain pools in the forest for young turtles. The exact situation of these hidden sheets of water is known only to a few practised huntsmen; we took one of these men with us from Ega, a mameluco named Pedro, and on our way called at Shimuní for Daniel to serve as an additional guide. We started from the praia at sunrise on the 7th in two canoes containing twenty-three persons, nineteen of whom were Indians. The morning was cloudy and cool, and a fresh wind blew from down river, against which we had to struggle with all the force of our paddles, aided by the current; the boats were tossed about most disagreeably, and shipped a great deal of water. On passing the lower end of Shimuní, a long reach of the river was before us, undivided by islands; a magnificent expanse of water stretching away to the south-east. The country on the left bank is not, however, terra firma, but a portion of the alluvial land which forms the extensive and complex delta region of the Japurá. It is flooded every year at the time of high water, and is traversed by many narrow and deep channels which serve as outlets to the Japurá, or at least are connected with that river by means of the interior water system of the Cupiyó. This inhospitable tract of country extends for several hundred miles, and contains in its midst an endless number of pools and lakes tenanted by multitudes of turtles, fishes, alligators, and water serpents. Our destination was a point on this coast situated about twenty miles below Shimuní, and a short distance from the mouth of the Ananá, one of the channels just alluded to as connected with the Japurá. After travelling for three hours in mid-stream we steered for the land, and brought to under a steeply-inclined bank of crumbly earth, shaped into a succession of steps or terraces, marking the various halts which the waters of the river make in the course of subsidence. The coast line was nearly straight for many miles, and the bank averaged about thirty feet in height above the present level of the river: at the top rose the unbroken hedge of forest. No one could have divined that pools of water existed on that elevated land. A narrow level space extended at the foot of the bank. On landing the first business was to get breakfast. Whilst a couple of Indian lads were employed in making the fire, roasting the fish, and boiling the coffee, the rest of the party moutned the bank, and with their long hunting knives commenced cutting a path through the forest; the pool, called the Aningal, being about half a mile distant. After breakfast a great number of short poles were cut and were laid crosswise on the path, and then three light montarias which we had brought with us were dragged up the bank by lianas, and rolled away to be embarked on the pool. A large net, seventy yards in length, was then disembarked and carried to the place. The work was done very speedily, and when Cardozo and I went to the spot at eleven o'clock we found some of the older Indians, including Pedro and Daniel, had begun their sport. They were mounted on little stages called moutás, made of poles and cross-pieces of wood secured with lianas, and were shooting the turtles as they came near the surface, with bows and arrows. The Indians seemed to think that netting the animals, as Cardozo proposed doing, was not lawful sport, and wished first to have an hour or two's old-fashioned practice with their weapons.

The pool covered an area of about four or five acres, and was closely hemmed in by the forest, which in picturesque variety and grouping of trees and foliage exceeded almost everything I had yet witnessed. The margins for some distance were swampy, and covered with large tufts of a fine grass called Matupá. These tufts in many places were overrun with ferns, and exterior to them a crowded row of arborescent arums, growing to a height of fifteen or twenty feet, formed a green palisade. Around the whole stood the taller forest trees; palmate-leaved Cecropiæ; slender Assai palms, thirty feet high, with their thin feathery heads crowning the gently-curving smooth stems; small fan-leaved palms; and as a background to all these airy shapes lay the voluminous masses of ordinary forest trees, with garlands, festoons, and streamers of leafy climbers hanging from their branches. The pool was nowhere more than five feet deep, one foot of which was not water, but extremely fine and soft mud.

Cardozo and I spent an hour paddling about. I was astonished at the skill which the Indians display in shooting turtles. They did not wait for their coming to the surface to breathe, but watched for the slight movements in the water which revealed their presence underneath. These little tracks on the water are called the Sirirí; the instant one was perceived an arrow flew from the bow of the nearest man, and never failed to pierce the shell of the submerged animal. When the turtle was very distant, of course the aim had to be taken at a considerable elevation, but the marksmen preferred a longish range, because the arrow then fell more perpendicularly on the shell, and entered it more deeply.

The arrow used in turtle shooting has a strong lancet-shaped steel point, fitted into a peg which enters the tip of the shaft. The peg is secured to the shaft by twine made of the fibres of pine-apple leaves, the twine being some thirty or forty yards in length, and neatly wound round the body of the arrow. When the missile enters the shell, the peg drops out, and the pierced animal descends with it towards the bottom, leaving the shaft floating on the surface. This being done, the sportsman paddles in his montaria to the place, and gently draws the animal by the twine, humouring it by giving it the rein when it plunges, until it is brought again near the surface, when he strikes it with a second arrow. With the increased hold given by the two cords he has then no difficulty in landing his game.

By mid-day the men had shot about a score of nearly full-grown turtles. Cardozo then gave orders to spread the net. The spongy, swampy nature of the banks made it impossible to work the net so as to draw the booty ashore; another method was therefore adopted. The net was taken by two Indians and extended in a curve at one extremity of the oval-shaped pool, holding it when they had done so by the perpendicular rods fixed at each end; its breadth was about equal to the depth of the water, its shotted side therefore rested on the bottom, whilst the floats buoyed it up on the surface, so that the whole, when the ends were brought together, would form a complete trap. The rest of the party then spread themselves around the swamp at the opposite end of the pool, and began to beat, with stout poles, the thick tufts of Matupá, in order to drive the turtles towards the middle. This was continued for an hour or more, the beaters gradually drawing nearer to each other, and driving the hosts of animals before them; the number of little snouts constantly popping above the surface of the water showing that all was going on well. When they neared the net, the men moved more quickly, shouting and beating with great vigour. The ends of the net were then seized by several strong hands and dragged suddenly forwards, bringing them at the same time together, so as to enclose all the booty in a circle. Every man now leapt into the inclosure, the boats were brought up, and the turtles easily captured by the hand and tossed into them. I jumped in along with the rest, although I had just before made the discovery that the pool abounded in ugly, red, four-angled leeches, having seen several of these delectable animals, which sometimes fasten on the legs of fishermen, although they did not, on this day, trouble us, working their way through cracks in the bottom of our montaria. Cardozo, who remained with the boats, could not turn the animals on their backs fast

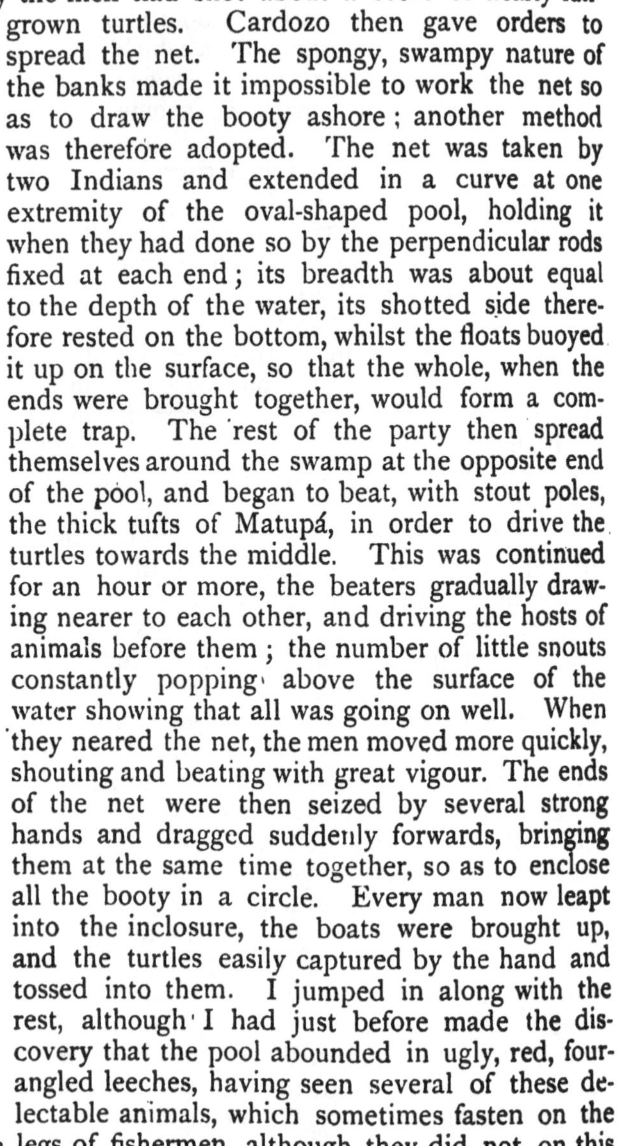

Arrow used in turtle shooting.

enough, so that a great many clambered out and got free again. However, three boat-loads, or about eighty, were secured in about twenty minutes. They were then taken ashore, and each one secured by the men tying the legs with thongs of bast.

When the canoes had been twice filled we desisted, after a very hard day's work. Nearly all the animals were young ones, chiefly, according to the statement of Pedro, from three to ten years of age: they varied from six to eighteen inches in length, and were very fat. Cardozo and I lived almost exclusively on them for several months afterwards. Roasted in the shell they form a most appetizing dish. These younger turtles never migrate with their elders on the sinking of the waters, but remain in the tepid pools, fattening on fallen fruits, and, according to the natives, on the fine nutritious mud. We captured a few full-grown mother-turtles, which were known at once by the horny skin of their breast-plates being worn, telling of their having crawled on the sands to lay eggs the previous year. They had evidently made a mistake in not leaving the pool at the proper time, for they were full of eggs, which, we were told, they would, before the season was over, scatter in despair over the swamp. We also found several male turtles, or Capitarís, as they are called by the natives. These are immensely less numerous than the females, and are distinguishable by their much smaller size, more circular shape, and the greater length and thickness of their tails. Their flesh is considered unwholesome, especially to sick people having external signs of inflammation. All diseases in these parts, as well as their remedies, and all articles of food, are classed by the inhabitants as "hot" and "cold," and the meat of the Capitarí is settled by unanimous consent as belonging to the "hot" list.

We dined on the banks of the river a little before sunset. The mosquitoes then began to be troublesome, and finding it would be impossible to sleep here, we all embarked and crossed the river to a sand-bank, about three miles distant, where we passed the night. Cardozo and I slept in our hammocks slung between upright poles, the rest stretching themselves on the sand round a large fire. We lay awake conversing until past midnight. It was a real pleasure to listen to the stories told by one of the older men; they were given with so much spirit. The tales always related to struggles with some intractable animal—jaguar, manatee, or alligator. Many interjections and expressive gestures were used, and at the end came a sudden "Pa! terra!" when the animal was vanquished by a shot or a blow. Many mysterious tales were recounted about the Bouto, as the large Dolphin of the Amazons is called. One of them was to the effect that a Bouto once had the habit of assuming the shape of a beautiful woman, with hair hanging loose to her heels, and walking ashore at night in the streets of Ega, to entice the young men down to the water. If any one was so much smitten as to follow her to the water-side, she grasped her victim round the waist and plunged beneath the waves with a triumphant cry. No animal in the Amazons region is the subject of so many fables as the Bouto; but it is probable these did not originate with the Indians, but with the Portuguese colonists. It was several years before I could induce a fisherman to harpoon Dolphins for me as specimens, for no one ever kills these animals voluntarily, although their fat is known to yield an excellent oil for lamps. The superstitious people believe that blindness would result from the use of this oil in lamps. I succeeded at length with Carepira, by offering him a high reward when his finances were at a very low point; but he repented of his deed ever afterwards, declaring that his luck had forsaken him from that day.

The next morning we again beat the pool. Although we had proof of there being a great number of turtles yet remaining, we had very poor success. The old Indians told us it would be so, for the turtles were "ladino" (cunning), and would take no notice of the beating a second day. When the net was formed into a circle, and the men had jumped in, an alligator was found to be inclosed. No one was alarmed, the only fear expressed being that the imprisoned beast would tear the net. First one shouted, "I have touched his head;" then another, "he has scratched my leg." One of the men, a lanky Miránha, was thrown off his balance, and then there was no end to the laughter and shouting. At last a youth of about fourteen years of age, on my calling to him, from the bank, to do so, seized the reptile by the tail, and held him tightly until, a little resistance being overcome, he was able to bring it ashore; the net was opened, and the boy slowly dragged the dangerous but cowardly beast to land through the muddy water, a distance of about a hundred yards. Meantime I had cut a strong pole from a tree, and as soon as the alligator was drawn to solid ground, gave him a smart rap with it on the crown of his head, which killed him instantly. It was a good-sized individual; the jaws being considerably more than a foot long, and fully capable of snapping a man's leg in twain. The

TURTLE FISHING AND ADVENTURE WITH ALLIGATOR.

species was the large cayman, the Jacaré-uassú of the Amazonian Indians (Jacare nigra).

On the third day we sent our men in the boats to net turtles in a larger pool, about five miles further down the river, and on the fourth returned to Ega.

It will be well to mention here a few circumstances relative to the large cayman, which, with the incident just narrated, afford illustrations of the cunning, cowardice, and ferocity of this reptile.

I have hitherto had but few occasions of mentioning alligators, although they exist by myriads in the waters of the Upper Amazons. Many different species are spoken of by the natives. I saw only three, and of these two only are common: one, the Jacaré-tinga, a small kind (five feet long when full-grown), having a long slender muzzle and a black-banded tail; the other, the Jacaré-uassú, to which these remarks more especially relate; and the third the Jacaré curúa, mentioned in a former chapter. The Jacaré-uassú, or large cayman, grows to a length of eighteen or twenty feet, and attains an enormous bulk. Like the turtles, the alligator has its annual migrations, for it retreats to the interior pools and flooded forests in the wet season, and descends to the main river in the dry season. During the months of high water, therefore, scarcely a single individual is to be seen in the main river. In the middle part of the Lower Amazons, about Obydos and Villa Nova, where many of the lakes with their channels of communication with the trunk stream dry up in the fine months, the alligator buries itself in the mud and becomes dormant, sleeping till the rainy season returns. On the Upper Amazons, where the dry season is never excessive, it has not this habit, but is lively all the year round. It is scarcely exaggerating to say that the waters of the Solimoens are as well stocked with large alligators in the dry season, as a ditch in England is in summer with tadpoles. During a journey of five days which I once made in the Upper Amazons steamer, in November, alligators were seen along the coast almost every step of the way, and the passengers amused themselves, from morning till night, by firing at them with rifle and ball. They were very numerous in the still bays, where the huddled crowds jostled together, to the great rattling of their coats of mail, as the steamer passed.

The natives at once despise and fear the great cayman. I once spent a month at Caiçara, a small village of semi-civilised Indians, about twenty miles to the west of Ega. My entertainer, the only white in the place, and one of my best and most constant friends, Senhor Innocencio Alves Faria, one day proposed a half-day's fishing with net in the lake,—the expanded bed of the small river on which the village is situated. We set out in an open boat with six Indians and two of Innocencio's children. The water had sunk so low that the net had to be taken out into the middle by the Indians, whence at the first draught two medium-sized alligators were brought to land. They were disengaged from the net and allowed, with the coolest unconcern, to return to the water, although the two children were playing in it not many yards off. We continued fishing, Innocencio and I lending a helping hand, and each time drew a number of the reptiles of different ages and sizes, some of them Jacaré-tingas; the lake, in fact, swarmed with alligators. After taking a very large quantity of fish we prepared to return, and the Indians, at my suggestion, secured one of the alligators with the view of letting it loose amongst the swarms of dogs in the village. An individual was selected about eight feet long: one man holding his head and another his tail, whilst a third took a few lengths of a flexible liana, and deliberately bound the jaws and the legs. Thus secured, the beast was laid across the benches of the boat, on which we sat during the hour and a half's journey to the settlement. We were rather crowded, but our amiable passenger gave us no trouble during the transit. On reaching the village we took the animal into the middle of the green, in front of the church, where the dogs were congregated, and there gave him his liberty, two of us arming ourselves with long poles to intercept him if he should make for the water, and the others exciting the dogs. The alligator showed great terror, although the dogs could not be made to advance, and made off at the top of its speed for the water, waddling like a duck. We tried to keep him back with the poles, but he became enraged, and seizing the end of the one I held, in his jaws, nearly wrenched it from my grasp. We were obliged, at length, to kill him to prevent his escape.

These little incidents show the timidity or cowardice of the alligator. He never attacks man when his intended victim is on his guard; but he is cunning enough to know when this may be done with impunity: of this we had proof at Caiçara, a few days afterwards. The river had sunk to a very low point, so that the port and bathing-place of the village now lay at the foot of a long sloping bank, and a large cayman made his appearance in the shallow and muddy water. We were all obliged to be very careful in taking our bath; most of the people simply using a calabash, pouring the water over themselves while standing on the brink. A large trading canoe, belonging to a Barra merchant named Soares, arrived at this time, and the Indian crew, as usual, spent the first day or two after their coming into port, in drunkenness and debauchery ashore. One of the men, during the greatest heat of the day, when almost every one was enjoying his afternoon's nap, took it into his head whilst in a tipsy state to go down alone to bathe. He was seen only by the Juiz de Paz, a feeble old man who was lying in his hammock, in the open verandah at the rear of his house on the top of the bank, and who shouted to the besotted Indian to beware of the alligator. Before he could repeat his warning the man stumbled, and a pair of gaping jaws, appearing suddenly above the surface, seized him round the waist and drew him under the water. A cry of agony, "Ai Jesús!" was the last sign made by the wretched victim. The village was aroused: the young men with praiseworthy readiness seized their harpoons and hurried down to the bank: but of course it was too late; a winding track of blood on the surface of the water was all that could be seen. They embarked, however, in montarias, determined on vengeance: the monster was traced, and when, after a short lapse of time, he came up to breathe—one leg of the man sticking out from his jaws—was dispatched with bitter curses.

The last of these minor excursions which I shall narrate, was made (again in company of Senhor Cardozo, with the addition of his housekeeper, Senhora Felippa) in the season when all the population of the villages turns out to dig up turtle eggs, and revel on the praias. Placards were posted on the church doors at Ega, announcing that the excavation on Shimuní would commence on the 17th of October, and on Catuá, sixty miles below Shimuní, on the 25th. We set out on the 16th, and passed on the road, in our well-manned igarité, a large number of people, men, women, and children, in canoes of all sizes, wending their way as if to a great holiday gathering. By the morning of the 17th some 400 persons were assembled on the borders of the sand-bank; each family having erected a rude temporary shed of poles and palm leaves to protect themselves from the sun and rain. Large copper kettles to prepare the oil, and hundreds of red earthenware jars, were scattered about on the sand.

The excavation of the taboleiro, collecting the eggs, and purifying the oil, occupied four days. All was done on a system established by the old Portuguese governors, probably more than a century ago. The commandante first took down the names of all the masters of households, with the number of persons each intended to employ in digging; he then exacted a payment of 140 reis (about fourpence) a head towards defraying the expense of sentinels. The whole were then allowed to go to the taboleiro. They ranged themselves round the circle, each person armed with a paddle, to be used as a spade, and then all began simultaneously to dig on a signal being given—the roll of drums—by order of the commandante. It was an animating sight to behold the wide circle of rival diggers throwing up clouds of sand in their energetic labours, and working gradually towards the centre of the ring. A little rest was taken during the great heat of mid-day, and in the evening the eggs were carried to the huts in baskets. By the end of the second day the taboleiro was exhausted: large mounds of eggs, some of them four to five feet in height, were then seen by the side of each hut, the produce of the labours of the family.

In the hurry of digging some of the deeper nests are passed over; to find these out the people go about provided with a long steel or wooden probe, the presence of the eggs being discoverable by the ease with which the spit enters the sand. When no more eggs are to be found, the mashing process begins. The egg, it may be mentioned, has a flexible or leathery shell; it is quite round, and somewhat larger than a hen's egg. The whole heap is thrown into an empty canoe and mashed with wooden prongs; but sometimes naked Indians and children jump into the mass and tread it down, besmearing themselves with yolk and making about as filthy a scene as can well be imagined. This being finished, water is poured into the canoe, and the fatty mess then left for a few hours to be heated by the sun, on which the oil separates and rises to the surface. The floating oil is afterwards skimmed off with long spoons, made by tying large mussel-shells to the end of rods, and purified over the fire in copper kettles.

The destruction of turtle eggs every year by these proceedings is enormous. At least 6,000 jars, holding each three gallons of the oil, are exported annually from the Upper Amazons and the Madeira to Para, where it is used for lighting, frying fish, and other purposes. It may be fairly estimated that 2,000 more jarfuls are consumed by the inhabitants of the villages on the river. Now, it takes at least twelve basketfuls of eggs, or about 6,000, by the wasteful process followed, to make one jar of oil. The total number of eggs annually destroyed amounts, therefore, to 48,000,000. As each turtle lays about 120, it follows that the yearly offspring of 400,000 turtles is thus annihilated. A vast number, nevertheless, remain undetected; and these would probably be sufficient to keep the turtle population of these rivers up to the mark, if the people did not follow the wasteful practice of lying in wait for the newly-hatched young, and collecting them by thousands for eating; their tender flesh and the remains of yolk in their entrails being considered a great delicacy. The chief natural enemies of the turtle are vultures and alligators, which devour the newly-hatched young as they descend in shoals to the water. These must have destroyed an immensely greater number before the European settlers began to appropriate the eggs than they do now. It is almost doubtful if this natural persecution did not act as effectively in checking the increase of the turtle as the artificial destruction now does. If we are to believe the tradition of the Indians, however, it had not this result; for they say that formerly the waters teemed as thickly with turtles as the air does now with mosquitoes. The universal opinion of the settlers on the Upper Amazons is, that the turtle has very greatly decreased in numbers, and is still annually decreasing.

We left Shimuní on the 20th, with quite a flotilla of canoes, and descended the river to Catuá, an eleven hours' journey by paddle and current. Catuá is about six miles long, and almost entirely encircled by its praia. The turtles had selected for their egg-laying a part of the sand-bank which was elevated at least twenty feet above the present level of the river; the animals, to reach the place, must have crawled up a slope. As we approached the island, numbers of the animals were seen coming to the surface to breathe, in a small shoaly bay. Those who had light montarias sped forward with bows and arrows to shoot them. Carepíra was foremost: having borrowed a small and very unsteady boat of Cardozo, and embarked in it with his little son. After bagging a couple of turtles, and whilst hauling in a third, he overbalanced himself; the canoe went over, and he with his child had to swim for their lives, in the midst of numerous alligators, about a mile from the land. The old man had to sustain a heavy fire of jokes from his companions for several days after this mishap. Such accidents are only laughed at by these almost amphibious people.

The number of persons congregated on Catuá was much greater than on Shimuní, as the population of the banks of several neighbouring lakes was here added. The line of huts and sheds extended half a mile, and several large sailing vessels were anchored at the place. The commandante was Senhor Macedo, the very worthy Indian blacksmith of Ega, who maintained excellent order during the fourteen days the process of excavation and oil manufacture lasted. There were also many primitive Indians here from the neighbouring rivers, amongst them a family of Shumánas, good-tempered, harmless people from the Lower Japurá. All of them were tattooed round the mouth, the bluish tint forming a border to the lips, and extending in a line on the cheeks towards the ear on each side. They were not quite so slender in figure as the Passés of Pedro-uassú's family; but their features deviated quite as much as those of the Passés from the ordinary Indian type. This was seen chiefly in the comparatively small mouth, pointed chin, thin lips, and narrow, high nose. One of the daughters, a young girl of about seventeen years of age, was a real beauty. The colour of her skin approached the light tanned shade of the Mameluco women; her figure was almost faultless, and the blue mouth, instead of being a disfigurement, gave quite a captivating finish to her appearance. Her neck, wrists, and ankles were adorned with strings of blue beads. She was, however, extremely bashful, never venturing to look strangers in the face, and never quitting, for many minutes together, the side of her father and mother. The family had been shamefully swindled by some rascally trader on another praia; and, on our arrival, came to lay their case before Senhor Cardozo, as the delegado of police of the district. The mild way in which the old man, without a trace of anger, stated his complaint in imperfect Tupí, quite enlisted our sympathies in his favour. But Cardozo could give him no redress; he invited the family, however, to make their rancho near to ours, and in the end gave them the highest price for the surplus oil which they manufactured.

It was not all work at Catuá; indeed there was rather more play than work going on. The people make a kind of holiday of these occasions. Every fine night parties of the younger people assembled on the sands, and dancing and games were carried on for hours together. But the requisite liveliness for these sports was never got up without a good deal of preliminary rum-drinking. The girls were so coy that the young men could not get sufficient partners for the dances, without first subscribing for a few flagons of the needful cashaça. The coldness of the shy Indian and Mameluco maidens never failed to give way after a little of this strong drink, but it was astonishing what an immense deal they could take of it in the course of an evening. Coyness is not always a sign of innocence in these people, for most of the half-caste women on the Upper Amazons lead a little career of looseness before they marry and settle down for life; and it is rather remarkable that the men do not seem to object much to their brides having had a child or two by various fathers before marriage. The women do not lose reputation unless they become utterly depraved, but in that case they are condemned pretty strongly by public opinion. Depravity is, however, rare, for all require more or less to be wooed before they are won. I did not see (although I mixed pretty freely with the young people) any breach of propriety on the praias. The merry-makings were carried on near the ranchos, where the more staid citizens of Ega, husbands with their wives and young daughters, all smoking gravely out of long pipes, sat in their hammocks and enjoyed the fun. Towards midnight we often heard, in the intervals between jokes and laughter, the hoarse roar of jaguars prowling about the jungle in the middle of the praia. There were several guitar players amongst the young men, and one most persevering fiddler, so there was no lack of music.

The favourite sport was the Pira-purasséya, or fish-dance, one of the original games of the Indians, though now probably a little modified. The young men and women, mingling together, formed a ring, leaving one of their number in the middle, who represented the fish. They then all marched round, Indian file, the musicians mixed up with the rest, singing a monotonous but rather pretty chorus, the words of which were invented (under a certain form) by one of the party who acted as leader. This finished, all joined hands, and questions were put to the one in the middle, asking what kind of fish he or she might be. To these the individual has to reply. The end of it all is that he makes a rush at the ring, and if he succeeds in escaping, the person who allowed him to do so has to take his place; the march and chorus then recommence, and so the game goes on hour after hour. Tupí was the language mostly used, but sometimes Portuguese was sung and spoken.

The details of the dance were often varied. Instead of the names of fishes being called over by the person in the middle, the name of some animal, flower, or other object was given to every fresh occupier of the place. There was then good scope for wit in the invention of nicknames, and peals of laughter would often salute some particularly good hit. Thus a very lanky young man was called the Magoary, or the gray stork; a moist, gray-eyed man, with a profile comically suggestive of a fish, was christened Jarakí (a kind of fish), which was considered quite a witty sally; a little Mameluco girl, with light-coloured eyes and brown hair, got the gallant name of Rosa branca, or the white rose; a young fellow who had recently singed his eyebrows by the explosion of fireworks was dubbed Pedro queimado (burnt Peter); in short, every one got a nickname, and each time the cognomen was introduced into the chorus as the circle marched round.

Our rancho was a large one, and was erected in a line with the others, near the edge of the sand-bank which sloped rather abruptly to the water. During the first week the people were all, more or less, troubled by alligators. Some half-dozen full-grown ones were in attendance off the praia, floating about on the lazily-flowing muddy water. The dryness of the weather had increased since we had left Shimuní, the currents had slackened, and the heat in the middle part of the day was almost insupportable. But no one could descend to bathe without being advanced upon by one or other of these hungry monsters. There was much offal cast into the river, and this of course attracted them to the place. One day I amused myself by taking a basketful of fragments of meat beyond the line of ranchos, and drawing the alligators towards me by feeding them. They behaved pretty much as dogs do when fed; catching the bones I threw them in their huge jaws, and coming nearer and showing increased eagerness after every morsel. The enormous gape of their mouths, with their blood-red lining and long fringes of teeth, and the uncouth shapes of their bodies, made a picture of unsurpassable ugliness. I once or twice fired a heavy charge of shot at them, aiming at the vulnerable part of their bodies, which is a small space situated behind the eyes, but this had no other effect than to make them give a hoarse grunt and shake themselves; they immediately afterwards turned to receive another bone which I threw to them.

Every day these visitors became bolder; at length they reached a pitch of impudence that was quite intolerable. Cardozo had

NIGHT ADVENTURE WITH ALLIGATOR.

a poodle dog named Carlito, which some grateful traveller whom he had befriended had sent him from Rio Janeiro. He took great pride in this dog, keeping it well sheared, and preserving his coat as white as soap and water could make it. We slept in our rancho in hammocks slung between the outer posts; a large wood fire (fed with a kind of wood abundant on the banks of the river, which keeps alight all night) being made in the middle, by the side of which slept Carlito on a little mat. Well, one night I was awoke by a great uproar. It was caused by Cardozo hurling burning firewood with loud curses at a huge cayman which had crawled up the bank and passed beneath my hammock (being nearest the water) towards the place where Carlito lay. The dog had raised the alarm in time; the reptile backed out and tumbled down the bank to the water, the sparks from the brands hurled at him flying from his bony hide. To our great surprise the animal (we supposed it to be the same individual) repeated his visit the very next night, this time passing round to the other side of our shed. Cardozo was awake, and threw a harpoon at him, but without doing him any harm. After this it was thought necessary to make an effort to check the alligators; a number of men were therefore persuaded to sally forth in their montarias and devote a day to killing them.

The young men made several hunting excursions during the fourteen days of our stay on Catuá, and I, being associated with them in all their pleasures, made generally one of the party. These were, besides, the sole occasions on which I could add to my collections, whilst on these barren sands. Only two of these trips afforded incidents worth relating.

The first, which was made to the interior of the wooded island of Catuá, was not a very successful one. We were twelve in number, all armed with guns and long hunting knives. Long before sunrise my friends woke me up from my hammock, where I lay, as usual, in the clothes worn during the day; and after taking each a cupful of cashaça and ginger (a very general practice in early morning on the sand-banks), we commenced our walk. The waning moon still lingered in the clear sky, and a profound stillness pervaded sleeping camp, forest, and stream. Along the line of ranchos glimmered the fires made by each party to dry turtle-eggs for food, the eggs being spread on little wooden stages over the smoke. The distance to the forest from our place of starting was about two miles, being nearly the whole length of the sand-bank, which was also a very broad one; the highest part, where it was covered with a thicket of dwarf willows, mimosas, and arrow grass, lying near the ranchos. We loitered much on the way, and the day dawned whilst we were yet on the road; the sand at this early hour feeling quite cold to the naked feet. As soon as we were able to distinguish things, the surface of the praia was seen to be dotted with small black objects. These were newly-hatched Aiyussá turtles, which were making their way in an undeviating line to the water, at least a mile distant. The young animal of this species is distinguishable from that of the large turtle and the Tracajá by the edges of the breast-plate being raised on each side, so that in crawling it scores two parallel lines on the sand. The mouths of these little creatures were full of sand, a circumstance arising from their having to bite their way through many inches of superincumbent sand, to reach the surface on emerging from the buried eggs. It was amusing to observe how constantly they turned again in the direction of the distant river, after being handled and set down on the sand with their heads facing the opposite quarter. We saw also several skeletons of the large cayman (some with the horny and bony hide of the animal nearly perfect) embedded in the sand: they reminded me of the remains of Ichthyosauri fossilized in beds of lias, with the difference of being buried in fine sand instead of in blue mud. I marked the place of one which had a well-preserved skull, and the next day returned to secure it. The specimen is now in the British Museum collection. There were also many footmarks of jaguars on the sand.

We entered the forest as the sun peeped over the tree-tops far away down river. The party soon after divided; I keeping with a section which was led by Bento, the Ega carpenter, a capital woodsman. After a short walk we struck the banks of a beautiful little lake, having grassy margins and clear dark water, on the surface of which floated thick beds of water-lilies. We then crossed a muddy creek or watercourse that entered the lake, and then found ourselves on a *restinga*, or tongue of land between two waters. By keeping in sight of one or the other of these there was no danger of our losing our way; all other precautions were therefore unnecessary. The forest was tolerably clear of underwood, and consequently easy to walk through. We had not gone far before a soft, long-drawn whistle was heard aloft in the trees, betraying the presence of Mútums (Curassow birds). The crowns of the trees, a hundred feet or more over our heads, were so closely interwoven that it was difficult to distinguish the birds; the practised eye of Bento, however, made them out, and a fine male was shot from the flock; the rest flying away and alighting at no great distance; the species was the one of which the male has a round red ball on its beak (Crax globicera). The pursuit of the others led us a great distance, straight towards the interior of the island, in which direction we marched for three hours, having the lake always on our right.

Arriving at length at the head of the lake, Bento struck off to the left across the restinga, and we then soon came upon a treeless space choked up with tall grass, which appeared to be the dried-up bed of another lake. Our leader was obliged to climb a tree to ascertain our position, and found that the clear space was part of the creek, whose mouth we had crossed lower down. The banks were clothed with low trees, nearly all of one species, a kind of araça (Psidium), and the ground was carpeted with a slender delicate grass, now in flower. A great number of crimson and vermilion coloured butterflies (Catagramma Peristera, male and female) were settled on the smooth white trunks of these trees. I had also here the great pleasure of seeing for the first time the rare and curious Umbrella Bird (Cephalopterus ornatus), a species which resembles in size, colour, and appearance our common crow, but is decorated with a crest of long curved hairy feathers having long bare quills, which, when raised, spread themselves out in the form of a fringed sun-shade over the head. A strange ornament, like a pelerine, is also suspended from the neck, formed by a thick pad of glossy steel-blue feathers, which grow on a long fleshy lobe or excrescence. This lobe is connected (as I found on skinning specimens) with an unusual development of the trachea and vocal organs, to which the bird doubtless owes its singularly deep, loud, and long-sustained fluty note. The Indian name of this strange creature is Uirá-mimbéu, or fife-bird,* in allusion to the tone of its voice. We had the good luck, after remaining quiet a short time, to hear its performance. It drew itself up on its perch, spread widely the umbrella-formed crest, dilated and waved its glossy breast-lappet, and then, in giving vent to its loud piping note, bowed its head slowly forwards. We obtained a pair, male and female: the female has only the rudiments of the crest and lappet, and is duller coloured alto-

* Mimbéu is the Indian name for a rude kind of pan-pipes used by the Caishánas and other tribes.

gether than the male. The range of this bird appears to be quite confined to the plains of the Upper Amazons (especially the Ygapó forests), not having been found to the east of the Rio Negro.

Bento and our other friends being disappointed in finding no more Curassows, or indeed any other species of game, now resolved to turn back. On reaching the edge of the forest we

Umbrella Bird.

sat down and ate our dinners under the shade; each man having brought a little bag containing a few handfuls of farinha, and a piece of fried fish or roast turtle. We expected our companions of the other division to join us at mid-day, but after waiting till past one o'clock without seeing anything of them (in fact, they had returned to the huts an hour or two previously), we struck off across the praia towards the encampment. An obstacle here presented itself on which we had not counted. The sun had shone all day through a cloudless sky untempered by a breath of wind, and the sands had become heated by it to a degree that rendered walking over them with our bare feet impossible. The most hardened footsoles of the party could not endure the burning soil. We made several attempts; we tried running: wrapped the cool leaves of Heliconiæ round our feet, but in no way could we step forward many yards. There was no means of getting back to our friends before night, except going round the praia, a circuit of about four miles, and walking through the water or on the moist sand. To get to the water-side from the place where we then stood was not difficult, as a thick bed of a flowering shrub, called tintarána, an infusion of the leaves of which is used to dye black, lay on that side of the sand-bank. Footsore and wearied, burthened with our guns, and walking for miles through the tepid shallow water under the brain-scorching vertical sun, we had, as may be imagined, anything but a pleasant time of it. I did not, however, feel any inconvenience afterwards. Every one enjoys the most lusty health whilst living this free and wild life on the rivers.

The other hunting trip which I have alluded so was undertaken in company with three friendly young half-castes. Two of them were brothers, namely, Joaõ (John) and Zephyrino Jabutí: Jabutí, or tortoise, being a nickname which their father had earned for his slow gait, and which, as is usual in this country, had descended as the surname of the family. The other was José Frazaõ, a nephew of Senhor Chrysostomo, of Ega, an active, clever, and manly young fellow, whom I much esteemed. He was almost a white, his father being a Portuguese and his mother a Mameluco. We were accompanied by an Indian named Lino, and a Mulatto boy, whose office was to carry our game.

Our proposed hunting-ground on this occasion lay across the water, about fifteen miles distant. We set out in a small montaria, at four o'clock in the morning, again leaving the encampment asleep, and travelled at a good pace up the northern channel of the Solimoens, or that lying between the island of Catuá and the left bank of the river. The northern shore of the island had a broad sandy beach reaching to its western extremity. We gained our destination a little after daybreak;

this was the banks of the Carapanatúba,* a channel some 150 yards in width, which, like the Ananá already mentioned, communicates with the Cupiyó. To reach this we had to cross the river, here nearly two miles wide. Just as the day dawned we saw a cayman seize a large fish, a Tambakí, near the surface; the reptile seemed to have a difficulty in securing its prey, for it reared itself above the water, tossing the fish in its jaws, and making a tremendous commotion. I was much struck also by the singular appearance presented by certain diving birds having very long and snaky necks (the Plotus Anhinga). Occasionally a long serpentine form would suddenly wriggle itself to a height of a foot and a half above the glassy surface of the water, producing such a deceptive imitation of a snake that at first I had some difficulty in believing it to be the neck of a bird; it did not remain long in view, but soon plunged again beneath the stream.

We ran ashore in a most lonely and gloomy place, on a low sand-bank covered with bushes, secured the montaria to a tree, and then, after making a very sparing breakfast on fried fish and mandioca meal, rolled up our trousers and plunged into the thick forest, which here, as everywhere else, rose like a lofty wall of foliage from the narrow strip of beach. We made straight for the heart of the land, John Jabutí leading, and breaking off at every few steps a branch of the lower trees, so that we might recognise the path on our return. The district was quite new to all my companions, and being on a coast almost totally uninhabited by human beings for some 300 miles, to lose our way would have been to perish helplessly. I did not think at the time of the risk we ran of having our canoe stolen by passing Indians; unguarded montarias being never safe even in the ports of the villages, Indians apparently considering them common property, and stealing them without any compunction. No misgivings clouded the lightness of heart with which we trod forwards in warm anticipation of a good day's sport.

The tract of forest through which we passed was Ygapó, but the higher parts of the land formed areas which went only a very few inches under water in the flood season. It consisted of a most bewildering diversity of grand and beautiful trees, draped, festooned, corded, matted, and ribboned with climbing

* Meaning in Tupi, the river of many mosquitoes: from carapaná, mosquito, and itúba, many.

plants, woody and succulent, in endless variety. The most prevalent palm was the tall Astryocaryum Jauarí, whose fallen spines made it necessary to pick our way carefully over the ground, as we were all barefoot. There was not much green underwood, except in places where bamboos grew; these formed impenetrable thickets of plumy foliage and thorny jointed stems, which always compelled us to make a circuit to avoid them. The earth elsewhere was encumbered with rotting fruits, gigantic bean-pods, leaves, limbs, and trunks of trees; fixing the impression of its being the cemetery as wel as the birthplace of the great world of vegetation overhead. Some of the trees were of prodigious height. We passed many specimens of the Moratinga, whose cylindrical trunks, I dare not say how many feet in circumference, towered up and were lost amidst the crowns of the lower trees, their lower branches, in some cases, being hidden from our view. Another very large and remarkable tree was the Assacú (Sapium aucuparium). A traveller on the Amazons, mingling with the people, is sure to hear much of the poisonous qualities of the juices of this tree. Its bark exudes, when hacked with a knife, a milky sap, which is not only a fatal poison when taken internally, but is said to cause incurable sores if simply sprinkled on the skin. My companions always gave the Assacú a wide berth when we passed one. The tree looks ugly enough to merit a bad name, for the bark is of a dingy olive colour, and is studded with short and sharp venomous-looking spines.

After walking about half a mile we came upon a dry water-course, where we observed, first, the old footmarks of a tapir, and, soon after, on the margin of a curious circular hole full of muddy water, the fresh tracks of a jaguar. This latter discovery was hardly made when a rush was heard amidst the bushes on the top of a sloping bank on the opposite side of the dried creek. We bounded forward; it was, however, too late, for the animal had sped in a few minutes far out of our reach. It was clear we had disturbed, on our approach, the jaguar, whilst quenching his thirst at the water-hole. A few steps further on we saw the mangled remains of an alligator (the Jacarétinga). The head, fore-quarters, and bony shell were the only parts which remained; but the meat was quite fresh, and there were many footmarks of the jaguar around the carcase; so that there was no doubt this had formed the solid part of the animal's breakfast. My companions now began to search for the alligator's nest, the presence of the reptile so far from the river being accountable for on no other ground than its maternal solicitude for its eggs. We found, in fact, the nest at the distance of a few yards from the place. It was a conical pile of dead leaves, in the middle of which twenty eggs were buried. These were of elliptical shape, considerably larger than those of a duck, and having a hard shell of the texture of porcelain, but very rough on the outside. They make a loud sound when rubbed together, and it is said that it is easy to find a mother alligator in the Ygapó forests by rubbing together two eggs in this way, she being never far off, and attracted by the sounds.

I put half a dozen of the alligator's eggs in my game-bag for specimens, and we then continued on our way. Lino, who was now first, presently made a start backwards, calling out "Jararáca!" This is the name of a poisonous snake (genus Craspedocephalus), which is far more dreaded by the natives than jaguar or alligator. The individual seen by Lino lay coiled up at the foot of a tree, and was scarcely distinguishable, on account of the colours of its body being assimilated to those of the fallen leaves. Its hideous flat triangular head, connected with the body by a thin neck, was reared and turned towards us: Frazaõ killed it with a charge of shot, shattering it completely, and destroying, to my regret, its value as a specimen. In conversing on the subject of Jararácas as we walked onwards, every one of the party was ready to swear that this snake attacks man without provocation, leaping towards him from a considerable distance when he approaches. I met, in the course of my daily rambles in the woods, many Jararácas, and once or twice very narrowly escaped treading on them, but never saw them attempt to spring. On some subjects the testimony of the natives of a wild country is utterly worthless. The bite of the Jararácas is generally fatal. I knew of four or five instances of death from it, and only of one clear case of recovery after being bitten; but in that case the person was lamed for life.

We walked over moderately elevated and dry ground for about a mile, and then descended (three or four feet only) to the dry bed of another creek. This was pierced in the same way as the former watercourse, with round holes full of muddy water. They occurred at intervals of a few yards, and had the appearance of having been made by the hand of man. The smallest were about two feet, the largest seven or eight feet in diameter. As we approached the most extensive of the larger ones, I was startled at seeing a number of large serpent-like heads bobbing above the surface. They proved to be those of electric eels, and it now occurred to me that the round holes were made by these animals working constantly round and round in the moist muddy soil. Their depth (some of them were at least eight feet deep) was doubtless due also to the movements of the eels in the soft soil, and accounted for their not drying up, in the fine season, with the rest of the creek. Thus, whilst alligators and turtles in this great inundated forest region retire to the larger pools during the dry season, the electric eels make for themselves little ponds in which to pass the season of drought.

My companions now cut each a stout pole, and proceeded to eject the eels in order to get at the other fishes, with which they had discovered the ponds to abound. I amused them all very much by showing how the electric shock from the eels could pass from one person to another. We joined hands in a line whilst I touched the biggest and freshest of the animals on the head with the point of my hunting-knife. We found that this experiment did not succeed more than three times with the same eel when out of the water; for the fourth time the shock was scarcely perceptible. All the fishes found in the holes (besides the eels) belonged to one species, a small kind of Acarí, or Loricaria, a group whose members have a complete bony integument. Lino and the boy strung them together through the gills with slender sipós, and hung them on the trees to await our return later in the day.

Leaving the bed of the creek, we marched onwards, always towards the centre of the land; guided by the sun, which now glimmered through the thick foliage overhead. About eleven o'clock we saw a break in the forest before us, and presently emerged on the banks of a rather large sheet of water. This was one of the interior pools of which there are so many in this district. The margins were elevated some few feet, and sloped down to the water, the ground being hard and dry to the water's edge, and covered with shrubby vegetation. We passed completely round this pool, finding the crowns of the trees on its borders tenanted by curassow birds, whose presence was betrayed as usual by the peculiar note which they emit. My companions shot two of them. At the further end of the lake lay a deep watercourse, which we traced for about half a mile, and found to communicate with another and smaller pool. This second one evidently swarmed with turtles, as we saw the snouts of many peering above the surface of the water: the same had not been seen in the larger lake, probably because we had made too much noise in hailing our discovery, on approaching its banks. My friends made an arrangement on the spot for returning to this pool, after the termination of the egg harvest on Catuá.

In recrossing the space between the two pools we heard the crash of monkeys in the crowns of trees overhead. The chase of these occupied us a considerable time. José fired at length at one of the laggards of the troop, and wounded him. He climbed pretty nimbly towards a denser part of the tree, and a second and a third discharge failed to bring him down. The poor maimed creature then trailed his limbs to one of the topmost branches, where we descried him soon after, seated and picking the entrails from a wound in his abdomen; a most heartrending sight. The height from the ground to the bough on which he was perched could not have been less than 150 feet, and we could get a glimpse of him only by standing directly underneath, and straining our eyes upwards. We killed him at last by loading our best gun with a careful charge, and resting the barrel against the tree trunk to steady the aim. A few shots entered his chin, and he then fell heels over head screaming to the ground. Although it was I who gave the final shot, this animal did not fall to my lot in dividing the spoils at the end of the day. I regret now not having preserved the skin, as it belonged to a very large species of Cebus, and one which I never met with afterwards.

It was about one o'clock in the afternoon when we again reached the spot where we had first struck the banks of the larger pool. We had hitherto had but poor sport, so after dining on the remains of our fried fish and farinha, and smoking our cigarettes, the apparatus for making which, including bamboo tinder-box and steel and flint for striking a light, being carried by every one always on these expeditions, we made off in another (westerly) direction through the forest to try to find better hunting-ground. We quenched our thirst with water from the pool, which I was rather surprised to find quite pure. These pools are, of course, sometimes fouled for a time by the movements of alligators and other tenants in the fine mud which settles at the bottom, but I never observed a scum of confervæ or traces of oil revealing animal decomposition on the surface of these waters, nor was there ever any foul smell perceptible. The whole of this level land, instead of being covered with unwholesome swamps emitting malaria, forms in the dry season (and in the wet also) a most healthy country. How elaborate must be the natural processes of self-purification in these teeming waters!

On our fresh route we were obliged to cut our way through a long belt of bamboo underwood, and not being so careful of my steps as my companions, I trod repeatedly on the flinty thorns which had fallen from the bushes, finishing by becoming completely lame, one thorn having entered deeply the sole of my foot. I was obliged to be left behind; Lino, the Indian, remaining with me. The careful fellow cleaned my wounds with his saliva, placed pieces of isca (the felt-like substance manufactured by ants) on them to staunch the blood, and bound my feet with tough bast to serve as shoes, which he cut from the bark of a Mongúba tree. He went about his work in a very gentle way and with much skill, but was so sparing of speech that I could scarcely get answers to the questions I put to him. When he had done, I was able to limp about pretty nimbly. An Indian, when he performs a service of this kind, never thinks of a reward. I did not find so much disinterestedness in negro slaves or half-castes. We had to wait two hours for the return of our companions; during part of this time I was left quite alone, Lino having started off into the jungle after a peccary (a kind of wild hog) which had come near to where we sat, but on seeing us had given a grunt and bounded off into the thickets. At length our friends hove in sight, loaded with game, having shot twelve curassows and two cujubíms (Penelope Pipile), a handsome black fowl with a white head, which is arboreal in its habits, like the rest of this group of gallinaceous birds inhabiting the South American forests. They had discovered a third pool containing plenty of turtles. Lino rejoined us at the same time, having missed the peccary, but in compensation shot a quandú, or porcupine. The mulatto boy had caught alive in the pool a most charming little water-fowl, a species of grebe. It was somewhat smaller than a pigeon, and had a pointed beak; its feet were furnished with many intricate folds or frills of skin instead of webs, and resembled very much those of the gecko lizards. The bird was kept as a pet in Jabuti's house at Ega for a long time afterwards, where it became accustomed to swim about in a common hand-basin full of water, and was a great favourite with everybody.

We now retraced our steps towards the water-side, a weary

walk of five or six miles, reaching our canoe by half-past five o'clock, or a little before sunset. It was considered by every one at Catuá that we had had an unusually good day's sport. I never knew any small party to take so much game in one day in these forests, over which animals are everywhere so widely and sparingly scattered. My companions were greatly elated, and on approaching the encampment at Catuá made a great commotion with their paddles to announce their successful return, singing in their loudest key one of the wild choruses of the Amazonian boatmen.

The excavation of eggs and preparation of the oil being finished, we left Catuá on the 3rd of November. Carepíra, who was now attached to Cardozo's party, had discovered another lake rich in turtles, about twelve miles distant, in one of his fishing rambles, and my friend resolved, before returning to Ega, to go there with his nets and drag it as we had formerly done the Aningal. Several mameluco families of Ega begged to accompany us to share the labours and booty; the Shumána family also joined the party; we therefore formed a large body, numbering in all eight canoes and fifty persons.

The summer season was now breaking up: the river was rising; the sky was almost constantly clouded, and we had frequent rains. The mosquitoes also, which we had not felt whilst encamped on the sand-banks, now became troublesome. We paddled up the north-westerly channel, and arrived at a point near the upper end of Catuá at ten o'clock p.m. There was here a very broad beach of untrodden white sand, which extended quite into the forest, where it formed rounded hills and hollows like sand dunes, covered with a peculiar vegetation: harsh, reedy grasses, and low trees matted together with lianas, and varied with dwarf spiny palms of the genus Bactris. We encamped for the night on the sands, finding the place luckily free from mosquitoes. The different portions of the party made arched coverings with the toldos or maranta-leaf awnings of their canoes to sleep under, fixing the edges in the sand. No one, however, seemed inclined to go to sleep, so after supper we all sat or lay around the large fires and amused ourselves. We had the fiddler with us, and in the intervals between the wretched tunes which he played, the usual amusement of story-telling beguiled the time: tales of hair-breadth escapes from jaguar, alligator, and so forth. There were amongst us a father and son who had been the actors, the previous year, in an alligator adventure on the edge of the praia we had just left. The son, whilst bathing, was seized by the thigh and carried under water; a cry was raised, and the father, rushing down the bank, plunged after the rapacious beast, which was diving away with his victim. It seems almost incredible that a man could overtake and master the large cayman in his own element; but such was the case in this instance, for the animal was reached and forced to release his booty by the man's thrusting his thumb into his eye. The lad showed us the marks of the alligator's teeth on his thigh. We sat up until past midnight listening to these stories and assisting the flow of talk by frequent potations of burnt rum. A large shallow dish was filled with the liquor and fired; when it had burnt for a few minutes, the flame was extinguished and each one helped himself by dipping a tea-cup into the vessel.

One by one the people dropped asleep, and then the quiet murmur of talk of the few who remained awake was interrupted by the roar of jaguars in the jungle about a furlong distant. There was not one only, but several of the animals. The older men showed considerable alarm, and proceeded to light fresh fires around the outside of our encampment. I had read in books of travels of tigers coming to warm themselves by the fires of a bivouac, and thought my strong wish to witness the same sight would have been gratified to-night. I had not, however, such good fortune, although I was the last to go to sleep, and my bed was the bare sand under a little arched covering open at both ends. The jaguars nevertheless must have come very near during the night, for their fresh footmarks were numerous within a score yards of the place where we slept. In the morning I had a ramble along the borders of the jungle, and found the tracks very numerous and close together on the sandy soil.

We remained in this neighbourhood four days, and succeeded in obtaining many hundred turtles, but we were obliged to sleep two nights within the Carapanatúba channel. The first night passed rather pleasantly, for the weather was fine, and we encamped in the forest, making large fires and slinging our hammocks between the trees. The second was one of the most miserable nights I ever spent. The air was close, and a drizzling rain began to fall about midnight, lasting until morning. We tried at first to brave it out under the trees. Several very large fires were made, lighting up with ruddy gleams the magnificent foliage in the black shades around our encampment.

The heat and smoke had the desired effect of keeping off pretty well the mosquitoes, but the rain continued until at length everything was soaked, and we had no help for it but to bundle off to the canoes with drenched hammocks and garments. There was not nearly room enough in the flotilla to accommodate so large a number of persons lying at full length; moreover the night was pitch dark, and it was quite impossible in the gloom and confusion to get at a change of clothing. So there we lay, huddled together in the best way we could arrange ourselves, exhausted with fatigue and irritated beyond all conception by clouds of mosquitoes. I slept on a bench with a sail over me, my wet clothes clinging to my body, and to increase my discomfort, close beside me lay an Indian girl, one of Cardozo's domestics, who had a skin disfigured with black diseased patches, and whose thick clothing, not having been washed during the whole time we had been out (eighteen days), gave forth a most vile effluvium.

We spent the night of the 7th of November pleasantly on the smooth sands, where the jaguars again serenaded us, and on the succeeding morning we commenced our return voyage to Ega. We first doubled the upper end of the island of Catuá, and then struck off for the right bank of the Solimoens. The river was here of immense width, and the current was so strong in the middle that it required the most strenuous exertions on the part of our paddlers to prevent us from being carried miles away down the stream. At night we reached the Juteca, a small river which enters the Solimoens by a channel so narrow that a man might almost jump across it, but a furlong inwards expands into a very pretty lake several miles in circumference. We slept again in the forest, and again were annoyed by rain and mosquitoes; but this time Cardozo and I preferred remaining where we were to mingling with the reeking crowd in the boats. When the grey dawn arose, a steady rain was still falling, and the whole sky had a settled leaden appearance, but it was delightfully cool. We took our net into the lake and gleaned a good supply of delicious fish for breakfast. I saw at the upper end of this lake the native rice of this country growing wild.

The weather cleared up at 10 o'clock a.m. At 3 p.m. we arrived at the mouth of the Cayambé, another tributary stream much larger than the Juteca. The channel of exit to the Solimoens was here also very narrow, but the expanded river inside is of vast dimensions; it forms a lake (I may safely venture to say) several score miles in circumference. Although prepared for these surprises, I was quite taken aback in this case. We had been paddling all day along a monotonous shore, with the dreary Solimoens before us, here three to four miles broad, heavily rolling onward its muddy waters. We come to a little gap in the earthy banks, and find a dark, narrow inlet with a wall of forest overshadowing it on each side; we enter it and at a distance of two or three hundred yards a glorious sheet of water bursts upon the view. The scenery of Cayambé is very picturesque. The land, on the two sides visible of the lake, is high and clothed with sombre woods, varied here and there with a white-washed house in the middle of a green patch of clearing, belonging to settlers. In striking contrast to these dark rolling forests is the vivid light-green and cheerful foliage of the woods on the numerous islets which rest like water-gardens on the surface of the lake. Flocks of ducks, storks, and snow-white herons inhabit these islets, and a noise of parrots with the tingling chorus of Tamburí-parás was heard from them as we passed. This had a cheering effect, after the depressing stillness and absence of life in the woods on the margins of the main river.

Cardozo and I took a small boat and crossed the lake to visit one of the settlers, and on our return to our canoe, whilst in the middle of the lake, a squall suddenly arose, in the direction towards which we were going, so that for a whole hour we were in great danger of being swamped. The wind blew away the awning and mats, and lashed the waters into foam, the waves rising to a great height. Our boat, fortunately, was excellently constructed, rising well towards the prow, so that with good steering we managed to head the billows as they arose, and escaped without shipping much water. We reached our igarité at sunset, and then made all speed to Curubarú, fifteen miles distant, to encamp for the night on the sands. We reached the praia at ten o'clock. The waters were now mounting fast upon the sloping beach, and we found on dragging the net next morning that fish were beginning to be scarce. Cardozo and his friends talked quite gloomily at breakfast time over the departure of the joyous *veraõ*, and the setting in of the dull, hungry winter season.

At nine o'clock in the morning of the 10th of November a light wind from down river sprang up, and all who had sails hoisted them. It was the first time during our trip that we had had occasion to use our sails; so continual is the calm on this upper river. We bowled along merrily, and soor entered the broad channel lying between Bariá and the mainland on the south bank. The wind carried us right into the mouth of the Teffé, and at four o'clock p.m. we cast anchor in the port of Ega.

CHAPTER XII.

ANIMALS OF THE NEIGHBOURHOOD OF EGA.

Scarlet-faced Monkeys—Parauacú Monkey—Owl-faced Night Apes—Marmosets—Jupurá—Bats—Birds—Cuvier's Toucan—Curl-crested Toucan—Insects—Pendulous Cocoons—Foraging Ants—Blind Ants.

As may have been gathered from the remarks already made, the neighbourhood of Ega was a fine field for a Natural History collector. With the exception of what could be learnt from the few specimens brought home, after transient visits, by Spix and Martius and the Count de Castelnau, whose acquisitions have been deposited in the public museums of Munich and Paris, very little was known in Europe of the animal tenants of this region; the collections that I had the opportunity of making and sending home attracted, therefore, considerable attention. Indeed, the name of my favourite village has become quite a household word amongst a numerous class of Naturalists, not only in England, but abroad, in consequence of the very large number of new species (upwards of 3,000) which they have had to describe, with the locality "Ega" attached to them. The discovery of new species, however, forms but a small item in the interest belonging to the study of the living creation. The structure, habits, instincts, and geographical distribution of some of the oldest-known forms supply inexhaustible materials for reflection. The few remarks I have to make on the animals of Ega will relate to the mammals, birds, and insects, and will sometimes apply to the productions of the whole Upper Amazons region. We will begin with the monkeys, the most interesting, next to man, of all animals.

Scarlet-faced Monkeys.—Early one sunny morning, in the year 1855, I saw in the streets of Ega a number of Indians, carrying on their shoulders down to the port, to be embarked on the Upper Amazons steamer, a large cage made of strong lianas, some twelve feet in length and five in height, containing a dozen monkeys of the most grotesque appearance. Their bodies (about eighteen inches in height, exclusive of limbs) were clothed from neck to tail with very long, straight, and shining whitish hair; their heads were nearly bald, owing to the very short crop of thin grey hairs, and their faces glowed with the most vivid scarlet hue. As a finish to their striking physiognomy, they had bushy whiskers of a sandy colour, meeting under the chin, and reddish-yellow eyes. These red-faced apes belonged to a species called by the Indians Uakarí, which is peculiar to the Ega district, and the cage with its contents was being sent as a present by Senhor Chrysostomo, the Director of Indians of the Japurá, to one of the Government officials at Rio Janeiro, in acknowledgment of having been made colonel of the new national guard. They had been obtained with great difficulty in the forests which cover the lowlands, near the principal mouth of the Japurá, about thirty miles from Ega. It was the first time I had seen this most curious of all the South American monkeys, and one that appears to have escaped the notice of Spix and Martius. I afterwards made a journey to the dsitrict inhabited by it, but did not then succeed in obtaining specimens; before leaving the country, however, I acquired two individuals, one of which lived in my house for several weeks.

The scarlet-faced monkey belongs, in all essential points of structure, to the same family (Cebidæ) as the rest of the large-sized American species; but it differs from all its relatives in having only the rudiment of a tail, a member which reaches in some allied kinds the highest grade of development known in the order. It was so unusual to see a nearly tailless monkey from America, that naturalists thought, when the first specimens arrived in Europe, that the member had been shortened artificially. Nevertheless, the Uakarí is not quite isolated from its related species of the same family, several other kinds, also found on the Amazons, forming a graduated passage between the extreme forms as regards the tail. The appendage reaches its perfection in those genera (the Howlers, the Lagothrix, and the Spider Monkeys) in which it presents on its under surface near the tip a naked palm, which makes it sensitive and useful as a fifth hand in climbing. In the rest of the genera of Cebidæ (seven in number, containing thirty-eight species), the tail is weaker in structure, entirely covered

SCARLET-FACED AND PARAUACÚ MONKEYS.

with hair, and of little or no service in climbing, a few species nearly related to our Uakarí having it much shorter than usual. All the Cebidæ, both long-tailed and short-tailed, are equally dwellers in trees. The scarlet-faced monkey lives in forests which are inundated during great part of the year, and is never known to descend to the ground; the shortness of its tail is therefore no sign of terrestrial habits, as it is in the Macaques and Baboons of the Old World. It differs a little from the typical Cebidæ in its teeth, the incisors being oblique and in the upper jaw converging, so as to leave a gap between the outermost and the canine teeth. Like all the rest of its family, it differs from the monkeys of the Old World, and from man, in having an additional grinding-tooth (premolar) on each side of both jaws, making the complete set thirty-six instead of thirty-two in number.

The white Uakarí (Brachyurus calvus), seems to be found in no other part of America than the district just mentioned, namely, the banks of the Japurá, near its principal mouth; and even there it is confined, as far as I could learn, to the western side of the river. It lives in small troops amongst the crowns of the lofty trees, subsisting on fruits of various kinds. Hunters say it is pretty nimble in its motions, but is not much given to leaping, preferring to run up and down the larger boughs in travelling from tree to tree. The mother, as in other species of the monkey order, carries her young on her back. Individuals are obtained alive by shooting them with the blow-pipe and arrows tipped with diluted Urarí poison. They run a considerable distance after being pierced, and it requires an experienced hunter to track them. He is considered the most expert who can keep pace with a wounded one and catch it in his arms when it falls exhausted. A pinch of salt, the antidote to the poison, is then put in its mouth, and the creature revives. The species is rare, even in the limited district which it inhabits. Senhor Chrysostomo sent six of his most skilful Indians, who were absent three weeks before they obtained the twelve specimens which formed his unique and princely gift. When an independent hunter obtains one, a very high price (thirty to forty milreis*) is asked, these monkeys being in great demand for presents to persons of influence down the river.

Adult Uakarís, caught in the way just described, very rarely

* Three pounds seven shillings to four pounds thirteen shillings.

become tame. They are peevish and sulky, resisting all attempts to coax them, and biting any one who ventures within reach. They have no particular cry, even when in their native woods; in captivity they are quite silent. In the course of a few days or weeks, if not very carefully attended to, they fall into a listless condition, refuse food and die. Many of them succumb to a disease which I suppose from the symptoms to be inflammation of the chest or lungs. The one which I kept as a pet died of this disorder, after I had had it about three weeks. It lost its appetite in a very few days, although kept in an airy verandah; its coat, which was originally long, smooth, and glossy, became dingy and ragged like that of the specimens seen in museums, and the bright scarlet colour of its face changed to a duller hue. This colour, in health, is spread over the features up to the roots of the hair on the forehead and temples, and down to the neck, including the flabby cheeks which hang down below the jaws. The animal in this condition looks at a short distance as though some one had laid a thick coat of red paint on its countenance. The death of my pet was slow; during the last twenty-four hours it lay prostrate, breathing quickly, its chest strongly heaving; the colour of its face became gradually paler, but was still red when it expired. As the hue did not quite disappear until two or three hours after the animal was quite dead, I judged that it was not exclusively due to the blood, but partly to a pigment beneath the skin, which would probably retain its colour a short time after the circulation had ceased.

After seeing much of the morose disposition of the Uakarí, I was not a little surprised one day at a friend's house to find an extremely lively and familiar individual of this species. It ran from an inner chamber straight towards me, after I had sat down on a chair, climbed my legs and nestled in my lap, turning round and looking up with the usual monkey's grin, after it had made itself comfortable. It was a young animal which had been taken when its mother was shot with a poisoned arrow; its teeth were incomplete, and the face was pale and mottled, the glowing scarlet hue not supervening in these animals before mature age; it had also a few long black hairs on the eyebrows and lips. The frisky little fellow had been reared in the house amongst the children, and allowed to run about freely, and take its meals with the rest of the household. There are few animals which the Brazilians of these villages have not succeeded in taming. I have even seen young jaguars running loose about a house, and treated as pets. The animals that I had, rarely became familiar, however long they might remain in my possession, a circumstance due no doubt to their being kept always tied up.

The Uakarí is one of the many species of animals which are classified by the Brazilians as "mortál," or of delicate constitution, in contradistinction to those which are "duro," or hardy. A large proportion of the specimens sent from Ega die before arriving at Pará, and scarcely one in a dozen succeeds in reaching Rio Janeiro alive. The difficulty it has of accommodating itself to changed conditions probably has some connection with the very limited range, or confined sphere of life, of the species in its natural state, its native home being an area of swampy woods, not more than about sixty square miles in extent, although no permanent barrier exists to check its dispersal, except towards the south, over a much wider space. When I descended the river in 1859, we had with us a tame adult Uakarí, which was allowed to ramble about the vessel, a large schooner. When we reached the mouth of the Rio Negro, we had to wait four days whilst the custom-house officials at Barra, ten miles distant, made out the passports for our crew, and during this time the schooner lay close to the shore, with its bowsprit secured to the trees on the bank. Well, one morning, scarlet-face was missing, having made his escape into the forest. Two men were sent in search of him, but returned after several hours' absence without having caught sight of the runaway. We gave up the monkey for lost, until the following day, when he re-appeared on the skirts of the forest, and marched quietly down the bowsprit to his usual place on deck. He had evidently found the forests of the Rio Negro very different from those of the delta lands of the Japurá, and preferred captivity to freedom in a place that was so uncongenial to him.

The Parauacú Monkey.—Another Ega monkey, nearly related to the Uakarís, is the Parauacú (Pithecia hirsuta), a timid inoffensive creature, with a long bear-like coat of harsh speckled-grey hair. The long fur hangs over the head, half concealing the pleasing diminutive face, and clothes also the tail to the tip, which member is well developed, being eighteen inches in length, or longer than the body. The Parauacú is found on the "terra firma" lands of the north shore of the Solimoens, from Tunantins to Peru. It exists also on the south side of the river, namely, on the banks of the Teffé, but there under a changed form, which differs a little from its type in colours. This form has been described by Dr. Gray as a distinct species, under the name of Pithecia albicans. The Parauacú is also a very delicate animal, rarely living many weeks in captivity; but any one who succeeds in keeping it alive for a month or two, gains by it a most affectionate pet. One of the specimens of Pithecia albicans now in the British Museum was when living the property of a young Frenchman, a neighbour of mine at Ega. It became so tame in the course of a few weeks, that it followed him about the streets like a dog. My friend was a tailor, and the little pet used to spend the greater part of the day seated on his shoulder, whilst he was at work on his board. It showed, nevertheless, great dislike to strangers, and was not on good terms with any other member of my friend's household than himself. I saw no monkey that showed so strong a personal attachment as this gentle, timid, silent little creature. The eager and passionate Cebi seem to take the lead of all the South American monkeys in intelligence and docility, and the Coaitá has perhaps the most gentle and impressible disposition; but the Parauacú, although a dull, cheerless animal, excels all in this quality of capability of attachment to individuals of our own species. It is not wanting, however, in intelligence as well as moral goodness, proof of which was furnished one day by an act of our little pet. My neighbour had quitted his house in the morning without taking Parauacú with him, and the little creature having missed its friend, and concluded, as it seemed, that he would be sure to come to me, both being in the habit of paying me a daily visit together, came straight to my dwelling, taking a short cut over gardens, trees, and thickets, instead of going the roundabout way of the street. It had never done this before, and we knew the route it had taken only from a neighbour having watched its movements. On arriving at my house and not finding its master, it climbed to the top of my table, and sat with an air of quiet resignation, waiting for him. Shortly afterwards, my friend entered, and the gladdened pet then jumped to its usual perch on his shoulder.

Owl-faced Night Apes.—A third interesting genus of monkeys, found near Ega, are the Nyctipitheci, or night Apes, called Ei-á by the Indians. Of these I found two species, closely related to each other, but nevertheless quite distinct, as both inhabit the same forests, namely, those of the higher and drier lands, without mingling with each other or intercrossing. They sleep all day long in hollow trees, and come forth to prey on insects and eat fruits only in the night. They are of small size, the body being about a foot long, and the tail fourteen inches, and are thickly clothed with soft grey and brown fur, similar in substance to that of the rabbit.' Their physiognomy reminds one of the owl, or tiger-cat; the face is round and encircled by a ruff of whitish fur; the muzzle is not at all prominent; the mouth and chin are small; the ears are very short, scarcely appearing above the hair of the head; and the eyes are large and yellowish in colour, imparting the staring expression of nocturnal animals of prey. The forehead is whitish, and decorated with three black stripes, which in one of the species (Nyctipithecus trivirgatus) continue to the crown, and in the other (N. felinus) meet on the top of the forehead. N. trivirgatus was first described by Humboldt, who discovered it on the banks of the Cassiquiare, near the head waters of the Rio Negro.

I kept a pet animal of the N. trivirgatus for many months, a young one having been given to me by an Indian *compadre*, as a present from my newly-baptized godson. These monkeys, although sleeping by day, are aroused by the least noise; so that when a person passes by a tree in which a number of them are concealed, he is startled by the sudden apparition of a group of little striped faces crowding a hole in the trunk. It was in this way that my compadre discovered the colony from which the one given to me was taken. I was obliged to keep my pet chained up; it therefore never became thoroughly familiar. I once saw, however, an individual of the other species (N. felinus) which was most amusingly tame. It was as lively and nimble as the Cebi, but not so mischievous and far more confiding in its disposition, delighting to be caressed by all persons who came into the house. But its owner, the Municipal Judge of Ega, Dr. Carlos Mariana, had treated it for many weeks with the greatest kindness, allowing it to sleep with him at night in his hammock, and to nestle in his bosom half the day as he lay reading. It was a great favourite with every one, from the cleanliness of its habits and the prettiness of its features and ways. My own pet was kept in a box, in which was placed a broad-mouthed glass jar; into this it would dive, head-foremost, when any one entered the room, turning round inside, and thrusting forth its inquisitive face an instant afterwards to stare at the intruder. It was very active at night, venting at frequent intervals a hoarse cry, like the suppressed barking of a dog, and scampering about the room, to the length of its tether, after cockroaches and spiders. In climbing between the box and the wall, it straddled the space, resting its hands on the palms and tips of the out-stretched fingers, with the knuckles bent at an acute angle, and thus mounted to the top with the greatest facility. Although seeming to prefer insects, it ate all kinds of fruit, but would not touch raw or cooked meat, and was very seldom thirsty. I was told by persons who had kept these monkeys loose about the house, that they cleared the chambers of bats as well as insect vermin. When approached gently, my Ei-á allowed itself to be caressed; but when handled roughly, it always took alarm, biting severely, striking out its little hands, and making a hissing noise like a cat. As already related, my pet was killed by a jealous Caiarára monkey, which was kept in the house at the same time.

Barrigudo Monkeys.—Ten other species of monkeys were found, in addition to those already mentioned, in the forests of the Upper Amazons. All were strictly arboreal and diurnal in their habits, and lived in flocks, travelling from tree to tree, the mothers with their children on their backs; leading, in fact, a life similar to that of the Parárauáte Indians, and, like them, occasionally plundering the plantations which lie near their line of march. Some of them were found also on the Lower Amazons, and have been noticed in former chapters of this narrative. Of the remainder, the most remarkable is the Macaco barrigudo, or big-bellied monkey of the Portuguese colonists, a species of Lagothrix. The genus is closely allied to the Coaitás, or spider monkeys, having, like them, exceedingly strong and flexible tails, which are furnished underneath with a naked palm like a hand for grasping. The Barrigudos, however, are very bulky animals, whilst the spider monkeys are remarkable for the slenderness of their bodies and limbs. I obtained specimens of what have been considered two species, one (L. olivaceus of Spix?) having the head clothed with gray, the other (L. Humboldtii) with black fur. They both live together in the same places, and are probably only differently coloured individuals of one and the same species. I sent home a very large male of one of these kinds, which measured twenty-seven inches in length of trunk, the tail being twenty-six inches long; it was the largest monkey I saw in America, with the exception of a black Howler, whose body was twenty-eight inches in height. The skin of the face in the Barrigudo is black and wrinkled, the forehead is low, with the eyebrows projecting, and, in short, the features altogether resemble in a striking manner those of an old negro. In the forests the Barrigudo is not a very active animal; it lives exclusively on fruits, and is much persecuted by the Indians, on account of the excellence of its flesh as food. From information given me by a collector of birds and mammals, whom I employed, and who resided a long time amongst the Tucuna Indians, near Tabatinga, I calculated that one horde of this tribe, 200 in number, destroyed 1,200 of these monkeys annually for food. The species is very numerous in the forests of the higher lands, but, owing to long persecution, it is now seldom seen in the neighbourhood of the larger villages. It is not found at all on the Lower Amazons. Its manners in captivity are grave, and its temper mild and confiding, like that of the Coaitás. Owing to these traits the Barrigudo is much sought after for pets; but it is not hardy like the Coaitás, and seldom survives a passage down the river to Pará.

Marmosets.—It now only remains to notice the Marmosets, which form the second family of American monkeys. Our old friend Midas ursulus, of Pará and the Lower Amazons, is not found on the upper river, but in its stead a closely allied species presents itself, which appears to be the Midas rufoniger of Gervais, whose mouth is bordered with longish white hairs. The habits of this species are the same as those of the M. ursulus, indeed it seems probable that it is a form or race of the same stock, modified to suit the altered local conditions under which it lives. One day, whilst walking along a forest pathway, I saw one of these lively little fellows miss his grasp as he was passing from one tree to another along with his troop. He fell head-foremost from a height of at least fifty feet, but managed cleverly to alight on his legs in the pathway; quickly turning round, he gave me a good stare for a few moments, and then bounded off gaily to climb another tree. At Tunantins I shot a pair of very handsome species of marmoset, the M. rufiventer, I believe, of zoologists. Its coat was very glossy and smooth, the back deep brown, and the underside of the body of rich black and reddish hues. A third species (found at Tabatinga, 200 miles further west) is of a deep black colour, with the exception of a patch of white hair around its mouth. The little animal at a short distance looks as though it held a ball of snow-white cotton in its teeth. The last I shall mention is the Hapale pygmæus, one of the most diminutive forms of the monkey order, three full-grown specimens of which, measuring only seven inches in length of body, I obtained near St. Paulo. The pretty Lilliputian face is furnished with long brown whiskers, which are naturally brushed back over the ears. The general colour of the animal is brownish-tawny, but the tail is elegantly barred with black. I was surprised on my return to England to learn, from specimens in the British Museum, that the pigmy marmoset was found also in Mexico, no other Amazonian monkey being known to wander far from the great river plain. Thus the smallest and apparently the feeblest species of the whole order is one which has by some means become the most widely dispersed.

The Jupurá.—A curious animal, known to naturalists as the Kinkajou, but called Jupurá by the Indians of the Amazons, and considered by them as a kind of monkey, may be mentioned in this place. It is the Cercoleptes caudivolvus of zoologists, and has been considered by some authors as an intermediate form between the Lemur family of apes and the plantigrade Carnivora, or bear family. It has decidedly no close relationship to either of the groups of American monkeys, having six cutting teeth to each jaw, and long claws instead of nails, with extremities of the usual shape of paws instead of hands. Its muzzle is conical and pointed, like that of many lemurs of Madagascar; the expression of its countenance, and its habits and actions, are also very similar to those of lemurs. Its tail is very flexible towards the tip, and is used to twine round branches in climbing. I did not see or hear anything of this animal whilst residing on the Lower Amazons, but on the banks of the upper river, from the Teffé to Peru, it appeared to be rather common. It is nocturnal in its habits, like the owl-faced monkeys, although unlike them it has a bright, dark eye. I once saw it in considerable numbers when on an excursion with an Indian companion along the low Ygapó shores of the Teffé, about twenty miles above Ega. We slept one night at the house of a native family living in the thick of the forest, where a festival was going on, and there being no room to hang our hammocks under shelter, on account of the number of visitors, we lay down on a mat in the open air, near a shed which stood in the midst of a grove of fruit trees and pupunha palms. Past midnight, when all became still after the uproar of holiday-making, as I was listening to the dull, fanning sound made by the wings of impish hosts of vampire bats crowding round the Cajú trees, a rustle commenced from the side of the woods, and a troop of slender, long-tailed animals were seen against the clear moonlit sky, taking flying leaps from branch to branch through the grove. Many of them stopped at the pupunha trees, and the hustling, twittering, and screaming, with sounds of falling fruits, showed how they were employed. I thought at first they were Nyctipitheci, but they proved to be Jupurás, for the owner of the house early next morning caught a young one, and gave it to me. I kept this as a pet animal for several weeks, feeding it on bananas and mandioca-meal mixed with treacle. It became tame in a very short time, allowing itself to be caressed, but making a distinction in the degree of confidence it showed between myself and strangers. My pet was unfortunately killed by a neighbour's dog, which entered the room where it was kept. The animal is so difficult to obtain alive, its place of retreat in the daytime not being known to the natives, that I was unable to procure a second living specimen.

Bats.—The only other mammals that I shall mention are the bats, which exist in very considerable numbers and variety in the forest, as well as in the buildings of the villages. Many small and curious species living in the woods conceal themselves by day under the broad leaf-blades of Heliconiæ and other plants which grow in shady places; others cling to the trunks of trees. Whilst walking through the forest in the daytime, especially along gloomy ravines, one is almost sure to startle bats from their sleeping-places; and at night they are often seen in great numbers flitting about the trees on the shady margins of narrow channels. I captured altogether, without giving especial attention to bats, sixteen different species at Ega.

The Vampire Bat.—The little gray bloodsucking Phyllostoma, mentioned in a former chapter as found in my chamber at Caripí, was not uncommon at Ega, where every one believes it to visit sleepers and bleed them in the night. But the vampire was here by far the most abundant of the family of leaf-nosed bats. It is the largest of all the South American species, measuring twenty-eight inches in expanse of wing. Nothing in animal physiognomy can be more hideous than the countenance of this creature when viewed from the front; the large leathery ears standing out from the sides and top of the head, the erect spear-shaped appendage on the tip of the nose, the grin and the glistening black eye, all combining to make up a figure that reminds one of some mocking imp of fable. No wonder that imaginative people have inferred diabolical instincts on the part of so ugly an animal. The vampire, however, is the most harmless of all bats, and its inoffensive character is well known to residents on the banks of the Amazons. I found two distinct species of it, one having the fur of a blackish colour, the other of a ruddy hue, and ascertained that both feed chiefly on fruits. The church at Ega was the head-quarters of both kinds; I used to see them, as I sat at my door during the short evening twilights, trooping forth by scores from a large open window at the back of the altar, twittering cheerfully as they sped off to the borders of the forest. They sometimes enter houses; the first time I saw one in my chamber, wheeling heavily round and round, I mistook it for a pigeon, thinking that a tame one had escaped from the premises of one of my neighbours. I opened the stomachs of several of these bats, and found them to contain a mass of pulp and seeds of fruits, mingled with a few remains of insects. The natives say they devour ripe cajús and guavas on trees in the gardens, but on comparing the seeds taken from their stomachs with those of all cultivated trees of Ega, I found they were unlike any of them; it is therefore probable that they generally resort to the forest to feed, coming to the village in the morning to sleep, because they find it more secure from animals of prey than their natural abodes in the woods.

Birds.—I have already had occasion to mention several of the more interesting birds found in the Ega district. The first thing that would strike a new-comer in the forests of the Upper Amazons would be the general scarcity of birds; indeed, it often happened that I did not meet with a single bird during a whole day's ramble in the richest and most varied parts of the woods. Yet the country is tenanted by many hundred species, many of which are in reality abundant, and some of them conspicuous from their brilliant plumage. The cause of their apparent rarity is to be sought in the sameness and density of the thousand miles of forest which constitute their dwelling-place. The birds of the country are gregarious, at least during the season when they are most readily found; but the frugivorous kinds are to be met with only when certain wild fruits are ripe, and to know the exact localities of the trees requires months of experience. It would not be supposed that the insectivorous birds are also gregarious; but they are so, numbers of distinct species, belonging to many different families, joining together in the chase or search of food. The proceedings of these associated bands of insect hunters are not a little curious, and merit a few remarks.

Whilst hunting along the narrow pathways that are made through the forest in the neighbourhood of houses and villages, one may pass several days without seeing many birds; but now and then the surrounding bushes and trees appear suddenly to swarm with them. There are scores, probably hundreds, of birds, all moving about with the greatest activity—woodpeckers and Dendrocolaptidæ (from species no larger than a sparrow to others the size of a crow) running up the tree trunks; tanagers, ant-thrushes, humming-birds, fly-catchers, and barbets, flitting about the leaves and lower branches. The bustling crowd loses no time, and although moving in concert, each bird is occupied on its own account in searching bark, or leaf, or twig; the barbets visiting every clayey nest of termites on the trees which lie in the line of march. In a few minutes the host is gone, and the forest path remains deserted and silent as before. I became in course of time so accustomed to this habit of birds in the woods near Ega, that I could generally find the flock of associated marauders whenever I wanted it. There appeared to be only one of these flocks in each small district; and as it traversed chiefly a limited tract of woods of second growth, I used to try different paths until I came up with it.

The Indians have noticed these miscellaneous hunting parties of birds, but appear not to have observed that they are occupied in searching for insects. They have supplied their want of knowledge, in the usual way of half-civilised people, by a theory which has degenerated into a myth, to the effect that the onward moving bands are led by a little gray bird called the Uirá-pará, which fascinates all the rest, and leads them a weary dance through the thickets. There is certainly some appearance of truth in this explanation; for sometimes stray birds encountered in the line of march are seen to be drawn into the throng, and purely frugivorous birds are now and then found mixed up with the rest, as though led away by some will-o'-the-wisp. The native women, even the white and half-caste inhabitants of the towns, attach a superstitious value to the skin and feathers of the Uirá-pará, believing that if they keep them in their clothes chest the relics will have the effect of attracting for the happy possessors a train of lovers and followers. These birds are consequently in great demand in some places, the hunters selling them at a high price to the foolish girls, who preserve the bodies by drying flesh and feathers together in the sun. I could never get a sight of this famous little bird in the forest. I once employed Indians to obtain specimens for me; but after the same man (who was a noted woodsman) had brought me at different times three distinct species of birds as the Uirá-pará, I gave up the story as a piece of humbug. The simplest explanation appears to be this, that the birds associate in flocks from the instinct of self-preservation, and in order to be a less easy prey to hawks, snakes, and other enemies, than they would be if feeding alone.

Toucans.—Cuvier's Toucan.—Of this family of birds, so conspicuous from the great size and light structure of their beaks, and so characteristic of tropical American forests, five species inhabit the woods of Ega. The commonest is Cuvier's Toucan, a large bird, distinguished from its nearest relatives by the feathers at the bottom of the back being of a saffron hue instead of red. It is found more or less numerously throughout the year, as it breeds in the neighbourhood, laying its eggs in holes of trees at a great height from the ground. During most months of the year it is met with in single individuals or small flocks, and the birds are then very wary. Sometimes one of these little bands of four or five is seen perched for hours together amongst the topmost branches of high trees, giving vent to their remarkably loud, shrill, yelping cries, one bird mounted higher than the rest, acting apparently as leader of the inharmonious chorus; but two of them are often heard yelping alternately and in different notes. These cries have a vague resemblance to the syllables Tocáno, Tocáno, and hence the Indian name of this genus of birds. At these times it is difficult to get a shot at Toucans, for their senses are so sharpened that they descry the hunter before he gets near the tree on which they are perched, although he may be half concealed amongst the underwood 150 feet below them. They stretch their necks downwards to look beneath, and on espying the least movement amongst the foliage, fly off to the more inaccessible parts of the forest. Solitary Toucans are sometimes met with at the same season, hopping silently up and down the larger boughs, and peering into the crevices of tree trunks. They moult in the months from March to June, some individuals earlier, others later. This season of enforced quiet being passed, they make their appearance suddenly in the dry forest near Ega, in large flocks, probably assemblages of birds gathered together from the neighbouring Ygapó forests, which are then flooded and cold. The birds have now become exceedingly tame, and the troops travel with heavy laborious flight from bough to bough amongst the lower trees. They thus become an easy prey to hunters, and every one at Ega who can get a gun of any sort and a few charges of powder and shot, or a blow-pipe, goes daily to the woods to kill a few brace for dinner; for, as already observed, the people of Ega live almost exclusively on stewed and roasted Toucans during the months of June and July, the birds being then very fat, and the meat exceedingly sweet and tender.

No one on seeing a Toucan can help asking what is the use of the enormous bill, which, in some species, attains a length of seven inches, and a width of more than two inches. A few remarks on this subject may be here introduced. The early naturalists having seen only the bill of a Toucan, which was esteemed as a marvellous production by the *virtuosi* of the sixteenth and seventeenth centuries, concluded that the bird must have belonged to the aquatic and web-footed order, as this contains so many species of remarkable development of beak, adapted for seizing fish. Some travellers also related fabulous stories of Toucans resorting to the banks of rivers to feed on fish, and these accounts also encouraged the erroneous views of the habits of the birds, which for a long time prevailed. Toucans, however, are now well known to be eminently arboreal birds, and to belong to a group (including trogons, parrots, and barbets*), all of whose members are fruit-eaters. On the Amazons, where these birds are very common, no one pretends ever to have seen a Toucan walking on the ground in its natural state, much less acting the part of a swimming or wading bird. Professor Owen found, on dissection, that the gizzard in Toucans is not so well adapted for the trituration of food as it is in other vegetable feeders, and concluded, therefore, as Broderip had observed the habit of chewing the cud in a tame bird, that the great toothed bill was useful in holding and re-masticating the food. The bill

* Capitoninæ, G. R. Gray.

can scarcely be said to be a very good contrivance for seizing and crushing small birds, or taking them from their nests in crevices of trees, habits which have been imputed to Toucans by some writers. The hollow, cellular structure of the interior of the bill, its curved and clumsy shape, and the deficiency of force and precision when it is used to seize objects, suggest a want of fitness, if this be the function of the member. But fruit is undoubtedly the chief food of Toucans, and it is in reference to their mode of obtaining it that the use of their uncouth bills is to be sought.

Flowers and fruit on the crowns of the large trees of South American forests grow principally towards the end of slender twigs, which will not bear any considerable weight; all animals, therefore, which feed upon fruit, or on insects contained in flowers, must, of course, have some means of reaching the ends of the stalks from a distance. Monkeys obtain their food by stretching forth their long arms, and, in some instances, their tails, to bring the fruit near to their mouths. Humming-birds are endowed with highly-perfected organs of flight, with corresponding muscular development, by which they are enabled to sustain themselves on the wing before blossoms whilst rifling them of their contents. These strong-flying creatures, however, will, whenever they can get near enough, remain on their perches whilst probing neighbouring flowers for insects. Trogons have feeble wings, and a dull, inactive temperament. Their mode of obtaining food is to station themselves quietly on low branches in the gloomy shades of the forest, and eye the fruits on the surrounding trees, darting off as if with an effort every time they wish to seize a mouthful, and returning to the same perch. Barbets (Capitoninæ) seem to have no especial endowment, either of habits or structure, to enable them to seize fruits; and in this respect they are similar to the Toucans, if we leave the bill out of question, both tribes having heavy bodies, with feeble organs of flight, so that they are disabled from taking their food on the wing. The purpose of the enormous bill here becomes evident. It is to enable the Toucan to reach and devour fruit whilst remaining seated, and thus to counterbalance the disadvantage which its heavy body and gluttonous appetite would otherwise give it in the competition with allied groups of birds. The relation between the extraordinarily lengthened bill of the Toucan and its mode of obtaining food, is therefore precisely similar to that between the long neck and lips of the Giraffe and the mode of browsing of the animal. The bill of the Toucan can scarcely be considered a very perfectly formed instrument for the end to which it is applied, as here explained; but nature appears not to invent organs at once for the functions to which they are now adapted, but avails herself, here of one already-existing structure or instinct, there of another, according as they are handy when need for their further modification arises.

One day whilst walking along the principal pathway in the woods near Ega, I saw one of these toucans seated gravely on a low branch close to the road, and had no difficulty in seizing it with my hand. It turned out to be a runaway pet bird; no one, however, came to own it, although I kept it in my house for several months. The bird was in a half-starved and sickly condition, but after a few days of good living it recovered health and spirits, and became one of the most amusing pets imaginable. Many excellent accounts of the habits of tame toucans have been published, and therefore I need not describe them in detail, but I do not recollect to have seen any notice of their intelligence and confiding disposition under domestication, in which qualities my pet seemed to be almost equal to parrots. I allowed Tocáno to go free about the house, contrary to my usual practice with pet animals; he never, however, mounted my working-table after a smart correction which he received the first time he did so. He used to sleep on the top of a box in a corner of the room, in the usual position of these birds, namely, with the long tail laid right over on the back, and the beak thrust underneath the wing. He ate of everything that we eat, beef, turtle, fish, farinha, fruit, and was a constant attendant at our table—a cloth spread on a mat. His appetite was most ravenous, and his powers of digestion quite wonderful. He got to know the meal hours to a nicety, and we found it very difficult, after the first week or two, to keep him away from the dining-room, where he had become very impudent and troublesome. We tried to shut him out by inclosing him in the back-yard, which was separated by a high fence from the street on which our front door opened, but he used to climb the fence and hop round by a long circuit to the dining-room, making his appearance with the greatest punctuality as the meal was placed on the table. He acquired the habit afterwards of rambling about the street near our house, and one day he was stolen, so we gave him up for lost. But two days afterwards he stepped through the open doorway at dinner hour, with his old gait and sly magpie-like expression, having escaped from the house where he had been guarded by the person who had stolen him, and which was situated at the further end of the village.

The Curl-crested Toucan (*Pteroglossus Beauharnaisii*).—Of the four smaller toucans, or arassarís, found near Ega, the Pteroglossus flavirostris is perhaps the most beautiful in colours, its breast being adorned with broad belts of rich crimson and black; but the most curious species by far is the Curl-crested, or Beauharnais Toucan. The feathers on the head of this singular bird are transformed into thin horny plates, of a lustrous black colour, curled up at the ends, and resembling shavings of steel or ebony wood, the curly crest

Curl-crested Toucan.

being arranged on the crown in the form of a wig. Mr. Wallace and I first met with this species on ascending the Amazons, at the mouth of the Solimoens; from that point it continues as a rather common bird on the terra firma, at least on the south side of the river, as far as Fonte Boa, but I did not hear of its being found further to the west. It appears in large flocks in the forests near Ega in May and June, when it has completed its moult. I did not find these bands congregated at fruit trees, but always wandering through the forest, hopping from branch to branch amongst the lower trees, and partly concealed amongst the foliage. None of the arassarís to my knowledge make a yelping noise like that uttered by the larger toucans (Ramphastos); the notes of the curl-

MOBBED BY CURL-CRESTED TOUCANS.

crested species are very singular, resembling the croaking of frogs. I had an amusing adventure one day with these birds. I had shot one from a rather high tree in a dark glen in the forest, and entered the thicket where the bird had fallen to secure my booty. It was only wounded, and on my attempting to seize it, set up a loud scream. In an instant, as if by magic, the shady nook seemed alive with these birds, although there was certainly none visible when I entered the jungle. They descended towards me, hopping from bough to bough, some of them swinging on the loops and cables of woody lianas, and all croaking and fluttering their wings like so many furies. If I had had a long stick in my hand, I could have knocked several of them over. After killing the wounded one I began to prepare for obtaining more specimens and punishing the viragos for their boldness; but the screaming of their companion having ceased, they remounted the trees, and before I could reload every one of them had disappeared.

Insects.—Upwards of 7,000 species of insects were found in the neighbourhood of Ega. I must confine myself in this place to a few remarks on the order Lepidoptera, and on the ants, several kinds of which, found chiefly on the Upper Amazons, exhibit the most extraordinary instincts.

I found about 550 distinct species of butterflies at Ega. Those who know a little of Entomology, will be able to form some idea of the riches of the place in this department, when I mention that eighteen species of true papilio (the swallow-tail genus) were found within ten minutes' walk of my house. No fact could speak more plainly for the surpassing exuberance of the vegetation, the varied nature of the land, the perennial warmth and humidity of the climate. But no description can convey an adequate notion of the beauty and diversity in form and colour of this class of insects in the neighbourhood of Ega. I paid especial attention to them, having found that this tribe was better adapted than almost any other group of animals or plants, to furnish facts in illustration of the modifications which all species undergo in nature under changed local conditions. This accidental superiority is owing partly to the simplicity and distinctness of the specific characters of the insects, and partly to the facility with which very copious series of specimens can be collected and placed side by side for comparison. The distinctness of the specific characters is due probably to the fact that all the superficial signs of change in

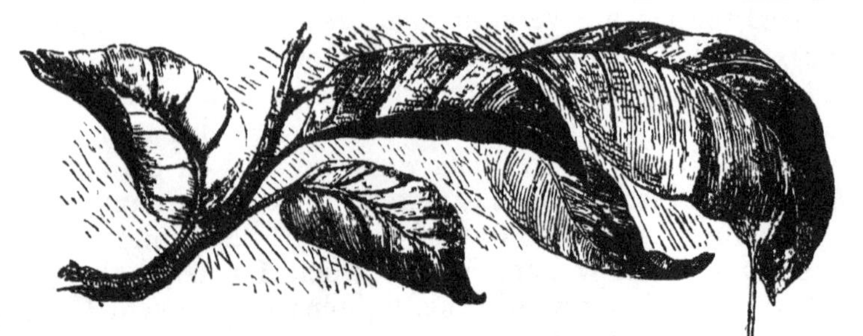

the organisation are exaggerated and made unusually plain by affecting the framework, shape, and colour of the wings, which, as many anatomists believe, are magnified extensions of the skin around the breathing orifices of the thorax of the insects. These expansions are clothed with minute feathers or scales, coloured in regular patterns, which vary in accordance with the slightest change in the conditions to which the species are exposed. It may be said, therefore, that on these expanded membranes nature writes, as on a tablet, the story of the modifications of species, so truly do all changes of the organisation register themselves thereon. Moreover, the same colour-patterns of the wings generally show, with great regularity, the degrees of blood-relationship of the species. As the laws of nature must be the same for all beings, the conclusions furnished by this group of insects must be applicable to the whole organic world; therefore the study of butterflies—creatures selected as the types of airiness and frivolity—instead of being despised, will some day be valued as one of the most important branches of Biological science.

Suspended cocoon of Moth.

Before proceeding to describe the ants, a few remarks may be made on the singular cases and cocoons woven by the caterpillars of certain moths found at Ega. The first that may be mentioned is one of the most beautiful examples of insect workmanship I ever saw. It is a cocoon, about the size of a sparrow's egg, woven by a caterpillar in broad meshes of either buff or rose-coloured silk, and is frequently seen in the narrow alleys of the forest, suspended from the extreme tip of an outstanding leaf by a strong silken thread five or six inches in length. It forms a very conspicuous object, hanging thus in mid-air. The glossy threads with which it is knitted are stout, and the structure is therefore not liable to be torn by the beaks of insectivorous birds, whilst its pendulous position makes it doubly secure against their attacks, the apparatus giving way when they peck at it. There is a small orifice at each end of the egg-shaped bag, to admit of the escape of the moth when it changes from the little chrysalis which sleeps tranquilly in its airy cage. The moth is

Sack-bearing Caterpillar (Saccophora).

of a dull slaty colour, and belongs to the Lithosiide group of the silk-worm family (Bombycidæ). When the caterpillar begins its work, it lets itself down from the tip of the leaf which it has chosen, by spinning a thread of silk, the thickness of which it slowly increases as it descends. Having given the proper length to the cord, it proceeds to weave its elegant bag, placing itself in the centre and spinning rings of silk at regular intervals, connecting them at the same time by means of cross threads; so that the whole, when finished, forms a loose web, with quadrangular meshes of nearly equal size throughout. The task occupies about four days: when finished, the enclosed caterpillar becomes sluggish, its skin shrivels and cracks, and there then remains a motionless chrysalis of narrow shape, leaning against the sides of its silken cage.

Many other kinds are found at Ega belonging to the same cocoon-weaving family, some of which differ from the rest in their caterpillars possessing the art of fabricating cases with fragments of wood or leaves, in which they live secure from all enemies whilst they are feeding and growing. I saw many species of these; some of them knitted together, with fine silken threads, small bits of stick, and so made tubes similar to those of caddice-worms; others (Saccophora) chose leaves for the same purpose, forming with them an elongated bag open at both ends, and having the inside lined with a thick web. The tubes of full-grown caterpillars of Saccophora are two inches in length, and it is at this stage of growth that I have generally seen them. They feed on the leaves of Melastomæ, and as in crawling the weight of so large a dwelling would be greater than the contained caterpillar could sustain, the insect attaches the case, by one or more threads, to the leaves or twigs near which it is feeding.

Foraging Ants.—Many confused statements have been published in books of travel and copied in Natural History works regarding these ants, which appear to have been confounded with the Saüba, a sketch of whose habits has been given in the first chapter of this work. The Saüba is a vegetable feeder, and does not attack other animals; the accounts that have been published regarding carnivorous ants which hunt in vast armies, exciting terror wherever they go, apply only to the Ecitons, or foraging ants, a totally different group of this tribe of insects. The Ecitons are called Tauóca by the Indians, who are always on the look-out for their armies when they traverse the forest, so as to avoid being attacked. I met with ten distinct species of them, nearly all of which have a different system of marching; eight were new to science when I sent them to England. Some are found commonly in every part of the country, and one is peculiar to the open campos of Santarem; but, as nearly all the species are found together at Ega, where the forest swarmed with their armies, I have left an account of the habits of the whole genus for this part of my narrative. The Ecitons resemble in their habits the Driver ants of Tropical Africa; but they have no close relationship with them in structure, and indeed belong to quite another sub-group of the ant tribe.

Like many other ants, the communities of Ecitons are composed, besides males and females, of two classes of workers, a large-headed (worker-major) and a small-headed (worker-minor) class; the large-heads have in some species greatly lengthened jaws, the small-heads have jaws always of the ordinary shape; but the two classes are not sharply defined in structure and function, except in two of the species. There is in all of them a little difference amongst the workers regarding the size of the head; but in some species this is not sufficient to cause a separation into classes, with division of labour; in others the jaws are so monstrously lengthened in the worker-majors, that they are incapacitated for taking part in the labours which the worker-minors perform; and again, in others the difference is so great that the distinction of classes becomes complete, one acting the part of soldiers, and the other that of workers. The peculiar feature in the habits of the Eciton genus is their hunting for prey in regular bodies or armies. It is this which chiefly distinguishes them from the genus of common red stinging ants, several species of which inhabit England, whose habit is to search for food in the usual irregular manner. All the Ecitons hunt in large organised bodies; but almost every species has its own special manner of hunting.

Eciton rapax.—One of the foragers, Eciton rapax, the giant of its genus, whose worker-majors are half an inch in length, hunts in single file through the forest. There is no division into classes amongst its workers, although the difference in size is very great, some being scarcely one-half the length of others. The head and jaws, however, are always of the same shape, and a gradation in size is presented from the largest to the smallest, so that all are able to take part in the common labours of the colony. The chief employment of the species seems to be plundering the nests of a large and defenceless ant of another genus (Formica), whose mangled bodies I have often seen in their possession as they were marching away. The armies of Eciton rapax are never very numerous.

Eciton legionis.—Another species, E. legionis, agrees with E. rapax in having workers not rigidly divisible into two classes; but it is much smaller in size, not differing greatly in this respect from our common English red ant (Myrmica rubra), which it also resembles in colour. The Eciton legionis lives in open places, and was seen only on the sandy campos of Santarem. The movement of its hosts were, therefore, much more easy to observe than those of all other kinds which inhabit solely the densest thickets; its sting and bite also were less formidable than those of other species. The armies of E. legionis consist of many thousands of individuals, and move in rather broad columns. They are just as quick to break line on being disturbed, and attack hurriedly and furiously any intruding object, as the other Ecitons. The species is not a common one, and I seldom had good opportunities of watching its habits. The first time I saw an army was one evening near sunset. The column consisted of two trains of ants, moving in opposite directions; one train empty-handed, the other laden with the mangled remains of insects, chiefly larvæ and pupæ of other ants. I had no difficulty in tracing the line to the spot from which they were conveying their booty; this was a low thicket; the Ecitons were moving rapidly about a heap of dead leaves; but as the short tropical twilight was deepening rapidly, and I had no wish to be benighted on the lonely campos, I deferred further examination until the next day.

On the following morning no trace of ants could be found near the place where I had seen them the preceding day, nor were there signs of insects of any description in the thicket; but at the distance of eighty or one hundred yards, I came upon the same army, engaged evidently on a razzia of a similar kind to that of the previous evening; but requiring other resources of their instinct, owing to the nature of the ground. They were eagerly occupied on the face of an inclined bank of light earth in excavating mines, whence, from a depth of eight or ten inches, they were extracting the bodies of a bulky species of ant of the genus Formica. It was curious to see them crowding round the orifices of the mines, some assisting their comrades to lift out the bodies of the Formicæ, and others tearing them in pieces, on account of their weight being too great for a single Eciton; a number of carriers seizing each a fragment, and carrying it off down the slope. On digging into the earth with a small trowel near the entrances of the mines, I found the nests of the Formicæ, with grubs and cocoons, which the Ecitons were thus invading, at a depth of about eight inches from the surface. The eager freebooters rushed in as fast as I excavated, and seized the ants in my fingers as I picked them out, so that I had some difficulty in rescuing a few entire for specimens. In digging the numerous mines to get at their prey, the little Ecitons seemed to be divided into parties, one set excavating and another set carrying away the grains of earth. When the shafts became rather deep, the mining parties had to climb up the sides each time they wished to cast out a pellet of earth; but their work was lightened for them by comrades, who stationed themselves at the mouth of the shaft, and relieved them of their burthens, carrying the particles, with an appearance of foresight which quite staggered me, a sufficient distance from the edge of the hole to prevent them from rolling in again. All the work seemed thus to be performed by intelligent co-operation amongst the host of eager little creatures; but still

Foraging Ants (Eciton drepanophora).

there was not a rigid division of labour, for some of them, whose proceedings I watched, acted at one time as carriers of pellets, and at another as miners, and all shortly afterwards assumed the office of conveyors of the spoil.

In about two hours all the nests of Formicæ were rifled, though not completely, of their contents, and I turned towards the army of Ecitons, which were carrying away the mutilated remains. For some distance there were many separate lines of them moving along the slope of the bank; but a short distance off these all converged, and then formed one close and broad column, which continued for some sixty or seventy yards, and terminated at one of those large termitariums or hillocks of white ants which are constructed of cemented material as hard as stone. The broad and compact column of ants moved up the steep sides of the hillock in a continued stream; many which had hitherto trotted along empty-handed, now turned to assist their comrades with their heavy loads, and the whole descended into a spacious gallery or mine opening on the top of the termitarium. I did not try to reach the nest, which I supposed to lie at the bottom of the broad mine, and therefore in the middle of the base of the stony hillock.

Eciton drepanophora.—The commonest species of foraging ants are the Eciton hamata and E. drepanophora, two kinds which resemble each other so closely that it requires attentive examination to distinguish them; yet their armies never intermingle, although moving in the same woods and often crossing each other's tracks. The two classes of workers look, at first sight, quite distinct, on account of the wonderful amount of difference between the largest individuals of the one, and the smallest of the other. There are dwarfs not more than one-fifth of an inch in length, with small heads and jaws, and giants half an inch in length, with monstrously enlarged head and jaws, all belonging to the same brood There is not, however, a distinct separation of classes, individuals existing which connect together the two extremes. These Ecitons are seen in the pathways of the forest at all places on the banks of the Amazons, travelling in dense columns of countless thousands. One or other of them is sure to be met with in a woodland ramble, and it is to them, probably, that the stories we read in books on South America apply, of ants clearing houses of vermin, although I heard of no instance of their entering houses, their ravages being confined to the thickest parts of the forest.

When the pedestrian falls in with a train of these ants, the first signal given him is a twittering and restless movement of small flocks of plain-coloured birds (ant thrushes) in the jungle. If this be disregarded until he advances a few steps farther, he is sure to fall into trouble, and find himself suddenly attacked by numbers of the ferocious little creatures. They swarm up his legs with incredible rapidity, each one driving its pincer-like jaws into his skin, and with the purchase thus obtained, doubling in its tail, and stinging with all its might. There is no course left but to run for it; if he is accompanied by natives, they will be sure to give the alarm, crying, "Tauóca!" and scampering at full speed to the other end of the column of ants. The tenacious insects who have secured themselves to his legs then have to be plucked off one by one, a task which is generally not accomplished without pulling them in twain, and leaving heads and jaws sticking in the wounds.

The errand of the vast ant-armies is plunder, as in the case of Eciton legionis; but from their moving always amongst dense thickets, their proceedings are not so easy to observe as in that species. Wherever they move, the whole animal world is set in commotion, and every creature tries to get out of their way. But it is especially the various tribes of wingless insects that have cause for fear, such as heavy-bodied spiders, ants of other species, maggots, caterpillars, larvæ of cockroaches, and so forth, all of which live under fallen leaves, or in decaying wood. The Ecitons do not mount very high on trees, and therefore the nestlings of birds are not much incommoded by them. The mode of operation of these armies, which I ascertained only after long-continued observation, is as follows. The main column, from four to six deep, moves forward in a given direction, clearing the ground of all animal matter, dead or alive, and throwing off here and there a thinner column to forage for a short time on the flanks of the main army, and re-enter it again after their task is accomplished. If some very rich place be encountered anywhere near the line of march, for example, a mass of rotten wood abounding in insect larvæ, a delay takes place, and a very strong force of ants is concentrated upon it. The excited creatures search every cranny and tear in pieces all the large grubs they drag to light. It is curious to see them attack wasps' nests, which are sometimes built on low shrubs. They gnaw away the papery covering to get at the larvæ, pupæ, and newly-hatched wasps, and cut everything to tatters, regardless of the infuriated owners which are flying about them. In bearing off their spoil in fragments, the pieces are apportioned to the carriers with some degree of regard to fairness of load, the dwarfs taking the smallest pieces, and the strongest fellows with small heads the heaviest portions. Sometimes two ants join together in carrying one piece, but the worker-majors, with their unwieldy and distorted jaws, are incapacitated from taking any part in the labour. The armies never march far on a beaten path, but seem to prefer the entangled thickets where it is seldom possible to follow them. I have traced an army sometimes for half a mile or more, but was never able to find one that had finished its day's course and returned to its hive. Indeed, I never met with a hive; whenever the Ecitons were seen they were always on the march.

I thought one day, at Villa Nova, that I had come upon a migratory horde of this indefatigable ant. The place was a tract of open ground near the river side, just outside the edge of the forest, and surrounded by rocks and shrubbery. A dense column of Ecitons was seen extending from the rocks on one side of the little haven, traversing the open space, and ascending the opposite declivity. The length of the procession was from sixty to seventy yards, and yet neither van nor rear was visible. All were moving in one and the same direction, except a few individuals on the outside of the column, which were running rearward, trotting along for a short distance, and then turning again to follow the same course as the main body. But these rearward movements were going on continually from one end to the other of the line, and there was every appearance of this being a means of keeping up a common understanding amongst all the members of the army, for the retrograding ants stopped very often for a moment to touch one or other of their onward-moving comrades with their antennæ, a proceeding which has been noticed in other ants, and supposed to be their mode of conveying intelligence. When I interfered with the column or abstracted an individual from it, news of the disturbance was very quickly communicated to a distance of several yards towards the rear, and the column at that point commenced retreating. All the small-headed workers carried in their jaws a little cluster of white maggots, which I thought at the time might be young larvæ of their own colony, but afterwards found reason to conclude were the grubs of some other species whose nests they had been plundering, the procession being most likely not a migration, but a column on a marauding expedition.

The position of the large-headed individuals in the marching column was rather curious. There was one of these extraordinary fellows to about a score of the smaller class; none of them carried anything in their mouths, but all trotted along empty-handed and outside the column, at pretty regular intervals from each other, like subaltern officers in a marching regiment of soldiers. It was easy to be tolerably exact in this observation, for their shining white heads made them very conspicuous amongst the rest, bobbing up and down as the column passed over the inequalities of the road. I did not see them change their position or take any notice of their small-headed comrades marching in the column, and when I disturbed the line they did not prance forth or show fight so eagerly as the others. These large-headed members of the community have been considered by some authors as a soldier class, like the similarly-armed caste in Termites; but I found no proof of this, at least in the present species, as they always seemed to be rather less pugnacious than the worker-minors, and their distorted jaws disabled them from fastening on a plane surface like the skin of an attacking animal. I am inclined, however, to think that they may act, in a less direct way, as protectors of the community, namely, as indigestible morsels to the flocks of ant-thrushes which follow the marching columns of these Ecitons, and are the most formidable enemies of the species. It is possible that the hooked and twisted jaws of the large-headed class may be effective weapons of annoyance when in the gizzards or stomachs of these birds, but I unfortunately omitted to ascertain whether this was really the fact.

The life of these Ecitons is not all work, for I frequently saw them very leisurely employed in a way that looked like recreation. When this happened, the place was always a sunny nook in the forest. The main column of the army and the branch columns, at these times, were in their ordinary relative positions; but instead of pressing forward eagerly, and plundering right and left, they seemed to have been all smitten with a sudden fit of laziness. Some were walking slowly about, others were brushing their antennæ with their fore feet; but the drollest sight was their cleaning one another. Here and there an ant was seen stretching forth first one leg and then another, to be brushed and washed by one or more of its comrades, who performed the task by passing the limb between the jaws and the tongue, finishing by giving the antennæ a friendly wipe. It was a curious spectacle, and one well calculated to increase one's amazement at the similarity between the instinctive actions of ants and the acts of rational beings, a similarity which must have been brought about by two different processes of development of the primary qualities of mind. The actions of these ants looked like simple indulgence in idle amusement. Have these little creatures, then, an excess of energy beyond what is required for labours absolutely necessary to the welfare of their species, and do they thus expend it in mere sportiveness, like young lambs or kittens, or in idle whims like rational beings? It is probable that these hours of relaxation and cleaning may be indispensable to the effective performance of their harder labours; but whilst looking at them, the conclusion that the ants were engaged merely in play was irresistible.

Eciton prædator.—This is a small dark-reddish species, very similar to the common red stinging ant of England. It differs from all other Ecitons in its habit of hunting, not in columns, but in dense phalanxes consisting of myriads of individuals, and was first met with at Ega, where it is very common. Nothing in insect movements is more striking than the rapid march of these large and compact bodies. Wherever they pass, all the rest of the animal world is thrown into a state of alarm. They stream along the ground and climb to the summits of all the lower trees, searching every leaf to its apex, and whenever they encounter a mass of decaying vegetable matter, where booty is plentiful, they concentrate, like other Ecitons, all their forces upon it, the dense phalanx of shining and quickly-moving bodies, as it spreads over the surface, looking like a flood of dark-red liquid. They soon penetrate every part of the confused heap, and then, gathering together again in marching order, onward they move. All soft-bodied and inactive insects fall an easy prey to them, and, like other Ecitons, they tear their victims in pieces for facility of carriage. A phalanx of this species, when passing over a tract of smooth ground, occupies a space of from four to six square yards; on examining the ants closely they are seen to move, not altogether in one straightforward direction, but in variously spreading contiguous columns, now separating a little from the general mass, now re-uniting with it. The margins of the phalanx spread out at times like a cloud of skirmishers from the flanks of an army. I was never able to find the hive of this species.

Blind Ecitons.—I will now give a short account of the blind species of Eciton. None of the foregoing kinds have eyes of the facetted or compound structure such as are usual in insects, and which ordinary ants (Formica) are furnished with, but all are provided with organs of vision composed each of a single lens. Connecting them with the utterly blind species of the genus, is a very stout-limbed Eciton, the E. crassicornis, whose eyes are sunk in rather deep sockets. This ant goes on foraging expeditions like the rest of its tribe, and attacks even the nests of other stinging species (Myrmica), but it avoids the light, always moving in concealment under leaves and fallen branches. When its columns have to cross a cleared space, the ants construct a temporary covered way with granules of earth, arched over, and holding together mechanically; under this the procession passes in secret, the indefatigable creatures repairing their arcade as fast as breaches are made in it.

Next in order comes the Eciton vastator, which has no eyes, although the collapsed sockets are plainly visible ; and, lastly, the Eciton erratica, in which both sockets and eyes have disappeared, leaving only a faint ring to mark the place where they are usually situated. The armies of E. vastator and E. erratica move, as far as I could learn, wholly under covered roads, the ants constructing them gradually but rapidly as they advance. The column of foragers pushes forward step by step under the protection of these covered passages, through the thickets, and on reaching a rotting log, or other promising hunting-ground, pour into the crevices in search of booty. I

Foraging ants (Eciton erratica) constructing a covered road—Soldiers sallying out on being disturbed.

have traced their arcades, occasionally, for a distance of one or two hundred yards; the grains of earth are taken from the soil over which the column is passing, and are fitted together without cement. It is this last-mentioned feature that distinguishes them from the similar covered roads made by Termites, who use their glutinous saliva to cement the grains together. The blind Ecitons, working in numbers, build up simultaneously the sides of their convex arcades, and contrive, in a surprising manner, to approximate them and fit in the key-stones without letting the loose uncemented structure fall to pieces. There was a very clear division of labour between the two classes of neuters in these blind species. The large-headed class, although not possessing monstrously lengthened jaws like the worker-majors in E. hamata and E. drepanophora, are rigidly defined in structure from the small-headed class, and act as soldiers, defending the working community (like soldier Termites) against all comers. Whenever I made a breach in one of their covered ways, all the ants underneath were set in commotion, but the worker-minors remained behind to repair the damage, whilst the large-heads issued forth in a most menacing manner, rearing their heads and snapping their jaws with an expression of the fiercest rage and defiance.

CHAPTER XIII.

EXCURSIONS BEYOND EGA.

Steamboat travelling on the Amazons—Passengers—Tunantins—Caishána Indians—The Jutahí—The Sapó—Marauá Indians—Fonte Boa—Journey to St. Paulo—Tucúna Indians—Illness—Descent to Pará—Changes at Pará—Departure for England.

November 7th, 1856.—Embarked on the Upper Amazons steamer, the "Tabatinga," for an excursion to Tunantins, a small semi-Indian settlement, lying 240 miles beyond Ega. The "Tabatinga" is an iron boat of about 170 tons burthen, built at Rio de Janeiro, and fitted with engines of fifty-horse power. The saloon, with berths on each side for twenty passengers, is above deck, and open at both ends to admit a free current of air. The captain or "commandante" was a lieutenant in the Brazilian navy, a man of polished, sailor-like address, and a rigid disciplinarian; his name, Senhor Nunes Mello Cardozo. I was obliged, as usual, to take with me a stock of all articles of food, except meat and fish, for the time I intended to be absent (three months); and the luggage, including hammocks, cooking utensils, crockery, and so forth, formed fifteen large packages. One bundle consisted of a mosquito tent, an article I had not yet had occasion to use on the river, but which was indispensable in all excursions beyond Ega; every person, man, woman, and child, requiring one, as without it existence would be scarcely possible. My tent was about eight feet long and five feet broad, and was made of coarse calico in an oblong shape, with sleeves at each end through which to pass the cords of a hammock. Under this shelter, which is fixed up every evening before sundown, one can read and write, or swing in one's hammock during the long hours which intervene before bed-time, and feel one's sense of comfort increased by having cheated the thirsty swarms of mosquitoes which fill the chamber.

We were four days on the road. The pilot, a mameluco of Ega, whom I knew very well, exhibited a knowledge of the river and powers of endurance which were quite remarkable. He stood all this time at his post, with the exception of three or four hours in the middle of each day, when he was relieved by a young man who served as apprentice; and he knew the breadth and windings of the channel, and the extent of all the yearly-shifting shoals from the Rio Negro to Loreto, a distance of more than a thousand miles. There was no slackening of speed at night, except during the brief but violent storms which occasionally broke upon us, and then the engines were stopped by the command of Lieutenant Nunes, sometimes against the wish of the pilot. The nights were often so dark that we passengers on the poop deck could not discern the hardy fellow on the bridge; but the steamer drove on at full speed, men being stationed on the look-out at the prow, to watch for floating logs, and one man placed to pass orders to the helmsman; the keel scraped against a sand-bank only once during the passage.

The passengers were chiefly Peruvians, mostly thin, anxious Yankee-looking men, who were returning home to the cities of Moyobamba and Chachapoyas, on the Andes, after a trading trip to the Brazilian towns on the Atlantic sea-board, whither they had gone six months previously, with cargoes of Panamá hats to exchange for European wares. These hats are made of the young leaflets of a palm tree, by the Indians and half-caste people who inhabit the eastern parts of Peru. They form almost the only article of export from Peru by way of the Amazons, but the money value is very great compared with the bulk of the goods, as the hats are generally of very fine quality, and cost from twelve shillings to six pounds sterling each; some traders bring down two or three thousands pounds' worth, folded into small compass in their trunks. The return cargoes consist of hardware, crockery, glass, and other bulky or heavy goods, but not of cloth, which, being of light weight, can be carried across the Andes from the ports on the Pacific to the eastern parts of Peru. All kinds of European cloth can be obtained at a much cheaper rate by this route than by the more direct way of the Amazons, the import duties of Peru being, as I was told, lower than those of Brazil, and the difference not being counterbalanced by increased expense of transit, on account of weight, over the passes of the Andes.

There was a great lack of amusement on board. The table was very well served, professed cooks being employed in these Amazonian steamers, and fresh meat insured by keeping on deck a supply of live bullocks and fowls, which are purchased whenever there is an opportunity on the road. The river scenery was similar to that already described as presented between the Rio Negro and Ega: long reaches of similar aspect, with two long, low lines of forest, varied sometimes with cliffs of red clay, appearing one after the other; an horizon of water and sky on some days limiting the view both up stream and down. We travelled, however, always near the bank, and, for my part, I was never weary of admiring the picturesque grouping and variety of trees, and the varied mantles of creeping plants which clothed the green wall of forest every step of the way. With the exception of a small village called Fonte Boa, retired from the main river, where we stopped to take in fire-wood, and which I shall have to speak of presently, we saw no human habitation the whole of the distance. The mornings were delightfully cool; coffee was served at sunrise, and a bountiful breakfast at ten o'clock; after that hour the heat rapidly increased until it became almost unbearable; how the engine-drivers and firemen stood it without exhaustion I cannot tell; it diminished after four o'clock in the afternoon, about which time dinner-bell rung, and the evenings were always pleasant.

November 11th to 30th.—The Tunantins is a sluggish black-water stream, about sixty miles in length, and towards its mouth from 100 to 200 yards in breadth. The vegetation on its banks has a similar aspect to that of the Rio Negro, the trees having small foliage of a sombre hue, and the dark piles of greenery resting on the surface of the inky water. The village is situated on the left bank, about a mile from the mouth of the river, and contains twenty habitations, nearly all of which are merely hovels, built of lath-work and mud. The short streets, after rain, are almost impassable, on account of the many puddles, and are choked up with weeds,—leguminous shrubs, and scarlet-flowered asclepias. The atmosphere in such a place, hedged in as it is by the lofty forest, and surrounded by swamps, is always close, warm, and reeking; and the hum and chirp of insects and birds cause a continual din. The small patch of weedy ground around the village swarms with plovers, sandpipers, striped herons, and scissor-tailed fly-catchers: and alligators are always seen floating lazily on the surface of the river in front of the houses.

On landing, I presented myself to Senhor Paulo Bitancourt, a good-natured half-caste, director of Indians of the neighbouring river Issá, who quickly ordered a small house to be cleared for me. This exhilarating abode contained only one room, the walls of which were disfigured by large and ugly patches of mud, the work of white ants. The floor was the bare earth, dirty and damp; the wretched chamber was darkened by a sheet of calico being stretched over the windows, a plan adopted here to keep out the Pium-flies, which float about in all shady places like thin clouds of smoke, rendering all repose impossible in the daytime whenever they can effect an entrance. My baggage was soon landed, and before the steamer departed I had taken gun, insect-net, and game-bag, to make a preliminary exploration of my new locality.

I remained here nineteen days, and, considering the shortness of the time, made a very good collection of monkeys, birds, and insects. A considerable number of the species (especially of insects) were different from those of the four other stations, which I examined on the south side of the Solimoens, and as many of these were "representative forms"* of others found on the opposite banks of the broad river, I concluded that there could have been no land connection between the two shores during, at least, the recent geological period. This conclusion is confirmed by the case of the Uakarí monkeys, described in the last chapter. All these strongly modified local races of insects confined to one side of the Solimoens (like the Uakarís), are such as have not been able to cross a wide treeless space such as a river. The acquisition which pleased me most, in this place, was a new species of butterfly (a Catagramma), which has since been named C. excelsior, owing to its surpassing in size and beauty all the previously-known species of its singularly beautiful genus. The upper surface of the wings is of the richest blue, varying in shade with the play of light, and on each side is a broad curved stripe of an orange colour. It is a bold flier, and is not confined, as I afterwards found, to the northern side of the river, for once I saw a specimen amidst a number of richly-coloured butterflies, flying about the deck of the steamer when we were anchored off Fonte Boa, 200 miles lower down the river.

* Species or races which take the place of other allied species or races.

With the exception of three mameluco families and a stray Portuguese trader, all the inhabitants of the village and neighbourhood are semi-civilized Indians of the Shumána and Passé tribes. The forests of the Tunantins, however, are inhabited by a tribe of wild Indians called Caishánas, who resemble much, in their social condition and manners, the debased Múras of the Lower Amazons, and have, like them, shown no aptitude for civilized life in any shape. Their huts commence at the distance of an hour's walk from the village, along gloomy and narrow forest-paths. My first and only visit to a Caishána dwelling was accidental. One day, having extended my walk further than usual, and followed one of the forest-roads until it became a mere *picada*, or hunters' track, I came suddenly upon a wel' trodden pathway, bordered on each side with Lycc-podia of the most elegant shapes, the tips of the fronds stretching almost like tendrils down the little earthy slopes which formed the edge of the path. The road, though smooth, was narrow and dark, and in many places blocked up by trunks of felled trees, which had been apparently thrown across by the timid Indians on purpose to obstruct the way to their habitations. Half-a-mile of this shady road brought me to a small open space on the banks of a brook or creek, on the skirts of which stood a conical hut with a very low doorway. There was also an open shed, with stages made of split palm stems, and a number of large wooden troughs. Two or three dark-skinned children, with a man and woman, were in the shed; but, immediately on espying me, all of them ran to the hut, bolting through the little doorway like so many wild animals scared into their burrows. A few moments after, the man put his head out with a look of great distrust; but on my making the most friendly gestures I could think of, he came forth with the children. They were all smeared with black mud and paint; the only clothing of the elders was a kind of apron made of the inner bark of the sapucaya tree, and the savage aspect of the man was heightened by his hair hanging over his forehead to the eyes. I stayed about two hours in the neighbourhood, the children gaining sufficient confidence to come and help me to search for insects. The only weapon used by the Caishánas is the blow-gun, and this is employed only in shooting animals for food. They are not a warlike people, like most of the neighbouring tribes on the Japurá and Issá.

The whole tribe of Cashánas does not exceed in number 400 souls. None of them are baptized Indians, and they do nct dwell in villages, like the more advanced sections of the Tupí stock; but each family has its own solitary hut. They are quite harmless, do not practise tattooing, or perforate their ears and noses in any way. Their social condition is of a low type, very little removed, indeed, from that of the brutes living in the same forests. They do not appear to obey any common chief, and I could not make out that they had Pajés, or medicine men, those rudest beginnings of a priest class. Symbolical or masked dances, and ceremonies in honour of the Juruparí, or demon, customs which prevail amongst all the surrounding tribes, are unknown to the Caishánas. There is amongst them a trace of festival keeping; but the only ceremony used is the drinking of cashirí beer, and fermented liquors made of Indian corn, bananas, and so forth. These affairs, however, are conducted in a degenerate style, for they do not drink to intoxication, or sustain the orgies for several days and nights in succession, like the Jurís, Passés, and Tucúnas. The men play a musical instrument, made of pieces of stem of the arrow-grass cut in different lengths and arranged like Pan-pipes. With this they while away whole hours, lolling in ragged bast hammocks slung in their dark, smoky huts. The Tunantins people say that the Caishánas have persecuted the wild animals and birds to such an extent near their settlements that there is now quite a scarcity of animal food. If they kill a toucan, it is considered an important event, and the bird is made to serve as a meal for a score or more persons. They boil the meat in earthenware kettles filled with Tucupí sauce, and eat it with beiju, or mandioca cakes. The women are not allowed to taste of the meat, but forced to content themselves with sopping pieces of cake in the liquor.

November 30th.—I left Tunantins in a trading schooner of eighty tons burthen belonging to Senhor Batalha, a tradesman of Ega, which had been out all the summer collecting produce, and was commanded by a friend of mine, a young Paraense, named Francisco Raiol. We arrived on the 3rd of December at the mouth of the Jutahí, a considerable stream about half a mile broad, and flowing with a very sluggish current. This is one of a series of six rivers, from 400 to 1,000 miles in length, which flow from the south-west through unknown lands lying between Bolivia and the Upper Amazons, and enter this latter river between the Madeira and the Ucayáli. We remained at anchor four days within the mouth of the Sapó, a small tributary of the Jutahí flowing from the south-east: Senhor Raiol having to send an igarité to the Cupatána, a large tributary some few miles farther up the river, to fetch a cargo of salt fish. During this time we made several excursions in the montaria to various places in the neighbourhood. Our longest trip was to some Indian houses, a distance of fifteen or eighteen miles up the Sapó, a journey made with one Indian paddler, and occupying a whole day. The stream is not more than forty or fifty yards broad; its waters are darker in colour than those of the Jutahí, and flow, as in all these small rivers, partly under shade between two lofty walls of forest. We passed, in ascending, seven habitations, most of them hidden in the luxuriant foliage of the banks; their sites being known only by small openings in the compact wall of forest, and the presence of a canoe or two tied up in little shady ports. The inhabitants are chiefly Indians of the Marauá tribe, whose original territory comprised all the small by-streams lying between the Jutahí and the Juruá, near the mouths of both these great tributaries. They live in separate families or small hordes; have no common chief, and are considered as a tribe little disposed to adopt civilised customs or be friendly with the whites. One of the houses belonged to a Jurí family, and we saw the owner, an erect, noble-looking old fellow, tattooed, as customary with his tribe, in a large patch over the middle of his face, fishing under the shade of a colossal tree in his port with hook and line. He saluted us in the usual grave and courteous manner of the better sort of Indians as we passed by.

We reached the last house, or rather two houses, about ten o'clock, and spent there several hours during the great heat of mid-day. The houses, which stood on a high clayey bank, were of quadrangular shape, partly open like sheds, and partly enclosed with rude mud-walls, forming one or more chambers. The inhabitants, a few families of Marauás, comprising about thirty persons, received us in a frank, smiling manner: a reception which may have been due to Senhor Raiol being an old acquaintance and somewhat of a favourite. None of them were tattooed; but the men had great holes pierced in their ear-lobes, in which they insert plugs of wood, and their lips were drilled with smaller holes. One of the younger men, a fine strapping fellow nearly six feet high, with a large aquiline nose, who seemed to wish to be particularly friendly with me, showed me the use of these lip-holes, by fixing a number of little white sticks in them, and then twisting his mouth about and going through a pantomime to represent defiance in the presence of an enemy. Nearly all the people were disfigured by dark blotches on the skin, the effect of a cutaneous disease very prevalent in this part of the country. The face of one old man was completely blackened, and looked as though it had been smeared with blacklead, the blotches having coalesced to form one large patch. Others were simply mottled; the black spots were hard and rough, but not scaly, and were margined with rings of a colour paler than the natural hue of the skin. I had seen many Indians and a few half-castes at Tunantins, and afterwards saw others at Fonte Boa, blotched in the same way. The disease would seem to be contagious, for I was told that a Portuguese trader became disfigured with it after cohabiting some years with an Indian woman. It is curious that, although prevalent in many places on the Solimoens, no resident of Ega exhibited signs of the disease: the early explorers of the country, on noticing spotted skins to be very frequent in certain localities, thought they were peculiar to a few tribes of Indians. The younger children in these houses on the Sapó were free from spots; but two or three of them, about ten years of age, showed signs of their commencement in rounded yellowish patches on the skin, and these appeared languid and sickly, although the blotched adults seemed not to be affected in their general health. A middle-aged half-caste at Fonte Boa told me he had cured himself of the disorder by strong doses of salsaparilla; the black patches had caused the hair of his beard and eyebrows to fall off, but it had grown again since his cure.

We left these friendly people about four o'clock in the afternoon, and in descending the umbrageous river, stopped, about half-way down, at another house, built in one of the most charming situations I had yet seen in this country. A clean, narrow, sandy pathway led from the shady port to the house, through a tract of forest of indescribable luxuriance. The buildings stood on an eminence in the middle of a level cleared space; the firm sandy soil, smooth as a floor, forming a broad terrace around them. The owner was a semi-civilised Indian, named Manoel; a dull, taciturn fellow, who, together with his wife and children, seemed by no means pleased at being intruded on in their solitude. The family must have been very industrious; for the plantations were very extensive, and included a little of almost all kinds of cultivated tropical productions: fruit trees, vegetables, and even flowers for ornament. The silent old man had surely a fine appreciation of the beauties of nature: for the site he had chosen commanded a view of surprising magnificence over the summits of the forest, and, to give finish to the prospect, he had planted a large quantity of banana trees in the foreground, thus concealing the charred and dead stumps which would otherwise have marred the effect of the rolling sea of greenery. The only information I could get out of Manoel was, that large flocks of richly-coloured birds came down in the fruit season and despoiled his trees. The sun set over the tree tops before we left this little Eden, and the remainder of our journey was made slowly and pleasantly, under the chequered shades of the river banks, by the light of the moon.

December 7th.—Arrival at Fonte Boa; a wretched, muddy, and dilapidated village, situated two or three miles within the mouth of a narrow by-stream called the Cayhiar-hy, which runs almost as straight as an artificial canal between the village and the main Amazons. The character of the vegetation and soil here was different from that of all other localities I had hitherto examined; I had planned, therefore, to devote six weeks to the place. Having written beforehand to one of the principal inhabitants, Senhor Venancio, a house was ready for me on landing. The only recommendation of the dwelling was its coolness. It was, in fact, decidedly damp; the plastered walls bore a crop of green mould, and a slimy moisture oozed through the black, dirty floor; the rooms were large, but lighted by miserable little holes in place of windows. The village is built on a clayey plateau, and the ruinous houses are arranged round a large square, which is so choked up with tangled bushes that it is quite impassable, the lazy inhabitants having allowed the fine open space to relapse into jungle. The stiff clayey eminence is worn into deep gullies which slope towards the river, and the ascent from the port in rainy weather is so slippery that one is obliged to crawl up to the streets on all-fours. A large tract of ground behind the place is clear of forest, but this, as well as the streets and gardens, is covered with a dense, tough carpet of shrubs, having the same wiry nature as our common heath. Beneath its deceitful covering the soil is always moist and soft, and in the wet season the whole is converted into a glutinous mud swamp. There is a very pretty church in one corner of the square, but in the rainy months of the year (nine out of twelve) the place of worship is almost inaccessible to the inhabitants on account of the mud, the only means of getting to it being by hugging closely the walls and palings, and so advancing sideways step by step.

I remained in this delectable place until the 25th of January, 1857. Fonte Boa, in addition to its other amenities, has the reputation throughout the country of being the head-quarters of mosquitoes, and it fully deserves the title. They are more annoying in the houses by day than by night, for they swarm in the dark and damp rooms, keeping, in the daytime, near the floor, and settling by half-dozens together on the legs. At night the calico tent is a sufficient protection; but this is obliged to be folded every morning, and in letting it down before sunset, great care is required to prevent even one or two of the tormentors from stealing in beneath, their insatiable thirst for blood, and pungent sting, making these enough to spoil all comfort. In the forest the plague is much worse; but the forest-mosquito belongs to a different species from that of the town, being much larger, and having transparent wings; it is a little cloud that one carries about one's person every step on a woodland ramble, and their hum is so loud that it prevents one hearing well the notes of birds. The town-mosquito has opaque speckled wings, a less severe sting, and a silent way of going to work; the inhabitants ought to be thankful the big noisy fellows never come out of the forest. In compensation for the abundance of mosquitoes, Fonte Boa has no piums; there was, therefore, some comfort outside one's door in the daytime; the comfort, however, was lessened by there being scarcely any room in front of the house to sit down or walk about; for, on our side of the square, the causeway was only two feet broad, and to step over the boundary, formed by a line of slippery stems of palms, was to sink up to the knees in a sticky swamp.

Notwithstanding damp and mosquitoes, I had capital health, and enjoyed myself much at Fonte Boa; swampy and weedy places being generally more healthy than dry ones on the Amazons, probably owing to the absence of great radiation of heat from the ground. The forest was extremely rich and picturesque, although the soil was everywhere clayey and cold, and broad pathways threaded it for many a mile over hill and dale. In every hollow flowed a sparkling brook, with perennial and crystal waters. The margins of these streams were paradises of leafiness and verdure; the most striking feature being the variety of ferns, with immense leaves, some terrestrial, others climbing over trees, and two, at least, arborescent. I saw here some of the largest trees I had yet seen: there was one especially, a cedar, whose colossal trunk towered up for more than a hundred feet, straight as an arrow; I never saw its crown, which was lost to view, from below, beyond the crowd of lesser trees which surrounded it. Birds and monkeys in this glorious forest were very abundant; the bear-like Pithecia hirsuta being the most remarkable of the monkeys, and the Umbrella Chatterer and Curl-crested Toucans amongst the most beautiful of the birds. The Indians and half-castes of the village have made their little plantations, and built huts for summer residence on the banks of the rivulets, and my rambles generally terminated at one or other of these places. The people were always cheerful and friendly, and seemed to be glad when I proposed to join them at their meals, contributing the contents of my provision-bag to the dinner, and squatting down amongst them on the mat.

The village was formerly a place of more importance than it now is, a great number of Indians belonging to the most industrious tribes, Shumánas, Passés, and Cambévas, having settled on the site and adopted civilised habits, their industry being directed by a few whites, who seem to have been men of humane views as well as enterprising traders. One of these old employers, Senhor Guerreiro, a well-educated Paraense, was still trading on the Amazons when I left the country, in 1859: he told me that forty years previously Fonte Boa was a delightful place to live in. The neighbourhood was then well cleared, and almost free from mosquitoes, and the Indians were orderly, industrious, and happy. What led to the ruin of the settlement was the arrival of several Portuguese and Brazilian traders of a low class, who in their eagerness for business taught the easy-going Indians all kinds of trickery and immorality. They enticed the men and women away from their old employers, and thus broke up the large establishments, compelling the principals to take their capital to other places. At the time of my visit there were few pure-blood Indians at Fonte Boa, and no true whites. The inhabitants seemed to be nearly all mamelucos, and were a loose-living, rustic, plain-spoken, and ignorant set of people. There was no priest or schoolmaster within 150 miles, and had not been any for many years: the people seemed to be almost without government of any kind, and yet crime and deeds of violence appeared to be of very rare occurrence. The principal man of the village, one Senhor Justo, was a big, coarse, energetic fellow, sub-delegado of police, and the only tradesman who owned a large vessel running directly between Fonte Boa and Pará. He had recently built a large house, in the style of middle-class dwellings of towns, namely, with brick floors and tiled roof, the bricks and tiles having been brought from Pará, 1,500 miles distant, the nearest place where they are manufactured in surplus. When Senhor Justo visited me, he was much struck with the engravings in a file of "Illustrated London News," which lay on my table. It was impossible to resist his urgent entreaties to let him have some of them "to look at," so one day he carried off a portion of the papers on loan. A fortnight afterwards, on going to request him to return them, I found the engravings had been cut out, and stuck all over the newly whitewashed walls of his chamber, many of them upside down. He thought a room thus decorated with foreign views would increase his importance amongst his neighbours, and when I yielded to his wish to keep them, was boundless in demonstrations of gratitude, ending by shipping a boat-load of turtles for my use at Ega.

These neglected and rude villagers still retained many religious practices which former missionaries or priests had taught them. The ceremony which they observed at Christmas, like that described as practised by negroes in a former chapter, was very pleasing for its simplicity, and for the heartiness with which it was conducted. The church was opened, dried, and swept clean a few days before Christmas-eve, and on the morning all the women and children of the village were busy decorating it with festoons of leaves and wild flowers. Towards midnight it was illuminated inside and out with little oil lamps, made of clay, and the image of the "menino Deus," or Child-God, in its cradle, was placed below the altar, which was lighted up with rows of wax candles, very lean ones, but the best the poor people could afford. All the villagers assembled soon afterwards, dressed in their best, the women with flowers in their hair, and a few simple hymns, totally irrelevant to the occasion, but probably the only ones known by them, were sung kneeling; an old half-caste, with black spotted face, leading off the tunes. This finished, the congregation rose, and then marched in single file up one side of the church and down the other, singing together a very pretty marching chorus, and each one, on reaching the little image, stooping to kiss the end of a ribbon which was tied round its waist. Considering that the ceremony was got up of their own freewill, and at considerable expense, I thought it spoke well for the good intentions and simplicity of heart of these poor neglected villagers.

I left Fonte Boa, for Ega, on the 25th of January, making the passage by steamer, down the middle of the current, in sixteen hours. The sight of the clean and neat little town, with its open spaces, close-cropped grass, broad lake, and white sandy shores, had a most exhilarating effect, after my trip into the wilder parts of the country. The district between Ega and Loreto, the first Peruvian village on the river, is, indeed, the most remote, thinly-peopled, and barbarous of the whole line of the Amazons, from ocean to ocean. Beyond Loreto, signs of civilisation, from the side of the Pacific, begin to be numerous; and from Ega, downwards, the improvement is felt from the side of the Atlantic.

September 5th, 1857.—Again embarked on the "Tabatinga," this time for a longer excursion than the last, namely, to St. Paulo de Olivença, a village higher up than any I had yet visited, being 260 miles distant in a straight line from Ega, or about 400 miles following the bends of the river.

The waters are now nearly at their lowest point; but this made no difference to the rate of travelling, night or day. Several of the Paraná-mirims, or by-channels, which the steamer threads in the season of full-water, to save a long circuit, were now dried up, their empty beds looking like deep sandy ravines in the midst of the thick forest. The large sand islands, and miles of sandy beach, were also uncovered; and these, with the swarms of large aquatic birds, storks, herons, ducks, waders, and spoon-bills, which lined their margins in certain places, made the river view much more varied and animated than it is in the season of the flood. Alligators of large size were common near the shores, lazily floating, and heedless of the passing steamer. The passengers amused themselves by shooting at them from the deck with a double-barrelled rifle we had on board. The sign of a mortal hit was the monster turning suddenly over, and remaining floating, with its white belly upwards. Lieutenant Nunes wished to have one of the dead animals on board, for the purpose of opening the abdomen, and, if a male, extracting a part which is held in great estimation amongst Brazilians as a "remedio," charm or medicine. The steamer was stopped, and a boat sent, with four strong men, to embark the beast; the body, however, was found too heavy to be lifted into the boat; so a rope was passed round it, and the hideous creature towed alongside, and hoisted on deck by means of the crane, which was rigged for the purpose. It had still some sparks of life, and when the knife was applied, lashed its tail, and opened its enormous jaws, sending the crowd of bystanders flying in all directions. A blow with a hatchet, on the crown of the head, gave him his quietus at last. The length of the animal was fifteen feet; but this statement can give but an imperfect idea of its immense bulk and weight. The number of turtles which were seen swimming in quiet shoaly bays passed on the road, also gave us much amusement. They were seen by dozens ahead, with their snouts peering above the surface of the water, and, on the steamer approaching, turning round to stare, but not losing confidence till the vessel had nearly passed, when they appeared to be suddenly smitten with distrust, diving like ducks under the stream.

The river scenery about the mouth of the Japurá is extremely grand, and was the subject of remark amongst the passengers. Lieutenant Nunes gave it as his opinion, that there was no diminution of width or grandeur in the mighty stream up to this point, a distance of 1,500 miles from the Atlantic; and yet we did not here see the two shores of the river on both sides at once; lines of islands or tracts of alluvial land, having by-channels in the rear, intercepting the view of the northern mainland, and sometimes also of the southern. Beyond the Issá, however, the river becomes evidently narrower, being reduced to an average width of about a mile; there were then no longer those magnificent reaches, with blank horizons, which occur lower down. We had a dark and rainy night after passing Tunantins, and the passengers were all very uneasy on account of the speed at which we were travelling, twelve miles an hour, with every plank vibrating with the force of the engines. Many of them could not sleep, myself amongst the number. At length, a little after midnight, a sudden shout startled us: "Back her!" (English terms being used in matters relating to steam-engines). The pilot instantly sprung to the helm, and in a few moments we felt our paddle-box brushing against the wall of forest into which we had nearly driven headlong. Fortunately the water was deep close up to the bank. Early in the morning of the 10th of September we anchored in the port of St. Paulo, after five days' quick travelling from Ega.

St. Paulo is built on a high hill, on the southern bank of the river. The hill is formed of the same Tabatinga clay, which occurs at intervals over the whole valley of the Amazons, but nowhere rises to so great an elevation as here, the height being about 100 feet above the mean level of the river. The ascent from the port is steep and slippery; steps and resting-places have been made to lighten the fatigue of mounting, otherwise the village would be almost inaccessible, especially to porters of luggage and cargo, for there are no means of making a circuitous road of more moderate slope, the hill being steep on all sides, and surrounded by dense forests and swamps. The place contains about 500 inhabitants, chiefly half-castes and Indians of the Tucúna and Collína tribes, who are very little improved from their primitive state. The streets are narrow, and in rainy weather inches deep in mud; many houses are of substantial structure, but in a ruinous condition, and the place altogether presents the appearance, like Fonte Boa, of having seen better days. Signs of commerce, such as meet the eye at Ega, could scarcely be expected in this remote spot, situate 1,800 miles, or seven months' round voyage by sailing-vessels, from Pará, the nearest market for produce. A very short experience showed that the inhabitants were utterly debased, the few Portuguese and other immigrants having, instead of promoting industry, adopted the lazy mode of life of the Indians, spiced with the practice of a few strong vices of their own introduction.

The head-man of the village, Senhor Antonia Ribeiro, half-white, half-Tucúna, prepared a house for me on landing, and introduced me to the principal people. The summit of the hill is grassy table-land, of two or three hundred acres in extent. The soil is not wholly clay, but partly sand and gravel; the village itself, however, stands chiefly on clay, and the streets, therefore, after heavy rains, become filled with muddy puddles. On damp nights the chorus of frogs and toads, which swarm in weedy back-yards, creates such a bewildering uproar, that it is impossible to carry on a conversation in-doors except by shouting. My house was damper even than the one I occupied at Fonte Boa, and this made it extremely difficult to keep my collections from being spoilt by mould. But the general humidity of the atmosphere in this part of the river was evidently much greater than it is lower down; it appears to increase gradually in ascending from the Atlantic to the Andes. It was impossible at St. Paulo to keep salt for many days in a solid state, which was not the case at Ega, when the baskets in which it is contained were well wrapped in leaves. Six degrees further westward, namely, at the foot of the Andes, the dampness of the climate of the Amazonian forest region appears to reach its acme, for Poeppig found at Chinchao that the most refined sugar in a few days, dissolved into syrup, and the best gunpowder became liquid, even when enclosed in canisters. At St. Paulo refined sugar kept pretty well in tin boxes, and I had no difficulty in keeping my gunpowder dry in canisters, although a gun loaded over-night could very seldom be fired off in the morning.

I remained at St. Paulo five months; five years would not have been sufficient to exhaust the treasures of its neighbourhood in Zoology and Botany. Although now a forest-rambler of ten years' experience, the beautiful forest which surrounds this settlement gave me as much enjoyment as if I had only just landed for the first time in a tropical country. The plateau on which the village is built extends on one side nearly a mile into the forest, but on the other side the descent into the lowland begins close to the streets; the hill sloping abruptly towards a boggy meadow surrounded by woods, through which a narrow winding path continues the slope down to a cool shady glen, with a brook of icy-cold water flowing at the bottom. At mid-day the vertical sun penetrates into the gloomy depths of this romantic spot, lighting up the leafy banks of the rivulet and its clean sandy margins, where numbers of scarlet, green, and black tanagers and brightly-coloured butterflies sport about in the stray beams. Sparkling brooks, large and small, traverse the glorious forest in almost every direction, and one is constantly meeting, whilst rambling through the thickets, with trickling rills and bubbling springs, so well provided is the country with moisture. Some of the rivulets flow over a sandy and pebbly bed, and the banks of all are clothed with the most magnificent vegetation conceivable. I had the almost daily habit, in my solitary walks, of resting on the clean banks of these swift-flowing streams, and bathing for an hour at a time in their bracing waters; hours which now remain amongst my most pleasant memories. The broad forest roads continue, as I was told, a distance of several days' journey into the interior, which is peopled by Tucúnas and other Indians, living in scattered houses and villages nearly in their primitive state, the nearest village lying about six miles from St. Paulo. The banks of all the streams are dotted with palm-thatched dwellings of Tucúnas, all half buried in the leafy wilderness, the scattered families having chosen the coolest and shadiest nooks for their abodes.

I frequently heard in the neighbourhood of these huts the "realejo" or organ bird (Cyphorhinus cantans), the most remarkable songster, by far, of the Amazonian forests. When its singular notes strike the ear for the first time, the impression cannot be resisted that they are produced by a human voice. Some musical boy must be gathering fruit in the thickets, and is singing a few notes to cheer himself. The tones become more fluty and plaintive; they are now those of a flageolet, and notwithstanding the utter impossibility of the thing, one is for the moment convinced that somebody is playing that instrument. No bird is to be seen, however closely the surrounding trees and bushes may be scanned, and yet the voice seems to come from the thicket close to one's ears. The ending of the song is rather disappointing. It begins with a few very slow and mellow notes, following each other like the commencement of an air; one listens expecting to hear a complete strain, but an abrupt pause occurs, and then the song breaks down, finishing with a number of clicking unmusical sounds like a piping barrel organ out of wind and tune. I never heard the bird on the Lower Amazons, and very rarely heard it even at Ega; it is the only songster which makes an impression on the natives, who sometimes rest their paddles whilst travelling in their small canoes along the shady by-streams, as if struck by the mysterious sounds.

The Tucúna Indians are a tribe resembling much the Shumánas, Passés, Jurís, and Mauhés in their physical appearance and customs. They lead like those tribes a settled agricultural life, each horde obeying a chief of more or less influence, according to his energy and ambition, and possessing its pajé or medicine man, who fosters its superstitions; but they are much more idle and debauched than other Indians belonging to the superior tribes. They are not so warlike and loyal as the Mundurucús, although resembling them in many respects, nor have they the slender figures, dignified mien, and gentle disposition of the Passés; there are, however, no trenchant points of difference to distinguish them from these highest of all the tribes. Both men and women are tattooed, the pattern being sometimes a scroll on each cheek, but generally rows of short straight lines on the face. Most of the older people wear bracelets, anklets, and garters of tapir-hide or tough bark; in their homes they wear no other dress except on festival days, when they ornament themselves with feathers or masked cloaks made of the inner bark of a tree. They were very shy when I made my first visits to their habitations in the forest, all scampering off to the thicket when I approached, but on subsequent days they became more familiar, and I found them a harmless, good-natured people.

A great part of the horde, living at the first Maloca or village dwell in a common habitation, a large oblong hut built and arranged inside with such a disregard of all symmetry, that it appeared as though constructed by a number of hands, each working independently, stretching a rafter or fitting in a piece of thatch, without reference to what his fellow-labourers were doing. The walls as well as the roof are covered with thatch of palm-leaves; each piece consisting of leaflets plaited and attached in a row to a lath many feet in length. Strong upright posts support the roof, hammocks being slung between them, leaving a free space for passage and for fires in the middle, and on one side is an elevated stage (*girao*) overhead, formed of split palm-stems. The Tucúnas excel most of the other tribes in the manufacture of pottery. They make broad-mouthed jars for Tucupí sauce, caysúma or mandioca beer, capable of holding twenty or more gallons, ornamenting them outside with crossed diagonal streaks of various colours. These jars, with cooking-pots, smaller jars for holding water, blow-guns, quivers, matirí bags* full of small articles, baskets, skins of animals, and so forth, form the principal part of the furniture of their huts, both large and small. The dead bodies of their chiefs are interred, the knees doubled up, in large jars under the floors of their huts.

The semi-religious dances and drinking bouts usual amongst the settled tribes of Amazonian Indians are indulged in to

* These bags are formed of remarkably neat twine made of Bromelia fibres elaborately knitted, all in one piece, with sticks; a belt of the same material, but more closely woven, being attached to the top to suspend them by. They afford good examples of the mechanical ability of these Indians. The Tucúnas also possess the art of skinning and stuffing birds, the handsome kinds of which they sell in great numbers to passing travellers.

MASKED DANCE AND WEDDING-FEAST OF TUCUNA INDIANS.

greater excess by the Tucúnas than they are by most other tribes. The Juruparí or Demon is the only superior being they have any conception of, and his name is mixed up with all their ceremonies, but it is difficult to ascertain what they consider to be his attributes. He seems to be believed in simply as a mischievous imp, who is at the bottom of all those mishaps of their daily life, the causes of which are not very immediate or obvious to their dull understandings. It is vain to try to get information out of a Tucúna on this subject; they affect great mystery when the name is mentioned, and give very confused answers to questions: it was clear, however, that the idea of a spirit as a beneficent God or Creator had not entered the minds of these Indians. There is great similarity in all their ceremonies and mummeries, whether the object is a wedding, the celebration of the feast of fruits, the plucking of the hair from the heads of their children, or a holiday got up simply out of a love of dissipation. Some of the tribe on these occasions deck themselves with the bright-coloured feathers of parrots and macaws. The chief wears a head-dress or cap made by fixing the breast-feathers of the Toucan on a web of Bromelia twine, with erect tail plumes of macaws rising from the crown. The cinctures of the arms and legs are also then ornamented with bunches of feathers. Others wear masked dresses: these are long cloaks reaching below the knee, and made of the thick whitish-coloured inner bark of a tree, the fibres of which are interlaced in so regular a manner, that the material looks like artificial cloth. The cloak covers the head; two holes are cut out for the eyes, a large round piece of the cloth stretched on a rim of flexible wood is stitched on each side to represent ears, and the features are painted in exaggerated style with yellow, red, and black streaks. The dresses are sewn into the proper shapes with thread made of the inner bark of the Uaissíma tree. Sometimes grotesque head-dresses, representing monkeys' busts or heads of other animals, made by stretching cloth or skin over a basket-work frame, are worn at these holidays. The biggest and ugliest mask represents the Juruparí. In these festival habiliments the Tucúnas go through their monotonous see-saw and stamping dances accompanied by singing and drumming, and keep up the sport often for three or four days and nights in succession, drinking enormous quantities of caysúma, smoking tobacco, and snuffing paricá powder.

I could not learn that there was any deep symbolical meaning in these masked dances, or that they commemorated any past event in the history of the tribe. Some of them seem vaguely intended as a propitiation of the Juruparí, but the masker who represents the demon sometimes gets drunk along with the rest, and is not treated with any reverence. From all I could make out, these Indians preserve no memory of events going beyond the times of their fathers or grandfathers. Almost every joyful event is made the occasion of a festival: weddings amongst the rest. A young man who wishes to wed a Tucúna girl has to demand her hand of her parents, who arrange the rest of the affair, and fix a day for the marriage ceremony. A wedding which took place in the Christmas week whilst I was at St. Paulo, was kept up with great spirit for three or four days, flagging during the heats of mid-day, but renewing itself with increased vigour every evening. During the whole time the bride, decked out with feather ornaments, was under the charge of the older squaws, whose business seemed to be sedulously to keep the bridegroom at a safe distance until the end of the dreary period of dancing and boosing. The Tucúnas have the singular custom, in common with the Collínas and Mauhés, of treating their young girls, on their showing the first signs of womanhood, as if they had committed some crime. They are sent up to the girao under the smoky and filthy roof, and kept there on very meagre diet, sometimes for a whole month. I heard of one poor girl dying under this treatment.

The only other tribe of this neighbourhood concerning which I obtained any information were the Majerónas, whose territory embraces several hundred miles of the western bank of the river Jauarí, an affluent of the Solimoens, 120 miles beyond St. Paulo. These are a fierce, indomitable, and hostile people, like the Aráras of the river Madeira; they are also cannibals. The navigation of the Jauarí is rendered impossible on account of the Majerónas lying in wait on its banks to intercept and murder all travellers, especially whites.

Four months before my arrival at St. Paulo, two young halfcastes (nearly white) of the village went to trade on the Jauarí; the Majerónas having shown signs of abating their hostility for a year or two previously. They had not been long gone, when their canoe returned with the news that the two young fellows had been shot with arrows, roasted, and eaten by the savages. José Patricio, with his usual activity in the cause of law and order, despatched a party of armed men of the National Guard to the place to make inquiries, and, if the murder should appear to be unprovoked, to retaliate. When they reached the settlement of the horde who had eaten the two men, it was found evacuated, with the exception of one girl, who had been in the woods when the rest of her people had taken flight, and whom the guards brought with them to St. Paulo. It was gathered from her, and from other Indians on the Jauarí, that the young men had brought their fate on themselves through improper conduct towards the Majeróna women. The girl, on arriving at St. Paulo, was taken care of by Senhor José Patricio, baptised under the name of Maria, and taught Portuguese. I saw a good deal of her, for my friend sent her daily to my house to fill the water jars, make the fire, and so forth. I also gained her goodwill by extracting the grub of an Œstrus fly from her back, and thus cured her of a painful tumour. She was decidedly the best-humoured and, to all appearance, the kindest-hearted specimen of her race I had yet seen. She was tall and very stout; in colour much lighter than the ordinary Indian tint, and her ways altogether were more like those of a careless, laughing country wench, such as might be met with any day amongst the labouring class in villages in our own country, than a cannibal. I heard this artless maiden relate, in the coolest manner possible, how she ate a portion of the bodies of the young men whom her tribe had roasted. But what increased greatly the incongruity of this business, the young widow of one of the victims, a neighbour of mine, happened to be present during the narrative, and showed her interest in it by laughing at the broken Portuguese in which the girl related the horrible story.

In the fourth month of my sojourn at St. Paulo I had a serious illness, an attack of the "sizoens," or ague of the country, which, as it left me with shattered health and damped enthusiasm, led to my abandoning the plan I had formed of proceeding to the Peruvian towns of Pebas and Moyobamba, 250 and 600 miles further west, and so completing the examination of the Natural History of the Amazonian plains up to the foot of the Andes. I made a very large collection at St. Paulo, and employed a collector at Tabatinga and on the banks of the Jauarí for several months, so that I acquired a very fair knowledge altogether of the productions of the country bordering the Amazons to the end of the Brazilian territory, a distance of 1,900 miles from the Atlantic at the mouth of the Pará; but beyond the Peruvian boundary I found now I should be unable to go. My ague seemed to be the culmination of a gradual deterioration of health, which had been going on for several years. I had exposed myself too much in the sun, working to the utmost of my strength six days a week, and had suffered much, besides, from bad and insufficient food. The ague did not exist at St. Paulo; but the foul and humid state of the village was, perhaps, sufficient to produce ague in a person much weakened from other causes. The country bordering the shores of the Solimoens is healthy throughout; some endemic diseases certainly exist, but these are not of a fatal nature, and the epidemics which desolated the Lower Amazons from Pará to the Rio Negro, between the years 1850 and 1856, had never reached this favoured land. Ague is endemic only on the banks of those tributary streams which have dark-coloured water.

I always carried a stock of medicines with me; and a small phial of quinine, which I had bought at Pará in 1851, but never yet had use for, now came in very useful. I took for each dose as much as would lie on the tip of a penknife-blade, mixing it with warm chamomile tea. The first few days after my first attack I could not stir, and was delirious during the paroxysms of fever; but the worst being over, I made an effort to rouse myself, knowing that incurable disorders of the liver and spleen follow ague in this country if the feeling of lassitude is too much indulged. So every morning I shouldered my gun or insect-net, and went my usual walk in the forest. The fit of shivering very often seized me before I got home, and I then used to stand still and brave it out. When the steamer ascended in January, 1858, Lieutenant Nunes was shocked to see me so much shattered, and recommended me strongly to return at once to Ega. I took his advice, and embarked with him, when he touched at St. Paulo on his downward voyage, on the 2nd of February. I still hoped to be able to turn my face westward again, to gather the yet unseen treasures of the marvellous countries lying between Tabatinga and the slopes of the Andes; but although, after a short rest in Ega, the ague left me, my general health remained in a state too weak to justify the undertaking of further journeys. At length I left Ega, on the 3rd of February, 1859, *en route* for England.

I arrived at Pará on the 17th of March, after an absence in the interior of seven years and a half. My old friends, English, American, and Brazilian, scarcely knew me again, but all gave me a very warm welcome, especially Mr. George Brocklehurst (of the firm of R. Singlehurst and Co., the chief foreign merchants, who had been my correspondents), who received me into his house, and treated me with the utmost kindness. I was rather surprised at the warm appreciation shown by many of the principal people of my labours; but, in fact, the interior of the country is still the "sertaõ" (wilderness),—a terra incognita to most residents of the seaport,—and a man who had spent seven years and a half in exploring it solely with scientific aims was somewhat of a curiosity. I found Pará greatly changed and improved. It was no longer the weedy, ruinous, village-looking place that it appeared when I first knew it in 1848. The population had been increased (to 20,000) by an influx of Portuguese, Madeiran, and German immigrants, and for many years past the provincial government had spent their considerable surplus revenue in beautifying the city. The streets, formerly unpaved or strewn with loose stones and sand, were now laid with concrete in a most complete manner; all the projecting masonry of the irregularly-built houses had been cleared away, and the buildings made more uniform. Most of the dilapidated houses were replaced by handsome new edifices, having long and elegant balconies fronting the first floors, at an elevation of several feet above the roadway. The large swampy squares had been drained, weeded, and planted with rows of almond and casuarina trees, so that they were now a great ornament to the city, instead of an eyesore as they formerly were. My old favourite road, the Monguba avenue, had been renovated and joined to many other magnificent rides lined with trees, which in a very few years had grown to a height sufficient to afford agreeable shade; one of these, the Estrada de Saò José, had been planted with coco-nut palms. Sixty public vehicles, light cabriolets (some of them built in Pará), now plied in the streets, increasing much the animation of the beautiful squares, streets, and avenues.

I found also the habits of the people considerably changed. Many of the old religious holidays had declined in importance, and given way to secular amusements; social parties, balls, music, billiards, and so forth. There was quite as much pleasure-seeking as formerly, but it was turned in a more rational direction, and the Paraenses seemed now to copy rather the customs of the northern nations of Europe, than those of the mother-country, Portugal. I was glad to see several new book-sellers' shops, and also a fine edifice devoted to a reading-room supplied with periodicals, globes, and maps, and a circulating library. There were now many printing-offices, and four daily newspapers. The health of the place had greatly improved since 1850, the year of the yellow fever, and Pará was now considered no longer dangerous to new-comers.

So much for the improvements visible in the place, and now for the dark side of the picture. The expenses of living had increased about fourfold, a natural consequence of the demand for labour and for native products of all kinds having augmented in greater ratio than the supply, through large arrivals of non-productive residents, and considerable importations of money on account of the steamboat company and foreign merchants. Pará, in 1848, was one of the cheapest places of residence on the American continent; it was now one of the dearest. Imported articles of food, clothing, and furniture were mostly cheaper, although charged with duties varying from 18 to 80 per cent., besides high freights and large profits, than those produced in the neighbourhood. Salt codfish was twopence per pound cheaper than the vile salt pirarucú of the country. Oranges, which could formerly be had almost gratis, were now sold in the streets at the rate of three for a penny; large bananas were a penny each; tomatos were from twopence to threepence each, and all other fruit in this fruit-producing country had advanced in like proportion. Mandioca-meal, the bread of the country, had become so scarce and dear and bad, that the poorer classes of natives suffered famine, and all who could afford it were obliged to eat wheaten bread at fourpence to fivepence per pound, made from American flour, 1,200 barrels of which were consumed monthly; this was now, therefore, a very serious item of daily expense to all but the most wealthy. House-rent was most exorbitant; a miserable little place of two rooms, without fixtures or conveniences of any kind, having simply blank walls, cost at the rate of £18 sterling a year. Lastly, the hire of servants was beyond the means of all persons in moderate circumstances; a lazy cook or porter could not be had for less than three or four shillings a day, besides his board and what he could steal. It cost me half-a-crown for the hire of a small boat and one man to disembark from the steamer, a distance of 100 yards.

In rambling over my old ground in the forests of the neighbourhood, I found great changes had taken place—to me, changes for the worse. The mantle of shrubs, bushes, and creeping plants which formerly, when the suburbs were undisturbed by axe or spade, had been left free to arrange itself in rich, full, and smooth sheets and masses over the forest borders, had been nearly all cut away, and troops of labourers were still employed cutting ugly muddy roads for carts and cattle, through the once clean and lonely woods. Houses and mills had been erected on the borders of these new roads. The noble forest trees had been cut down, and their naked half-burnt stems remained in the midst of ashes, muddy puddles, and heaps of broken branches. I was obliged to hire a negro boy to show me the way to my favourite path near Una, which I have described in the second chapter of this narrative; the new clearings having quite obliterated the old forest roads. Only a few acres of the glorious forest near Una now remained in their natural state. On the other side of the city, near the old road to the rice mills, several scores of woodsmen were employed, under government, in cutting a broad carriage-road through the forest to Maranham, the capital of the neighbouring province, distant 250 miles from Pará, and this had entirely destroyed the solitude of the grand old forest path. In the course of a few years, however, a new growth of creepers will cover the naked tree-trunks on the borders of this new road, and luxuriant shrubs form a green fringe to the path: it will then become as beautiful a woodland road as the old one was. A naturalist will have, henceforward, to go farther from the city to find the glorious forest scenery which lay so near in 1848, and work much more laboriously than was formerly needed, to make the large collections which Mr. Wallace and I succeeded in doing in the neighbourhood of Pará.

June 2, 1859.—At length, on the second of June, I left Pará, probably for ever; embarking in a North American trading-vessel, the "Frederick Demming," for New York, the United States route being the quickest as well as the pleasantest way of reaching England. My extensive private collections were divided into three portions, and sent by three separate ships, to lessen the risk of loss of the whole. On the evening of the third of June, I took a last view of the glorious forest for which I had so much love, and to explore which I had devoted so many years. The saddest hours I ever recollect to have spent were those of the succeeding night, when, the mameluco pilot having left us free of the shoals and out of sight of land, though within the mouth of the river, at anchor, waiting for the wind, I felt that the last link which connected me with the land of so many pleasing recollections was broken. The Paraenses, who are fully aware of the attractiveness of their country, have an alliterative proverb, "Quem vai para (o) Pará para," "He who goes to Pará stops there," and I had often thought I should myself have been added to the list of examples. The desire, however, of seeing again my parents and enjoying once more the rich pleasures of intellectual society, had succeeded in overcoming the attractions of a region which may be fittingly called a Naturalist's Paradise. During this last night on the Pará river a crowd of unusual thoughts occupied my mind. Recollections of English climate, scenery, and modes of life came to me with a vividness I had never before experienced during the eleven years of my absence. Pictures of startling clearness rose up of the gloomy winters, the long grey twilights, murky atmosphere, elongated shadows, chilly springs, and sloppy summers; of factory chimneys and crowds of grimy operatives, rung to work in early morning by factory bells; of union workhouses, confined rooms, artificial cares, and slavish conventionalities. To live again amidst these dull scenes I was quitting a country of perpetual summer, where my life had been spent like that of three-fourths of the people in gipsy fashion, on the endless streams or in the boundless forests. I was leaving the equator, where the well-balanced forces of Nature maintained a land-surface and climate that seemed to be typical of mundane order and beauty, to sail towards the North Pole, where lay my home under crepuscular skies somewhere about fifty-two degrees of latitude. It was natural to feel a little dismayed at the prospect of so great a change; but now, after three years of renewed experience of England, I find how incomparably superior is civilised life, where feelings, tastes, and intellect find abundant nourishment, to the spiritual sterility of half-savage existence, even though it be passed in the garden of Eden. What has struck me powerfully is the immeasurably greater diversity and interest of human character and social conditions in a single civilised nation, than in equatorial South America, where three distinct races of man live together. The superiority of the bleak north to tropical regions, however, is only in their social aspect; for I hold to the opinion that, although humanity can reach an advanced state of culture only by battling with the inclemencies of nature in high latitudes, it is under the equator alone that the perfect race of the future will attain to complete fruition of man's beautiful heritage, the earth.

The following day, having no wind, we drifted out of the mouth of the Pará with the current of fresh water that is poured from the mouth of the river, and in twenty-four hours advanced in this way seventy miles on our road. On the 6th of June, when in 7° 55′ N. lat. and 52° 30′ W. long., and therefore about 400 miles from the mouth of the main Amazons, we passed numerous patches of floating grass mingled with tree-trunks and withered foliage. Amongst these masses I espied many fruits of that peculiarly Amazonian tree the Ubussú palm; this was the last I saw of the Great River.

Watson and Hazell, Printers, London and Aylesbury.

INDEX.

THE END.